Models in Geography

Models in GEOGRAPHY

Majid Husain

RAWAT PUBLICATIONS
Jaipur • New Delhi • Bangalore • Guwahati • Kolkata

ISBN 978-81-316-0135-8 (hardcover)

Reprinted, 2025

Published by
Prem Rawat for **Rawat Publications**
Satyam Apts., Sector 3, Jawahar Nagar, Jaipur - 302 004 (India)
Phones: 0141 265 1748 / 265 7006
E-mail : info@rawatbooks.com
Website: rawatbooks.com

New Delhi Office
4858/24, Ansari Road, Daryaganj, New Delhi 110 002
Phone: 011-23263290

Also at *Bangalore, Guwahati* and *Kolkata*

Typeset by Rawat Computers, Jaipur
Printed at Nice Printing Press, New Delhi

To My Grandchildren

Contents

Preface

Models and theories in Human Geography is an important component in the course structures of the Indian Administrative Services. The prescribed models and theories are from the Agricultural Geography, Industrial Geography, Urban Geography, Population Geography, Political Geography and Settlement Geography. Being from different branches of Human Geography, literature on the prescribed models is not available at one place. The students face a lot of strain in consulting the different books and volumes to master the topics of models and theories. The aim of the present textbook is to provide literature in a concise and cogent manner on the models and theories of Human Geography at one place. This will facilitate the candidates who prepare themselves for the various competitive examinations and will economise on their precious time.

Some of the important theories and models presented in the volume are system analysis, Mathusian, Marxian and Demographic Transitional Models, Central Place Theory of Christaller and Losch, von Thunen Model of Agricultural Land Use, Weber's Theory of Industrial Location, Rostov's Stages of Growth, Heartland and Rimland theories, Laws of International Boundaries and Frontiers. In addition to these, the Limits to Growth, and Laws of Migration have also been systematically examined. A full chapter has been 'devoted to types and patterns of rural settlements, primate city, rank-size rule and models of urban land. Documented with the latest data, maps and diagrams one chapter has been written on climatic change. It is hoped the book will be of immense help to the students and teachers alike. Constructive criticism is invited to improve the subsequent editions of the book.

Planning and writing a book is always a cumbersome task. The task becomes even more tedious, if the contents pertain to different branches of Human Geography. For the collection of relevant literature, data, maps and diagrams, the author owes much to the work of previous scholars, authors and cartographers, who have dedicated their lives to exploring, probing, and writing on different models in Geography. A bibliography of the books consulted has been given at the end of the book. The author expresses his most sincere gratitude to all the authors and publishers whose books, research papers and maps and diagrams have been utilized in the preparation of the present book.

The constant demand from the students of the different universities as well as from the candidates who appear in the various competitive examinations, was a great motivating force behind this venture. I thank each one of them for the confidence they put in the abilities of the author.

Writing of a book demands a congenial environment at home as well as at the place of work. I am thankful to all the members of my family who assisted me throughout to complete this assignment. Dr. Ramesh Singh, Director, Civils India and his staff deserves special thank for their perennial support. My special thanks to Rawat Publications who completed the book with great dexterity in a record time.

March 2007 **Majid Husain**
New Delhi

List of Figures

List of Tables

1

Introduction to Geography

Geography is eclectic, integrating a whole range of subject matter from diverse fields; virtually any subject can be examined geographically. The word 'geography' has been derived from the Greek *geo*, the earth, *graphein*, to write (describe). It was Eratosthenes (3rd century B.C.) who was the first to use the word *geography*. The Greek geographers credit the beginning of geographical writings to Homer, but the earliest known map was made by the Sumerians (present valley of Tigris and Euphrates, Iraq) who are credited with the first extraction of copper and bronze, about 2700 B.C. There had not been a consensus among the geographers about the definition of geography. Throughout its history, geography has been defined differently by different geographers. Some of the important definitions of the discipline of geography have been concisely presented in this chapter. These will help in understanding the objectives and social relevance of the subject.

"Geography is the study of spatial variation on the earth's surface." This is a conventional definition of geography, which has been criticized as it does not give any stress on the explanation of the geographical phenomena and the changing character of the earth's surface.

In the opinion of Haggett (1960), geography is the study of the earth's surface within which the human population lives.

Tuan (1991) a strong supporter of the humanistic approach to geography, defined geography as the study of the earth's surface as the home of people.

The best known formal definition of geography was provided by Richard Hartshorne, a leading American geographer, in his books, *The Nature of Geography* (Chicago, 1939) and *Perspective on the Nature of Geography* (Chicago, 1959). According to Hartshorne, "Geography is concerned to provide accurate, orderly, and rational description and interpretation of the variable character of the earth surface." In this definition two words have special significance, i.e., 'variable character' and 'earth surface'.

For geographers, the variable character means the spatial (regional) variation that can occur on the earth from one region to another and from one location to another. The variation in character of the earth may occur in all the physical variables (landforms, drainage, temperature, precipitation, natural vegetation, soils, minerals etc.), and cultural variables (ethnicity, language, religion, customs, traditions, occupations, cultural ethos, and modes of life). These variations, in fact, can be observed at all scales of maps, i.e., global, national, regional, local, down to a very local level, say between the districts, sub-divisions (tehsils), blocks, villages, and within an urban area. In the definition of Hartshorne 'earth surface' is meant the upper thin shell of the crust, which may be only about one-thousandth of the planet's (earth's) circumference and forms the environment (habitat) within which the human population is able to survive.

Despite all these definitions, geography occupies a puzzling position within the traditional disciplines of knowledge. It is neither a purely natural science nor a purely social science. Its intellectual origins as a distinctive field of study date back to classical Greece; people were viewed then as an integral part of nature. The major thrust of Greeks was on the description of the living and non-living things. This approach was also known as chorology or chorography. Chorology is the areal differentiation of the earth's surface. It was first set by Hecataeus of Miletus in the 6th century B.C. The same point of view was confided most elegantly by Strabo in his *Geography* written in the form of seventeen volumes, sometime between 8 B.C. and 18 A.D. Strabo declared, a geographer is "the person who attempts to describe the parts of the earth". The two key words were 'describe' and 'parts'. In effect, Strabo was recommending what would now be called *Regional Geography* as the core of Geography. In his opinion, "If there is one science which is worthy to be practised by a 'philosopher', he argued, then it is geographical science."

For Strabo, geography describes those worthwhile things from which one could learn about truth, nobility and virtue: it was a complement to political and ethical philosophy. For this reason Strabo's geography was fundamentally concerned with human activities. It was directed towards social and political ends, and paid considerable attention to the interests of chorography that was not supposed to provide a comprehensive gazetteer or a regional inventory. He began with the chorography of 'Europe' because it is admirably adapted to nature for the development of excellence in man and government.

In opposition to Strabo, Claudius Ptolemaeus or Ptolemy, around A.D. 150, opined that the purpose of geography was to provide "a view of the whole, analogous to the drawing (map-making) of the whole head", and this meant that he separated geography from chorography which, so he said, "has the purpose of describing the parts, as if one were to draw only an ear or eye". As this passage implies, for Ptolemy, *graphein* (graphy) did not mean describing but drawing and, specifically mapping.

Ptolemy's 'geography' is geodetic and cartography and he preferred to leave out everything which had no direct connection with that aim.

The distance between Strabo and Ptolemy could not be plainer, and it is indelibly present in the constitution of a distinctively modern geography too. As late as the 17th century, Strabo and Ptolemy continued to provide the main models for European Geography. The usual distinction was between *special (particular/regional) geography*, devoted to the description of particular regions (including the study of human population), and a *general geography (systematic or universal geography)* mathematically oriented and concerned with the globe as a whole. The premier illustration is provided by Bernhard Varenius, who published both studies in special geography and his famous *Geography Generalis*. After the publication of *Geography Generalis*, for the first time, geography sought to engage with the ideas of Descartes, Bacon and Galilieo (Bowen, 1981).

The modern case for geography as a 'geographic science' was argued most forcefully by Hartshorne in the *Nature of Geography* (1939), and ever since the subsequent debate over exceptionalism in geography—and despite the nuances registered—chorology has been used in polemical opposition to spatial science (Sack, 1974). But, the temper of the original version, with its acknowledgement of the importance of political power and philosophical reflection, is a forceful reminder of the continuing need to attend to the politics of geographical enquiry.

Hartshorne argued forcefully that the focus of geography is areal differentiation, the mosaic of separate landscape on the earth's surface. Thus, geography, according to Hartshorne, is:

> ... a science that interprets the realities of areal differentiation of the world as they found, not only in terms of the differences in certain things from place to place, but also in terms of total combination of phenomena in each place, different from those at every other place.

According to Hartshorne, the main purpose of geography is synthesis or the integration of relevant characteristics to provide a total description of a place or region, which is identifiable by its peculiar combination of those characteristics. According to Hartshorne, there is a close analogy between geography and history; the latter provides for temporal section of reality whereas the former performs a similar task for spatial section of the earth's surface.

In opposition to Hartshorne's areal differentiation, Schaefer, who was essentially a regional economist, considered geography as a social science. In his opinion, geography is the science concerned with the formulation of the general laws governing the distribution of certain features on the surface of the earth. Using analogies from physics and economics, Schaefer argued that geography is not peculiar in focusing on unique phenomena: all sciences deal with unique events which can only be accounted for an integration of laws from various systematic sciences.

Immanuel Kant (University of Koningsberg, Germany) was a staunch supporter of 'physical geography' though he also gave some weight to political, commercial, moral and theological geography. Kant recognized that specialization in geography like other subjects is not possible, since its single object (the earth) is unique in the universe. In his opinion the main area of geography is 'general geography'. He also propounded the concept of the deflection of wind direction resulting from the earth's rotation. Kant described the field of physical geography as the world. It was Wilhelmi who limited the field of geography to the earth's surface.

The geography of Kant's period included both 'general studies' and 'particular studies' of the earth's surface. The distinction between the two directions (general and particular) was first perhaps clearly made by Varenius (a German, Bernhard Varen, living in Amsterdam) in 1650. Varenius divided geography into 'general or universal geography' and 'special geography' or 'chorology'. The general geography at present is called as 'systematic geography' and the special geography as 'regional geography'.

In the second half of the 18th century, the geographers tried to make geography as *'reine Geographie'* or 'pure geography'. It was during this period when geographers increasingly stressed to divide the earth into durable natural divisions. In other words the objective of geography became to delineate 'natural boundaries' of 'lands' rather than political boundaries of states. Kant's work led rapid growth in various physical sciences.

Humboldt and Ritter, who are known as the founders of modern geography, took into consideration all kinds of phenomena whose differences in different parts of the world made them appear significant to acknowledge the world. They echoed the famous statement of Strabo, "Geography, in addition to its vast importance to social life and the art of government ... acquaints us with the occupants of land and ocean and the vegetation, fruits and peculiarities of the various parts of the earth".

Essential Characteristics of Geography

Considered as the mother of all natural and social sciences, the salient characteristics of geography are as under:

1. The first characteristic of geography is the emphasis on *location*. Geography is concerned with the location or spatial variation in both

physical (landforms, relief, topography, drainage, temperature, precipitation, natural vegetation, soils, minerals etc.) and human (race, ethnicity, faith, religion, language, customs, traditions, occupations and gamut of life) phenomena at the earth's surface. Geography tries to establish locations accurately, to represent them effectively and economically. The geographers are also interested in ascertaining the factors that led to particular spatial patterns of economic, agricultural, industrial, social and political activities. The von Thunen Model (1826) and the Central Place Theory by Christaller (1933) may be cited as the good examples of location of certain patterns.

In human geography it may also propose alternative spatial patterns which are more equitable (welfare geography approach to human geography that emphasizes questions of inequality and social justice) or more efficient (economic geography, which deals with people's struggle to make living). It is significant that many of the techniques developed within geography for the study of such spatial variation are general in character, and not specific to phenomena studied in either physical or human geography.

The establishment of accurate location is possible with the help of cartographic techniques. Cartography is a body of theoretical and practical knowledge that map-makers employ to construct maps as a distinct mode of representation. The geographers in the past concentrated on the technical and professional practices of making maps.

2. The second characteristic of geography is ecological emphasis on society-land relations. (Ecology: the study of interrelations between organisms and their biotic and abiotic environments.) The geographers give especial stress on the man and environment relationship, i.e., the impact of land (environment) on people and the impact of people on land or environment.
3. The third characteristic of geography is regional analysis, in which the spatial and ecological approaches described above are fused. The homogenous earth's surface (regions) are identified, their internal (intra-regional) morphology and ecological linkages are traced, and their external (extra-regional) relations are established (chorology and regional geography).

For some geographers the process of region building, sometimes also termed as the areal differentiation (delineation of regions) of the earth's surface, represented the core area of the discipline, whereas for others, the most significant aspect of geography is the systematic study of the discipline to frame universal laws in association with the sister disciplines. Thus, there are two approaches to study geography: (i) either we can study geography, region by region, or (ii) we can consider a significant theme or element and trace it systematically over whole earth's surface.

The first approach is termed as regional geography, and the second, as systematic geography.

Nature of Geography

The nature of geography had been and is a topic of great debate among the geographers and other natural and social scientists. Some of the academicians, including geographers, assume that geography is some kind of knowledge that concerned with the earth. Some geographers consider geography as a physical science, while others take it as a subject of social sciences. Whether geography is a science or an art, or in what sense a science or an art, or both, are the questions which need objective assessment. Whatsoever the geography may be, it is not an infant subject, born out of the womb of American geology a few decades ago. The change in the concept of geography and its nature is neither desirable nor easy. It is also difficult to define precisely what geography ought to be, though it is easy to discuss what it contains.

Of essential concern to the geographers are spatial patterns and interrelationships of climate, landforms, vegetation, soils, population, economic activities and political units, either on global scale or in a more limited area. The geographers seek to explain the location of various elements in the environment (space) and to establish the spatial patterns of these elements. Besides, the processes that form and change these patterns are also analysed. One principal means to this end is the map, which for the geographers is an indispensable research tool as well as visual representation.

Systematic or Scientific Geography

Frobel, a German geographer, was perhaps the first of many students of Ritter who attempted to construct a science of geography based on a study of geography itself, in the form of natural science. He assumed that geography as *Erdkunde* (science of the earth) must correspond to the literal translation of its name and that it can only be science be natural science.

In the opinion of Frobel, geography can be divided into two parts: (i) Geography as a natural science or 'systematic geography', consisting of separate studies of each of the features of the earth, classified by kind. In the systematic geography, he included the study of man as a feature of the earth. (ii) Those aspects of geography that do not fit into systematic (scientific) geography including what we call 'aesthetic geography', i.e., the relations of historical phenomena to geography, and apparent regional geography—he recognized as a form of 'applied geography' or 'historical-philosophical geography'. The two points of view are to be developed by their own methods, the former following the analytic, the latter the synthetic method. Each can be built into a separate system, but the two apparently cannot be combined. Geography is, therefore, not merely 'dualistic' but, one would suppose, is completely separated into two kinds.

During the decade following 1832, Frobel taught geography and mineralogy in Zurich and devoted a considerable part of his time to the preparation of a full study of systematic geography that was to lead to the theory of earth science, but this was never to be completed.

In the early part of the 19th century there was no professional geographer having a university training in the discipline of geography. Most of the experts who developed the field of geography in the latter part of the 19th century had come into geography after being trained in mathematics, geology, zoology, botany or history (Penck, 90, I, 38). Gerland, in his essay, in 1887 appreciated the work of Kant and suggested the geographers that they should take to heart the word of great philosopher, namely, "physical geography—a general abstract of nature is the foundation of history and all other possible geographies". It was in this background that Gerland planned the (i) geography of inorganic earth, (ii) geography of plants, (iii) geography of animals, and (iv) geography of man.

If we examine the Gerland's thesis in detail, we find that he focused his attention on (i) nature of a science, and (ii) nature of the earth, which is the main field of enquiry of geographers. The question of nature of science is a philosophical, rather than a scientific one; that scientists, as such, are not qualified to answer it, is reflected in the wide range of answers that different scientists provide. In the opinion of Leighly, during that period, natural science was thought to include the objective study of man, 50 years later the concept of 'natural' had been narrowed to exclude man. Natural science requires fixed conclusions and laws of certainties, and therefore geography that could only deal with 'probabilities' was excluded from the category of science.

As stated above, Frobel and Gerland reduced the field of geography to *Erdkunde* (science of the earth). The essential question at this stage arose, what is earth? Earth is like millions of other heavenly bodies, a great complex of materials in state of change, bound together and interrelated by various forces to form a unit whole, though acted upon also external forces, notably from the sun and the atmosphere. The object of geography is therefore to study the interrelations of these forces and the resultant changes in earth materials.

The content of geography is thus the *earth as its subject of study*. The rocks, water and air are in the state of interrelated development and therefore the forces that affect these elements are also the subject matter of geography. Equally, the materials of which human animals are made, and those which they transform into cultural products, are earth materials and therefore their study is also the subject matter of geography. All the materials and forces of the earth represent a unit whole, and we cannot exclude some of these materials and forces still have a unit whole left.

Contrary to this, Frobel asserted that human phenomena could not be included in a science and must therefore be thrown out of geography. He also opined that scientific laws cannot be formulated about the human behaviour. Similarly, the biological sciences are not the same as the physical sciences. In brief, Gerland opined that plant and animal geography as well as human and regional geography must be excluded from the science of geography.

Leighly presumed geography as 'geo-physics' (physical science of the earth). He stressed on the study of landforms, weathering, erosion, meteorology, seismology, vulcanism, geodesy, hydrology, climatic types and their characteristics, etc. The sum of these branches of sciences does not form a 'physical earth science' but rather many physical earth sciences. Gerland also opposed the idea of Ratzel to develop a geographic law concerning the separation of peoples by mountains. According to him, "the appropriate law here is not found in the connection of men with mountains but in certain psychophysical reactions that are released by the mountains, hence does not lie in the field of geographic thought, but completely in psychology and physiology".

In opposition to Forbel and Gerland, Ratzel created great interest in the field of human and regional geography by publishing his book *Anthropogeography.* This work was of great significance in establishing systematic (general) human geography. His regional studies have been highly regarded, particularly by those interested in aesthetic aspect of the subject. Relatively speaking, his importance was greatest in political geography. In the opinion of L. Gumplovicz, "Ratzel's works contain more and more important knowledge concerning the state, than the entire theoretical political science literature of the last hundred years." Ratzel is universally recognized as having established the foundation of this particular branch of modern geography.

Areal Differentiation

The study of areal variation of human and physical phenomena as they relate to other spatially proximate and causally linked phenomena is known as areal differentiation. The term 'areal differentiation' was coined and used by Hartshorne in his classic work, *The Nature of Geography*, published in 1939. Drawing from Hettner, Hartshorne's central claim about geography is its integrative or synthetic purpose. The areal differentiation is also known as 'chorology' or 'chorography'. Chorology is the study of the areal differentiation of the earth's surface. Geography, on this definition, is solely concerned with the unique character of different areas of the earth's surface. Areal differentiation may be termed as "idiographic as it is concerned with the unique and particular".

In the 1980s areal differentiation was reinstated as a central perspective in human geography. Intellectual inspiration has come from three general directions, none of which is directly connected to older positions in debates about areal differentiation or uses the same terminology as the others.

The first is from the streams of thought referred to collectively as humanistic geography, which gives central and active role to human awareness and human agency, human consciousness and human creativity. The humanistic method (iconographic technique) seeks to explore the composition of landscapes, interpreting their symbolic content and reimbursing landscaping in their social and historical contexts. Successful iconographic interpretation allows us to see human landscapes as both shaped by and themselves shaping broader social and cultural processes, and thus having ideological significance.

The second focus of revival has been the analysis of uneven development and changing spatial division of labour. Some of the geographers tried to explain spatial variation in economic activity and well-being with a Marxist approach.

The third influence has come from attempts in geography to create Contextual Theory, in which the place or region is viewed as geographically mediating the interpolation of human agency and social structure and is thereby implicated directly in the production of geographical sameness and differences.

The third direction could be seen as potentially integrating the other two, but this would be a superficial view. There are important philosophical differences between all the three. For example, the first direction tends either to privilege or to emphasize the human objective experience of place—more often than not that of the scholar engaged in writing about it—whereas the second and third view the division of space in terms of objective socio-spatial processes with, for the third view, sense of place arising out of the conditions created by such processes.

In the areal differentiation the most challenging dilemma is the difficulty of neat boundary delimitation between places and regions when the territoriality of social groups is dynamic and irreducible to a singular and temporally fixed set of spatial units. Areal differentiation has also been criticized that it is incapable of contributing towards effective generalization.

Areal differentiation helped in the reconstruction of regional geography. It emphasized, however, that regions must not be studied solely as separate entities. Thus, regional geography must focus on the unique characteristics of the place being studied, but must not express them as if they were singular.

Despite all these criticisms, areal differentiation is being considered the rational and scientific definition of the discipline of geography.

Exceptionalism in Geography

The father of exceptionalism is Immanuel Kant. Kant claimed exceptionalism not only for geography but also for history. According to him, history and geography find themselves in an exceptional position different from that of the so-called systematic sciences. This grouping of geography with history has tempted many subsequent writers to elaborate the alleged similarity in order to obtain some insight into the nature of geography. This is one of the roots of historical variant of the claim to uniqueness. It is in Kant's work that one finds the statement on geography and history that has been quoted so reverently time and again by those who make it the cornerstone of geographical method. Humboldt and Ritter used it, so did Hettner and eventually Hartshorne.

The term 'exceptionalism' in geography is, however, usually identified with Schaefer. Schaefer was originally an economist. He joined the group of geographers teaching in the economics department at the University of Iowa after his escape from Nazi Germany. He made a critical analysis of the book of Hartshorne, *The Nature of Geography*, and published a paper 'Exceptionalism in Geography: A Methodological Examination' in the *Annals of Association of American Geographers* (1953, Vol. 43). Schaefer's essay was widely regarded as a rallying point for a new generation of human geographers.

Schaefer claimed that his paper was the first to challenge Hartshorne's interpretation of the works of Hettner and others. He criticized Hartshorne's exceptionalist claims for regional geography, and presented an alternative case for geography adopting the philosophy and methods of the positivist school of science. He first outlined the nature of science and then defined the peculiar characteristics of geography as a social science. He argued that a claim for geography as the integrating science, which puts together the findings of individual systematic sciences, was arrogant, and that in any case its products were somewhat lacking in "starting new and deeper insights". A science is characterized by its explanations, and explanations require laws. "To explain the phenomena, one has described means always to recognize them as instances of laws." In geography, in the opinion of Schaefer, the major regularities which are described refer to spatial patterns. "Hence, geography has to be conceived as the science concerned with the formulation of laws governing the spatial distribution of certain features on the surface of the earth", and these spatial arrangements of phenomena, not the phenomena themselves, should be the subject of geographers' search for law like statements. Geographical procedures would then not differ from those employed in other sciences, both natural and social: observation would lead to hypothesis—about the interrelationship between two spatial patterns, for example, and this would be tested against larger number of cases, to provide the material for a law if it were thereby verified. Thus, Schaefer conceived geography as a systematic social science in which universal laws need to be made to interpret the distributional patterns of phenomena.

The argument against this definition of geography as the science of spatial arrangements was termed 'exceptionalism'. It claims that geography does not share the methodology of other sciences because of the peculiar nature of its subject matter—the study of unique places, or regions, and compares geography with history, which studies unique periods of time. Using analogies from physics and economics, Schaefer argued that geography is not peculiar in focusing on unique phenomena, all sciences deal with unique events which can only be accounted for by an integration of laws from various systematic sciences, but this does not prevent (although undoubtedly makes more difficult) the development of laws. It is, therefore, absurd to maintain that geographers are distinguished among the scientists through the integration of heterogeneous phenomena that they achieve. There is nothing extraordinary about geography in that respect.

Thus, Schaefer argued that there is nothing exceptional in the philosophy (nature) and methodology of geography as advocated by Hartshorne. As stated above, he positioned human geography in the category of social sciences rather than in humanities or natural sciences. Geography, he argued, must be like other social sciences and not 'exceptional'.

The belief that geography and history are methodologically distinct from other fields of enquiry because they are peculiar and concerned with the study of unique and particular. Schaefer rejected the idiographic orthodoxy enshrined in Hartshorne's *The Nature of Geography*. In other words, he argued for *nomethetic geography* declaring geography as general or systematic science which aims to furnish general and universal 'morphological laws' about spatial patterns. He declared geography as the science of earth's surface in which the general and universal laws about spatial patterns are to be formulated instead of the regional and local laws.

Earlier, Humboldt and Ritter recognized that the major concern of geography is to examine the manner in which the natural phenomena, including man, are distributed in space. This implies that geographers must describe and explain the manner in which things combine "to fill an area". These combinations change, of course, from area to area. These differences either in the combination of factors or in their arrangement from place to place underlie the common sense notion that areas differ. Following the Greek geographers, this viewpoint is called chorographic or chorological one, depending on the level of abstraction. Geography, thus, must pay attention to the spatial arrangement of the phenomena in an area and not so much to the phenomena themselves. Spatial patterns are the one's that matter in geography, and no others. Non-spatial relations found among the phenomena in an area are the subject matter of other specialists such as ecologist, anthropologist or economist.

Subsequently, Kraft, while discussing Humboldt and Ritter, agrees with them that geography is a science trying to discover laws; that it is limited to the earth's surface; and that it is essentially chorological. Incidentally, he also feels that this suffices to set geography logically apart as an exceptional discipline.

Hettner, one of the leading German geographers, declared that "both history and geography are essentially chronological". History arranges phenomena in time, geography in space. Both, in contrast to other disciplines, integrate phenomena heterogeneous among themselves. Also, these phenomena are unique. No historical event and historical period is like any other. In geography no two phenomena and no two regions are alike. Thus, both fields face the task of explaining the unique (exceptional). Hettner calls history 'time-Wissenschaft' and geography 'space-Wissenschaft'. Hartshorne translated them into 'time science' and 'space science'.

Hartshorne's views were in fact more naunced than Schaefer maintained, and he never accepted any clear division between the idiographic (regional or particular) and the nomothetic (general or systematic) because both are "present in all branches of science". But, he did insist that any general concept used in geography should be directed towards the analysis of specific 'regions', and that its essential task was to study areal differentiation rather than (as Schaefer preferred) the elucidation of laws of location that were supposed to underpin these regional configurations.

It may be summarized that both Hartshorne and Schaefer relied upon mainly German sources such as Kant, Humboldt and Hettner, and both directed geographers towards spatial patterns. Even so Schaefer was among the first geographers to teach the ideas of Christaller, von Thunen and Losch, which Hartshorne has not acknowledged but which were to become the mainstay of 1960s' human geography. He also positioned human geography in the social sciences rather than the humanities or natural sciences. Geography, he argues, must be like other social sciences and not exceptional.

Spatial Analysis

The quantitative techniques employed in locational analysis is known as 'spatial analysis'. It is sometimes used as a synonym for the locational analysis. Unwin (1981) presents spatial analysis as the study of the arrangements of points, lines, areas and surfaces on a map.

The followers and advocates of spatial science consider human geography as that component of social sciences which focuses on the role of space as a fundamental variable, influencing both society's organization and operation and the behaviour of its individual members. Spatial analysis got popularity during the period of quantitative revolution. It is closely associated with the philosophy of positivism.

The goal of spatial analysis was "building accurate generalizations with predictive power by precise quantitative description of spatial distribution, spatial structure and organization and spatial relationships".

The generalizations arrived at with the application of spatial analysis could be based on three fundamental spatial concepts: (i) direction, (ii) distance, and (iii) connection (or relative position).

In the spatial analysis some of the geographers merely apply techniques derived from the General Linear Model to geographical example, while others have argued that spatial data analysis poses particular statistical problems (such as spatial auto-correlation) which means developing procedures specifically designed to counter them.

Spatial analysis for man and environment relationship has been criticized on several counts. One of the main criticisms is that spatial analysis focused on spatial determinism, and the logical impossibility of defining spatial variables independent of the context within which they were supposed to operate.

The second weakness of spatial analysis is that it does not take into consideration the cultural values and normative questions while attempting to establish the man and environment relationship.

The Marxist and radicalists consider spatial analysis as a device to promote capitalism and exploit workers and environment (see quantitative revolution).

Locational Analysis

Locational analysis is an approach to human geography which focuses on the spatial arrangement of phenomena. Its usual methodology is that of spatial science. The main objective of locational analysis was expressed as building accurate generalizations, models and theories with productive power (Berry and Marble, 1968).

Locational analysis is based on the philosophy of positivism. The philosophy of positivism underpins the approach, which concentrates on the identification of theories of spatial arrangements and so is closely linked to the discipline's quantitative revolution.

A number of geographers in USA advocated the cause of locational analysis in the 1950s, although it has much deeper roots in the work of pioneers who were later adopted by geographers. Bunge (1966), for example, wrote a thesis on *Theoretical Geography* based on the premises who stated that geography is the 'science of locations'. Others, such as McCarty, were strongly influenced by developments in the field of economics, to which they introduced the spatial variable. These links led to the close interrelationship between geographers and regional scientists in the 1960s and 1970s, and illustrated by attempts to build economic geography theories of spatial arrangements (Smith, 1981).

Locational analysis is based on empiricism. Empiricism is a philosophy which accords special privilege to empirical observations over theoretical statements. Specifically, it assumes that observational statements are the only ones which make direct reference to phenomena in the real world, and that they can be declared true or false without reference to the truth or falsity of the theoretical statements. In empirical inquiry, it is assumed that its facts 'speak for themselves'. They presented a strong case for using geometry as the language for the study of spatial form.

Haggett, in his book *Locational Analysis in Human Geography* (1965), appealed to adopt the geometrical tradition to explain order, location order and patterns in human geography. Such a focus needed (1) to adopt a system approach which concentrates on the patterns and linkages within a whole assemblage; (2) to employ models as to understand man and environment relationship; and (3) to use quantitative techniques to make precise statements (generalizations) about locational order. For the spatial analysis they suggested to adopt 'linear model', spatial auto-correlation and regression.

Other geographers, who contributed substantially to the field of locational analysis, are Morril, Col, Chorley, Cox, Harvey, Johnston, Pooler, Sack and Smith.

Morril was strongly influenced by the geometrical traditions adopted by Bunge and Haggett. In his book, *The Spatial Organisation of Society*, Morril argued that people seek to maximize spatial interaction at minimum cost and so bring related activities into proximity—the result is that human society is surprisingly alike from place to place... [because of] the predictable, organized pattern of locations and interactions.

The locational approach in human geography has been criticized on philosophical and methodological grounds by the behaviouralists and humanists. Some of the main criticisms against locational analysis are as under:

1. The locational analysis based on positivism ignores the normative questions to explain the man and environment relationship. It was their mistaken belief that "positive theory would lead to normative insight". The cultural values are quite important in any decision-making process. The ideal location for any economic activity may not be acceptable to individuals and the society (see quantitative revolution).
2. The locational analysis did not reflect the reality of decision-making processes and so was of little value in predicting locational arrangement.
3. The models developed with the help of locational analysis conceal the complexities of the real world.
4. At present, there is economic interdependence of societies at the global level, which means that spatial interdependence has become much more important and "locally experienced environmental dependencies lost their rationale".
5. Locational analysis has also been criticized on the ground that it encourages the social order of capitalism in which the owners of the means of production become rich and the poor becomes poorer.
6. The locational analysis has given a chance to the capitalists to optimize their profits. It gives an uncontrolled liberty and licence for plunder and miscalled profit.
7. Owing to locational analysis, there is over-production and the economy enters the era of over-industrialization.
8. It is mainly because of the locational analysis and capitalism that there is a total newness—new technology, new means of transportation, new education, new art, new morals, new media, new amusement, new weapons, new violence, new terrorism, new war and new mode of exploitation.
9. The followers of spatial science (positivists) treat people as dots on a map, statistics (data) on a graph, or numbers in an equation. They consider humans as non-living and other livings (plants and animals).

It is because of the inadequacies of the locational analysis that the 'behaviouralism' and 'humanism' achieved much significance in human geography.

Whatever the reason for its origin, there is little doubt that locational analysis substantially changed the nature of human geography from the mid-1960s, although there is some doubt that it ever dominated the discipline (Mikesell, 1984). It presented geography as a positivist social science, concerned to develop precise, quantitatively stated generalization about pattern of spatial organization, thereby enriching and being enriched by Location Theory, and to offer models and procedures which could be used in physical planning. By 1978, therefore, Haggett could write that "the spatial economy is more carefully defined than before, we know a little more about its organization, the way it responds to shocks, and the way some regional sections are tied into others. There now exist theoretical bridges, albeit incomplete and shaky, which span from pure spaceless economics to a more spatial reality".

Twelve years later, he continued to promote the search for 'scientific generalization' (Haggett, 1990), while accepting, that in the search for spatial order "the answer largely depends on what we are prepared to look for and what we accept as order": for only a minority of geographers can now claim that order is the focus of their quest.

Geography as a Chorographic or Chorologic (Regional) Science

As stated earlier, chorography or chorology is the study of the areal differentiation of the earth's surface. It represents the oldest tradition of geographical inquiry. It was first set forth and codified most elegantly by Strabo in his 17 books on geography written sometime between 8 B.C. and A.D. 18. The geographer, he declared, is the person who attempts to describe the parts of the earth (in Greek, *chorographein*). The two key words were 'describe' and 'parts'. In effect, Strabo was recommending what would now be called regional geography as the core of geographical study. For Strabo, geography described those worthwhile and unique things which one could learn about truth nobility and virtue. Strabo's geography was fundamentally concerned with human activities. It was directed towards social, political and military ends. The modern case for geography as a 'chorographic' science was argued most forcefully by Hartshorne in the *Nature of Geography* (1939).

Geography, as a chorographic study, has always found justification in the widespread desire of many people to know what other parts of the world are like. Further, since the ordinary person actually knows but partially the area of what is happening in the world in his lifetime, 'home geography' is as necessary as history.

In the opinion of Volz, "the purpose and significance of geography is that it teaches us to know the space (area/region)—the surface of the earth". According to Sauer, geography assumes the responsibility for the study of the areas (regions), because there exists a common curiosity about that subject. The fact that every school child knows that geography provides information about different countries is enough to establish the validity of regional geography. No other subject attempts the study of areas.

From the point of view of each specialized science, while each subject deals with a restricted class of phenomena, geography deals with heterogeneity of phenomena. The study of numerous types of phenomena helps to break the field up into separate specialized parts (economic geography, social geography, cultural geography, political geography, urban geography, agricultural geography, industrial geography, settlement geography, etc.).

Vidal de Lablache also justified the regional geography as the core of geography. He observed that man is interested above all in the totality of conditions in an area. There are many different ways of studying the world, but since men as individuals, and as individual groups, do not in any very full sense live in the entire world, but each in relatively restricted area of the world, one of the most significant methods of studying the world is to study it by areas (regions). Geography, thus, attempts to provide complete, accurate and organized knowledge to satisfy man's curiosity about how things differ in the different parts of the world. In order to achieve this objective, geography must consider the world in terms of limited areas (regions) within which things are closely associated.

Geography as the Science of the Planet Earth

In the opinion of some of the geographers like Gerland, Ricthofen, Lehmann and Penck, geography is the science of the planet earth. There had been a puzzling question in the mind of geographers, whether the field of geography includes the earth as a whole or whether it is limited to that part of the earth which we know directly—the earth surface. While in some of the textbooks they take the earth planet as the sphere of geography, others opine that only that part of the earth is the sphere of geography with which we are more intimately familiar, i.e., the earth surface. In practice, geography is to study the world, as that term is most commonly understood.

Some of the geographers have arbitrarily restricted geography as a science which studies the outer shell of the earth. Richthofen, in his presidential address at Leipzig, in 1883, declared that the geographer's main concern is the outer shell of the earth. This concept was, however, declared illogical by Hettrier.

Gerland viewed the earth as the subject of study in geography but he made a distinction between geography and geophysics. About the question what is the earth, Gerland discarded not only the teleological and anthropocentric views, but also, by implication, the geocentric viewpoint. Looking at the earth from the cosmic point of view, he saw it as but one of many similar units in the universe. From this point of view, the earth body would be included in an earth science, and thus the interior of the earth would appear vastly to outweigh the surface. Since we cannot make a direct observation on the interior of the earth, the geographers should concentrate on that part of the earth which can be observed directly, i.e., external surface, which is the result of the internal and external forces. The very fact that we have the science of geography, but no science of Mars-Geography, reflects that geography is the study of the outer shell of the earth. Our earth is a unique planet. It is the one body in the universe on whose surface temperatures fall within that very narrow range in which different materials exist simultaneously in solid, liquid and gaseous states. Consequently, earth is the body whose surface shows the greatest multiplicity of varying interactions of factors and differential results. Moreover, on it alone, life is found in millions of generic and individual variations. All these variations are, however, near the surface of the earth and, therefore, the unique interest of geographers is in the surface of the earth—the thin shell of outer earth in which are found the unique phenomena of the planet earth.

Lehmann took different point of view about the earth as the sphere of geography. He explained that outside of the learned circles the word 'world' particularly in the many combinations in which it is commonly used 'world tour', 'world map', 'world trade', 'world-wide', 'world war', etc. In all of these cases, there is no uncertainty as to what physical extent of space is included: they, however, do not include the centre of the earth, nor do they include the moon or other 'heavenly' bodies, but simply that outer shell of our planet as high in the atmosphere and as deep in the ground as man can experience.

If earth surface is regarded as the domain of geographic study, it is also the meeting place for all sciences that are concerned with earth (meteorology, climatology, geomorphology, geology, geophysics and environmental studies), history, economics and sociology. But each of these is interested in the earth surface from a different point of view, just as mineralogy and paleontology may study the same rocks from entirely distinct point of view.

It may be concluded from the above discussion that the domain of geography is not the entire planet but only the outer shell of the earth, where we may see the convergence of the lithosphere, hydrosphere, atmosphere and biosphere. Geography, in fact, is concerned to provide accurate, orderly and rational description and interpretation of the variable character of the earth surface.

Geography: An Idiographic or Nomethetic Discipline

Geographical thoughts, right from its inception, have been confronted with the problems of dichotomies and dualism. The idiographic and nomothetic dichotomy has also been the point of discussion among the geographers. According to some scholars, geography is an idiographic science, while others consider it as a nomethetic science. Kant, Hettner and Hartshorne considered geography as an idiographic science. In the words of Hartshorne, geography deals with the spatial differentiations of phenomena. He conceives geography as the study of peculiar phenomena of regions. Since geography deals with peculiar phenomena, generalization in the form of laws is useless, and any prediction in geography is of insignificant value. For Kant, geography is description; for Hartshorne, it is a 'naive science'. If we accept this meaning of science, geography is 'naive description'.

Nomothetic sciences deal with general laws. Systematic geography or general geography comes under this category. Humboldt, Sauer and many of their followers laid stress on systematic geography and the formulation of general laws. James, while discussing the idiographic and nomothetic approaches, emphatically declared that there is no such thing as a 'real region'. The region exists only as an intellectual concept which is useful for a particular purpose.

Despite many upheavals, crises, developments and revolutions in geography, it did not succeed in the formulation of nomothetic laws. It was only after the Second World War that geographers concentrated on theoretical issues and prepared diffusion models, location theory and gravity models and many other geometrical models to explain geographical patterns. This process has passed the take-off stage and it is hoped that new nomothetic (general) laws will be formulated which shall put geography on a sound footing and will bring it greater recognition in sister disciplines.

Geography: A Discipline of Synthesis

It has often been alleged that geography does not have its own area of study. It does not have its own laws and thus encroaches upon the spheres of other disciplines. Some geographers argue that the subject matter of geography is largely shared with other disciplines. Some others are of the opinion that while the subject matter of geography is exclusive, geographers alone study places. There is, however, a consensus on the definition given by Ackerman. Ackerman asserted that the fundamental approach in geography "is the differentiation of the content of space on the earth's surface and the analysis of space within the same universe".

In order to overcome the controversy of the subject matter of geography, Fenneman, in 1919, prepared a model showing the circumference of geography. Fenneman's model has been presented in Figure 1.1.

It may be noted from Figure 1.1 that sciences overlap and that each one of the overlapping zones, which also represent specialized, systematic branch of geography, belongs equally to some other science. For example, the geologists deal with rocks, the botanists with plants, the

Figure 1.1 *The Circumference of Geography (adopted from Fenneman)*

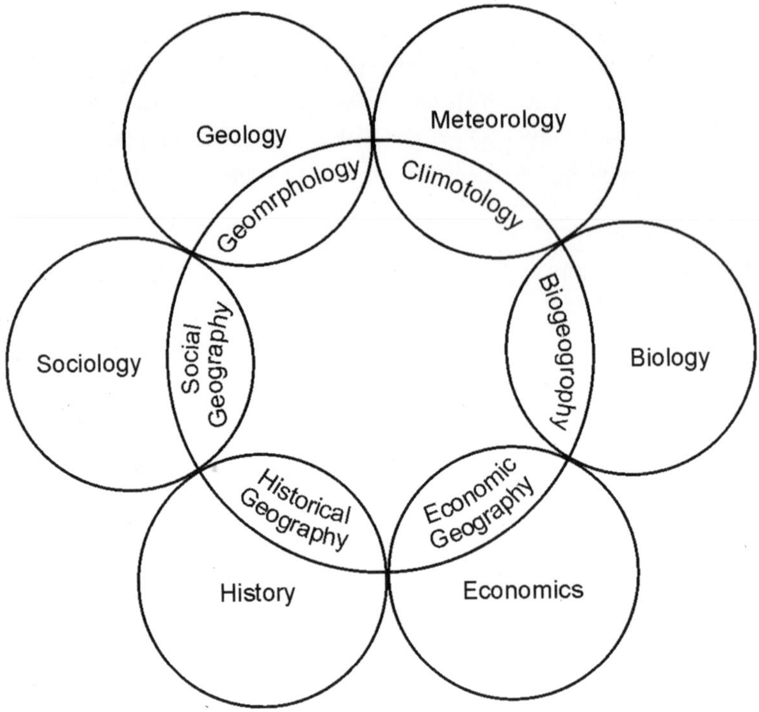

meteorologists with climates, the demographers with population conditions, and the economists with economic conditions. Such studies, unfortunately, fail to consider the overall interaction between phenomena. It is geography which deals with an overall interaction of the biotic and abiotic phenomena in time and space with an anthropocentric approach. The sphere where atmosphere, lithosphere, hydrosphere and biosphere converge is the region where geographers study the man-nature relationship. Thus, geography has its own area of study which may appear to be in the nature of an encroachment upon other disciplines. It is because of this peculiarity that geography is regarded as a *science of synthesis*. The geographers attempt a synthesis of the physical and human phenomena within an area or region. The models and laws formulated in general geography are tested by them in the regional studies.

The concept of geography, as a science of regional synthesis, is not very old. The British and American geographers did rigorous exercises to define the objectives, scope and nature of geography. Before the advent of 'quantitative revolution' geography used to be considered as "short a theory and long a fact". Notwithstanding this definition, formulation of theory in geography is essential. In the words of Burton, theory "provides the sieve through which myriads of facts are sorted, and without it the facts remain a meaningless jumble".

The Geographic Continuum

As already stated, geography is eclectic, integrating a wide range of subject matter from diverse fields; virtually any subject can be examined geographically. Figure 1.2 shows a continuous distribution—a continuum —along which the content of geography is arranged. Disciplines in the physical and life sciences are listed on the continuum's left; those in the cultural human sciences are listed on the right. Specialties within geography that draw from these disciplines are listed at either end of the continuum.

This continuum reflects a basic duality within geography—*physical geography* versus *human cultural geography*. This duality is paralleled in society by the tendency of those who live in developed countries to distance themselves from their supportive life-sustaining environment and to think of themselves as exempt from the physical functions of earth. In contrast, many people in Third World countries live close to nature and are acutely aware of its importance in their daily lives. Regardless of our philosophies toward earth, we all depend on earth's systems, which provide oxygen, water, nutrients, energy, and materials.

Our modern world requires that we shift our study of geographic processes, and perhaps our philosophies, towards the centre of the continuum to attain a more holistic, or complete, perspective. Modern environmental problems call for geographers to examine relations among earth, society and cultures throughout space and time. Critical environmental concerns that are integral to physical geography include:

1. *Global ozone depletion* in the upper atmosphere that allows increasing amounts of ultraviolet radiation to reach earth's surface.
2. *Possible global warming* through human-caused increases of carbon dioxide in the atmosphere.
3. *Worsening air pollution*, particularly in metropolitan areas.
4. *Identification of natural hazards* that threaten society, such as hurricanes, earthquakes, tsunamis, landslides, droughts and floods.
5. *Delineate destruction of earth's forests.*
6. *Increasing losses of plant and animal diversity* as habitats disappear.
7. *Spatial analysis of disasters* such as the 1986 Chornobyl Nuclear Power Station explosion and the 1989 *Exon Valdez* oil tanker catastrophe.
8. *Accounting for natural resources* as environmental assets on national economic balance sheets.

Geography is in a unique position to synthesize the environmental, spatial and human aspects of these concerns.

Figure 1.2 *Distribution of Geographic Content Along a Continuum*

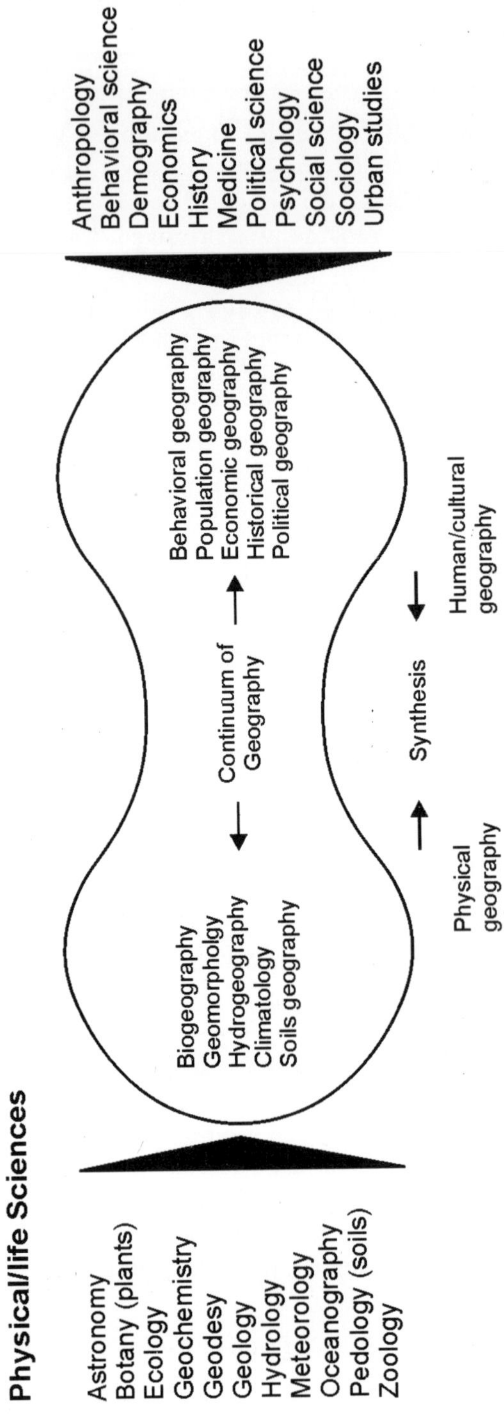

Geography as the Science of Relationship

In the latter part of the 19th century and in the beginning of the 20th century, in Germany, France, Britain, America and Japan, geography was considered as a science of relationship between man and nature (environment). Even, prior to Charles Darwin (1859), geographers tried to understand the complex relationship between terrain, topography, climate, natural vegetation, soils, minerals and the distribution of population and the mode of life of the people. The publication of *Anthropogeography* by Ratzel (1872-1899) created great interest in the study of man and environment relationship. Ratzel, a strong supporter of environmental determinism, declared that "similar locations lead to similar mode of life" and in support of his argument presented the examples of Britain and Japan, then world powers. These two countries, in the opinion of Ratzel, lie in temperate latitudes, have insular locations, and mountainous topography. Their history, political systems and levels of economic development are almost identical.

The geographers and historians of France were also greatly influenced by the works of Ratzel. The French scholars tried to develop a close relationship between history and geography and advocated the philosophical approach of *possibilism.* The possibilism is a belief which asserts that nature provides options, the number of which increases as the knowledge and technology of the cultural group develop. Febvre, a leading French historian, coined the famous phrase: "There are no necessities, but everywhere possibilities." Brunhes published a short outline of geography *La geographie humaine* and defined human geography as the science of man and nature (environment) relationship. In brief, the thrust of the French geographers was that human geography is the study of relationships between active man and his changing environment. The concept of *pays* (region) also dominated the works of French geographers.

In America, the philosophy of geography as the study of relationship was preached by Miss Semple—a committed student of Ratzel. After extensive field studies in Europe, Asia and America, she published her monumental work *Influences of Geographic Environment.*

Hettner published a discussion in 1907 explaining that the natural factors influence human (cultural) phenomena. Barrow, in his presidential address of 1922, stressed that relationships in geography should be studied "from man's adjustment to environment, rather than the reeography verse". Ellsworth Huntington, an American geographer, who authored the *Principles of Human Geography* (1945) also considered geography as the discipline of interrelationship. In his opinion, the supreme achievements of civilization in any region were always bound up with a particular type of climate and variations in climate led to 'pulzation' in the history of culture. He suggested that the best climatic conditions for mental and physical work are found in the countries of North-West

Europe. Robert Mill, Sauer, Herbertson, Taylor, Geddes, Lord Boyd Orr, Gilkh, etc., were also the believers in geography as the discipline of interrelationship.

Geography, as a discipline of relationship, was, however, not been accepted by all the geographers. It was criticized by the positivists and followers of quantitative revolution. In fact, similar locations may not lead to similar history and similar mode of life.

Geography as the Science of Distribution

It was during the period of Varenius and Kant (pre-classical period of modern geography) when some of the experts interested in geography declared this subject as the science of distribution. Carl Sauer was also a strong supporter of "geography as the study of distribution of geographical phenomena". In his opinion, geography is a subject which deals with the distribution of different physical and cultural phenomena. Marthe was a strong supporter of this point of view who described geography as the study of "thegs where of thin". The same point of view was adopted by De Geer.

If the study of distribution is fundamental to the character of geography, it must form a characteristic distinguishing geography from other fields. This necessary conclusion has led to many to suppose that when a botanist determines the areal distribution of a particular plant, or geologist determines the location of volcano, or the sociologist maps the distribution of population in a country, the botanist, geologist or sociologist becomes thereby geographer—or, at least, is working in geography. Each of these scholars is trying to understand the particular type of phenomena that he is trying from the point of view, not of geography, but of his own science. In the opinion of Sauer, "if history may be considered as the science of *when*, geography would be logically the science of the *where*". The historians always seek the answer of *when*, the geographer is always concerned to know *where* his phenomena are.

This approach was, however, criticized by many geographers. One of the main objections is that the geographers may start describing and explaining the phenomena in which they are not very competent; for example, the causes and distribution of diseases, the distributional patterns of electoral behaviour of people in a democracy, etc. Hettner (1905), therefore, criticized this approach of geography as the study of distribution.

At present, most of the geographers will not agree on the concept of geography as the study of distribution, it is extremely difficult to avoid the apparently logical conclusion that because the geographer cannot study any phenomena without knowing where they are, the study of the *where* is itself a part, at least, of our field. In brief, all the systematic sciences that are concerned with phenomena located within the earth surface must also answer the same question, the problem of determining and explaining the *where* of things is not the distinctive function of geography. Geography is thus more than the mere description of the distributional patterns.

Geography as a Chorographic Science

Geography as a chorographic or (chorologic) study has always found its justification in the widespread desire of many people to know that other parts of the world are like, just as history finds ample justification in the common desire to know what happened and what things were like in the past time. Further, since the ordinary person actually knows but partially the area in which he lives, just as he understands but little of what is happening in the world in his life time, 'home geography' is as necessary as current history.

Geography teaches us to know the space we live in, the surface of the earth. "Geography assumes the responsibility for the study of areas", as Sauer says, "because there exists a common curiosity about that subject. The fact that every school child knows that geography provides information about different countries is enough to establish the validity of such a discipline. No other subject deals with the study of area."

Right from the inception of the word 'geography', the word 'area' is inseparable from the discipline of geography. All the ancient geographers described the salient characteristics of the different countries. They described and provided ample details about the mountains, rivers, deserts, lakes, forests and ethnic groups. According to Hartshorne's definition of geography, a geographer is supposed to provide complete, detailed, accurate, sound, explanatory and organized answers to such general questions about the variable phenomena on the changing face of the earth. In other words, geography demands serious attention if it thrives to provide complete, accurate and organized knowledge to satisfy man's curiosity about how things differ in the different parts of the world, just as history in similar fashion strives to satisfy his curiosity about what things were like in the past.

History and Evolution of Geography

The historical developments of geography as a discipline are shrouded in mystery. Right from the periods of Sumerians, Egyptians, Babylonians, Assyrians, Greeks, Romans, Indians, and Arabs, there had been numerous informative and scholarly writings about the various parts of the world. To promote the subject of geography, a number of Geographical Societies were established in the beginning of the early 19th century onwards. Geography, however, established itself as a university discipline rather late.

Early Geographers

Many consider geography as one of the oldest academic disciplines. Its antecedents may be traced to ancient Greek geographers, such as the Thales of Miletus (6th century B.C.), Hecataeus and Herodotus (5th century B.C.). Much of the geographic writings of this period described their physical environment. The Greeks also observed the planet as a whole. Aristotle (4th century B.C.) is most often credited with concerning the "earth as a sphere", a conclusion he arrived by using philosophical reasoning and astronomical observations. His speculations about earth's shape were seconded by Eratosthenes based on a limited, measured arc of longitude. The astronomer Hipparchus of Rhode (22nd century B.C.) developed a location system of lines on the surface of the earth—the forerunner of latitudes and longitudes. Strabo (c 63 B.C.-A.D. 21) wrote geography as a 17-volume work that describes first-hand observations of the world known to him. He is considered the father of regional geography, because he substituted divisions based on natural boundaries (such as mountains, rivers, deserts etc.) for the less permanent and artificially drawn political units. Ptolemy of Alexandria (A.D. 121-51), perhaps the most famous ancient geographer, further systematized the geographical study of areas by classing regions in terms of size and details. He drew a map of the then known world that is remarkably accurate. His eight- volume *Guide to Geography* consisted of a list of all known places tabulated according to latitude and longitude—a system that he devised herself.

After the collapse of the Roman Empire there occurred a Dark Age in Europe. The geographical knowledge developed by the Greeks and Romans was forgotten and numerous wrong concepts were developed in Christian Europe. In their opinion, the world was peopled with fantastic creatures conjured up by imaginations unfetted by facts. Thus, the word 'geography' disappeared from the ordinary vocabulary. In A.D. 326, Helena, the mother of Emperor Constantine, made one of the earliest pilgrimages from Rome to the Holy Land. Silvia of Aquitaine, a Roman lady, was one of the earliest women geographers. She travelled overland to Jerusalem and then on to Egypt, Arabia and Mesopotamia; eventually, she wrote an account of her travels. As the number of pilgrims increased, itineraries were complied to guide them on routes to Jerusalem.

The Greek heritage was preserved in the Arab world. The followers of Prophet Mohammad, between 8th and 13th century, made significant contribution to the field of geography. In 762, the Muslims founded the new city of Baghdad and for more than a century it remained the centre of the intellectual world. With the patronage of Caliph Harun-al-Rashid, an academy, called 'Baitul-Hikma' was established. In this academy, scholars were invited from all over the world. The Arabs made significant contribution to the fields of astronomy, physical geography and geography of different countries. Versatile scholars like Ibne-Hawqal, Al-Masudi, Al-Biruni, Al-Idrisi, Ibn-Battuta and Ibn-Khaldun were great explorers, traders, travellers, cartographers and versatile writers. In the 12th century Al-Idrisi devised a refined system of climatic classification. The great 14th century explorer, Ibn-Battuta, found on his travels through Africa and Asia, concrete evidence that refuted Aristotle's thesis that hot regions of the world would be too warm for human habitation. In the same century, Ibn-Khaldun wrote an important historical geographic treatise. Thus, the explorations, discoveries, scientific inventions and scholarly works of the Arabs provided a new realistic picture of the then known world.

Age of Discoveries

The revival of knowledge in Europe took place in the 15th century. The Europeans started moving through land and sea in different directions to preach Christianity and to obtain metals and minerals from the different parts of the world. From 1391 to 1492 the Spaniards waged continuous warfare with the Muslims, and the last Muslim stronghold (Granada) was taken by the Christians just before Columbus set sail for America.

The first initiative towards wider exploration came from Portugal. The individual leader was Prince Henery, called 'The Navigator', the third son of the Portuguese king. In 1418 Prince Henery established the first research centre of geography at Sagres where he built an observatory. He brought to Sagres scholars (geographers, cartographers, mathematicians, astronomers, and experts in reading of manuscripts in many languages) of all faiths from all around the Mediterranean region. Thus, there was a concomitant renewal of interest in world exploration, mapping and geographical description.

Marco Polo (1254-1324) through the arduous land route reached Chinas. In 1492 Columbus discovered North America, while in 1498 Vasco de Gama reached India through the Cape of Good Hope.

In 1507, German cartographer Martin Waldseemuller (1470-1521) produced a map of the world that clearly indicated both North and South America. Fifteen years later Ferdinand Magellan's party circumnavigated the world. In 1569, Mercator produced the Mercator projection to show the Great Circle routes to the navigators.

The explorations and discoveries had a great impact on the development of geographical concepts. The geocentric concept of Ptolemy was rejected and was replaced by the heliocentric universe (the sun lies in the centre of the solar system, not the earth). Galilio, for the first time, formulated the concept of universal mathematical order, a universe that could be described in terms of mathematical law rather than in the verbal and logical terms of Aristotle. Apart from astronomy, the geomorphology

(Davis), political geography (Ratzel), human geography (Ratzel and Semple) and regional geography (Lablache) as well as the geographical accounts of the different parts of the world and countries were closely influenced by the Great Age of Discoveries.

Geography, as an academic discipline, was rejuvenated by Varenius (1622-1650 A.D.), who in his *Geography Generalis* (1650), established the concept of general (topical geography) and special (regional) geography. Thus, Varenius was the first scholar who laid the foundation of the dichotomy of systematic vs. regional geography. Subsequently, the great philosopher Immanuel Kant (1724-1804), who was mainly interested in physical geography, stressed that there are two different ways of grouping or classifying empirical phenomena for the purpose of studying them systematically. These phenomena may be classified either in accordance with nature, or in accordance with their position in time and space. The former is a logical classification and the latter a physical one. Logical classification lays the foundation of systematic science—the study of animals in zoology, the study of plants in botany, the study of rocks in geology, etc. The physical classification gives the scientific basis for history and geography. History studies the phenomena which follows one after the other in time (chronological science), while geography studies phenomena which lie side by side in space (chorological science). In his opinion, both history and geography are essential sciences, standing alongside the systematic sciences. Without them man cannot understand the world fully.

Founders of Modern Geography

Although the roots of geography, as a field of study, reach back to classical antiquity, its establishment as a modern science was essentially the work of the century from 1750 to 1850. The second half of this period (1800 to 1850), the time of Humboldt and Ritter, is commonly known as the *classical period of geography*. In fact, the great contributions of these scholars in the form of 'Cosmos' and 'Erdkunde' greatly influenced the works of the subsequent scholars. Most of the fundamental concepts of geography, especially pertaining to physical geography, may be found in the writings of German geographers immediately preceding Ritter and Humboldt.

During the 19th century there were renewed efforts to develop geography as a descriptive science (chorological science). Contributions towards that end were made by the German geographer Alexander von Humboldt (1790-1859) and Carl Ritter (1779-1859). These two scholars are often been called as the *founders of modern scientific geography,* although neither of them was trained in geography. Both of them traveled widely, conducted field studies and made original contributions in geomorphology, climatology, oceanography, plant geography and regional geography. Humboldt laid the foundation of physical geography and plant geography, while Ritter is credited with introducing humankind to geographical studies, particularly humanity in relation to environment. Ratzel took Ritter's human geography still further, sub-dividing it into anthropogeography and political geography. Ratzel, a staunch environmental determinist, is best known for his organic theory of state (*Lebensraum*), in which he compares the evolution of state to that of living organism. Ratzel's disciples like Miss Ellen Churchill Semple and Huntington, followed and preached Ratzel's philosophy of environmental determinism, and produced the classical works in the form of Influences of Geographic Environment and Principles of Human Geography respectively.

In USA geography was not pursued by university scholars until the end of the 19th century, and the beginning of the 20th century. About the same time, the USA emerged as a world power. Many of the early American geographers, such as Miss Semple, were the former students of Ratzel.

The 19th century witnessed a fundamental change in the role of universities. The traditional university system (training the students in Greek and Latin classics, on theology and law, or on logic and rhetoric) was replaced by professional fields or disciplines. In fact, the universities assumed the responsibility for the advanced training of younger generations of scholars. This kind of education required courses of advanced study in graduate schools.

The German geographers, especially Ratzel and his disciples, were the strong supporters of *environmental determinism*. Environmental determinism lie in the work of Charles Darwin, whose seminal book *On The Origin of Species* was first published in 1859. His notions regarding evolution were taken up by an American geographer, William Morris Davis, in his cycle of erosion model of landform development. Ratzel applied his notion in the form of Lebensraum, taking state as a living organism.

Environmental determinism implies that environment controls the course of human action. In other words, the belief that variation in human behaviour around the world can be explained by the differences in the natural environment. The essence of the deterministic philosophy is that history, culture, living style, stage of development and the overall gamut of life are exclusively or largely governed by the physical factors of environment, i.e., topography, climate, natural vegetation, soils, etc. The determinists consider man as a passive agent and nature (environment) as an active agent, which decides the destiny of mankind.

The deterministic approach was criticized by Febvre and Vidal de Lablache. The French geographers supported and preached the philosophy of *possibilism*. In the opinion of the possibilists, the physical environment provides the opportunity for a range of human responses and that people have considerable discretion to choose between them. In the opinion of Lablache, "nature is never more than an advisor". Vidal's programme was endorsed by the historian Lucian Febvre in a famous phrase: "there are not necessities but everywhere possibilities, and man as the master of these possibilities is the judge of their use" (Febvre, 1932).

Quantitative Revolution in Geography

Traditionally, geography was considered as chorography (description of the earth surface). In order to put the subject at a scientific footing, after the Second World War, the American and British geographers applied statistical and mathematical techniques in geography. The introduction of these techniques in the subject is known as 'Quantitative Revolution in Geography' (Burton, 1963). Apart from application of statistical techniques, there was also an attempt by geographers at hypothesis testing. It was during this period when I. Burton published his book *The Quantitative Revolution and Theoretical Geography*. Based on the philosophy of positivism, the quantitative revolution in geography was criticized on several counts. It followed the methodology of spatial science, stressing on locational analysis and reduced the subject to space geometry. In reality, the man and environment relationship cannot be established by the mechanistic models designed with the help of statistical techniques.

As a criticisms to quantitative geography, some new approaches were introduced in geography. Geographers applied the behaviouralistic, humanistic, welfare and radical approaches to resolve the man and environmental issues.

The hallmark of an academic discipline, according to one of geography's chroniclers (James, 1972), is an educational organization which provides specialist training in the subject. James dates the beginning of such an organization of geography around 1874, when the first university geography departments were established in Germany. Britain and United States followed a little later, with the main development coming in the 20th century. Before 1874, geography was investigated either by ametures or by scientists trained in other fields (Geology, Geomorphology, Mathematics, Climatology, Botany, History, etc.). With the inauguration of its own specialized institutional training, geography left its classical age and entered what James termed its modern period, which lasted for about eighty years, being superseded after 1945 by what James calls its contemporary period.

The Shifting Viewpoint in the Second Half of the 19th Century

The classical period in the development of geography is considered to be terminated with the death of Humboldt and Ritter in 1859. After the death of Ritter, his students and followers tended to emphasize the 'historical' aspect of geography. This was a drift from the systematic geography to regional geography. Paschel, one of the laudable students of Ritter, extended his interest into political problems. Arnold Guyot (French-Swiss), another follower, if not student of Ritter, who held the first chair of geography in USA at Princeton, also explained the man and environment relationship. Another student of Ritter, Elisee Reclus, stressed on general systematic geography before proceeding to a complete regional survey of the world.

In opposition to Ritter, Humboldt had no university post but he inspired a number of scholars to conduct scientific travellings and expeditions. Following the death of Ritter there was no professor of geography in any German university. The general scientific atmosphere during the latter part of the 19th century was far from receptive to the philosophical concepts of the earlier period, whether those of Ritter or Humboldt.

The shift in geographic work is generally regarded by German geographers to have been due primarily to the work of Peschel. Peschel led geographers to study primarily the morphology of landforms. He attempted to show the influence of landforms on human history. Richthofen and Hettner, Penck (a geologist) and Schmidt also stressed on the importance of geomorphology as the main focus of geography. Similarly, in America, Davis developed the school of geomorphology as the major field of geography.

Human geography, on the other hand, including ethnography, agricultural land use, trade and human migration, was considered in relation to landforms, or was confined to studies in regional geography. This led to the confusion of dualism of physical (now meaning non-human) geography, and a regional, human geography. This division of the subject into the physical geography and human geography pointed logically to two opposite directions of development. It was at this stage that declared geography as a pure natural science.

In opposition to the school of physical geography, Ratzel in his monumental work *Anthropogeography* (1882) suggested geography of man in terms of individuals and ethnic groups and races. He was a strong believer in environmental determinism stating that the man is the product of his environment. The environmental deterministic approach dominated geography for several decades.

Richthofen, another leading German geographer, also attempted to establish the relationship between systematic and regional geography. In his opinion, the actual purpose of systematic geography is to lead to an understanding of the causal relations of phenomena in areas, an understanding which may be expressed in principles than can be applied in the interpretation of individual regions, i.e., chorology (regional geography).

The major distinction in geography, between systematic and regional geography (chorology), is therefore not a difference in materials studied. Richthofen recognized that it is not possible to limit the materials to be studied in geography—but a difference in method of study. Because of the heterogeneous character of these materials (geographical phenomena), they must be studied by classes, in systematic geography. In the chorological studies (regional geography), the causal relation between all the groups of phenomena unites them in a single unitary study. Not bound by any limited concept of science, Richthofen, like Humboldt, saw no objection to a single science considering different kinds of things that exist together and are bound together by causal connections. Hettner, who was influenced by the works of Humboldt, Ritter and Richthofen, tried to declare geography as the outgrowth of historical development.

The French geographers, like Vidal de Lablache, Brunhes and Demangeon, also helped in the development of regional geography.

Contrary to this, the British geographers have shown perhaps the least interest in attempting to determine the nature and scope of their field. It was Mackinder who sought to state the unity of geography in terms of natural regions based primarily on the hydrosphere rather than lithosphere—a conception that no doubt owes much to Herbertson. Bryan's book on *Cultural Geography* shows no acquaintance with the developments of Germany since Ratzel. Stamp, Crowe and Dickinson attempted to examine the philosophy and methodology of German geography.

The American geographers' interest in discussing and arguing over the nature of their field than the studying the problem is well known. Carl Sauer developed the cultural geography as the main focus of human geography. In his opinion, "the modern expression of most ancient geography"; it is appropriate therefore to use the term common in ancient geography, namely, chorology, the science of region.

The historical development of geography has been attempted by scholars like Freeman (1961), Stoddart (1986), who belonged to the school of human geography. Modern geography, in the opinion of Sauer is, "the modern expression of the most ancient geography" (Gaile and Willmott, 1989; Livingstone, 1992; Martin and James, 1992). In the opinion of Freeman, there are six main trends in the geographical literature as under:

1. *The Encyclopaedic Trend*: This trend is associated with the collection of new information about the world, particularly areas little known to Western Europeans and North Americans. Although the Great Age of Discovery was over, by the late 19th century, much of the world had

been visited by Anglo-Saxon explorers, vast tracts remained, notably in Africa, which if not *terra incognita* (unknown land) were almost empty on contemporary maps. Indeed, at the beginning of geography's modern period, much of the North American continent itself remained to be settled by permanent farms.

Thus, it may be said that in the latter half of the 19th century, under the influence of the development of the specialized natural sciences, geography for a time appeared to be changing into a field quite different in character from that which Humboldt and Ritter had inherited and developed. The emphasis on systematic studies appeared to divide geography into two halves, one a natural science, the other a social science, united only in a study of regions that hardly appeared to be a science at all, in the sense in which the time conceived that term. By the end of the period, however, reaction had set in, so that the direction of geography was again essentially that which it had been before. Geography, as Richthofen and Hettner conceived, throughout most of its history, was concerned to study the areas of the earth according to their causally related differences—the science of areal differentiation of the earth surface.

2. *The Educational Trend:* This trend, in the opinion of James, was characterized by the opening of separate geography departments in several of the West European universities. Much work was undertaken to achieve a solid foundation of geographical work in schools, colleges, and universities, involving both proselytizers and architects of curricula and course structures.
3. *The Colonial Trend*: The colonial trend reflects a major environmental preoccupation during the early decades of the modern period, especially in Britain, whose empire was being consolidated and developed into a spatial division of labour based on its metropolitan hub and covering a considerable proportion of the earth surface. Organization of the commercial world required a great deal of information about the various countries.
4. *The Generalizing Trend*: The generalizing trend describes the use to which data collected in the encyclopaedic and colonial traditions were increasingly employed. An academic study involved more than collecting and collating facts: these defined the early paradigms of the disciplines development.
5. *The Political Trend*: The political trend was reflected in the contemporary uses made of geographical expertise. Isaiah Bowman was the chief advisor to Woodrow Wilson at the conference which redrew the map of the world after the First World War, for example, and the

work of geopoliticians such as Haushofer was influential on the *Lebensraum* ideology of Nazi Germany.

6. *The Specialization Trend*: The specialization trend was a reaction to the growth of knowledge and the inability of any one individual to master it all, even within the single discipline of geography. Prior to the modern period, scientists and other academics could be extremely catholic in their interests and expertise, investigation demanded longer and more rigorous training so it became necessary for the individual to specialize, first as a geographer and then within geography, focusing either on one substantive area or on a particular region of the earth's surface.

Some of these trends represent philosophies, some methodologies, and some ideologies with regard to the purpose of academic geography. From them it is possible to identify three paradigms which characterized the modern period in human geography as a whole and of its component parts, such as urban geography.

References

Alber, R.F., 1971, Distance, Intercommunications, and Geography, *Proceedings, Association of American Geographers*, 3, pp. 1-5.

Alber, R.F., Adams, J.S. and Gould, P.R., 1971, *Spatial Organization: The Geographers's View of the World,* Englewood Cliff, NJ: Prentice Hall.

Beazley, C.R., 1897-1906, *The Dawn of Modern Geography*, 3 volumes, Oxford: Dietrich Reimer Verlag.

Grolier, 1993, *Encyclopaedia of Knowledge*, Danbury: Connecticut.

Haggett, P., 1990, *The Geographer's Art*, Oxford: Blackwell.

—, 1983, *Geography a Modern Synthesis*, New York: Harper and Row.

Hartshorne, R., 1939, *The Nature of Geography*, Chicago.

—, 1955, Exceptionalism in Geography Re-examined, *Ann. Ass. Am. Geogr.*, 45, pp. 205-44.

—, 1959, *Perspective on the Nature of Geography*, Chicago, Rand, McNally/London; John Murry.

Harvey, M.E. and B.P. Holly, 1989, *Themes in Geographic Thought*, Jaipur: Rawat Publications.

Holt-Jenesen, A., 1988, *Geography: History and Concepts*, London: Paul Chapman.

Husain, M., *Evolution of Geographical Thought,* 5th ed. Jaipur: Rawat Publications.

James, P.E. and Martin, G.J., 1981, *All Possible Worlds: A History of Geographical Ideas*, New York: John Wiley.

Johnston, R.J., 1991, *Geography and Geographers: Anglo-American Human Geography since 1945,* London: Edward Arnold.

Kuhn, T.S., 1962, *The Structure of Scientific Revolution*, Chicago, University of Chicago Press.

Larkin, R.P. and Peters, G.L., 1993, *Biographical Dictionary of Geography*, Westport, CT: Greenwood Press.

Montefiore, A.G. and Williams, W.M., 1955, Determinism and Possibilism: A Search for Clarification, *Geogrl: Stud.* 2, pp. 1-11.

Sack, R., 1974, Chorology and Spatial Analysis, *Ann. Ass. Am. Geogr.*, 64, pp. 439-52.

Spate, O.H.K., 1957, How Determined is Possibilism? *Geographical Studies*, 4, pp. 3-12.

Stodart, D.R., 1986, *On Geography and its History*, Oxford: Blackwell.

Taylor, G., 1951, *Geography in the Twentieth Century*, New York.

Taylor, P. J., 1985, The Value of Geographical Perspective. In Johnston, R.J. (ed.), *The Future of Geography*, London, Methuen, pp. 92-108.

Tuan, Y.F., 1991, A View of Geography, *Geogrl. Rev.*, 81, pp. 99-107.

Von Paassen, e *Classicalk Tradition of Geography*, Groningen: J.B. Walters.

2

Models and Theories in Geography

Model has been defined as an idealized and structured representation of the real world. A number of definitions of models, have, however, been given by geographers. In the opinion of Skilling (1964), a model is "either a theory, a law, a hypothesis, or a structured idea. Most important, from the geographical point of view, it can also include reasoning about the real world (physical and cultural landscape) by means of relation in space or time. It can be a role, a relation or an equation".

In the opinion of Ackoff, "a model may be regarded as a formal presentation of a theory or law using the tools of logic, set theory and mathematics".

According to Young and Petch, "any device or mechanism which generates a prediction is a model". Accordingly, modelling, like experimentation and observation, is simply an activity which enables theories to be tested and examined critically.

Thus, most of the geographers of the post-Second World War period have widely conceived models as idealized simplified representation of geographical reality (physical and cultural landscapes).

Significance of Geographical Models

Geography can be formally defined as the study of the earth's surface within which the human population lives (Haggett, 1990). The word is derived from the Greek *geo*, the earth, and *graphein*, to describe.

Geography is a discipline which deals with the interpretation of man-nature relationship. The earth—the real document of geographical studies—is, however, quite complex and cannot be comprehended easily. The surface of the earth has great physical and cultural diversity. In geography, we examine location, landforms, climate, natural vegetation, soils and minerals, their spatial distribution and utilization by mankind which leads to the development of cultural landscape. Moreover, geography is a dynamic subject as the geographical phenomena (created by nature and by man) change in space and time. The subject matter of geography, i.e., the complex relationship of man and environment, can be examined and studied scientifically by means of hypotheses, models and theories. This basic aim of all models is to simplify a complex situation and thus render it more amenable to investigations. In fact, models are tools which allow theories to be tested. A more restricted view of models is that they are predictive devices.

Utility of Modelling in Geography

For the last few decades geographers have been increasingly concentrating on the development of models and theories (both in the physical geography and human geography) like that of in the physical, biological and other social sciences (Economics, Sociology, Psychology, etc.) Since models are a device of understanding the vast interacting system comprising all humanity and its natural environment on the surface of the earth, they are of immense importance for understanding the real document of the earth surface. Modelling in geography is therefore done to achieve the following objectives:

1. A model-based approach is often the only possible means for arriving at any kind of quantification or formal measurement of unobserved phenomena. Models help in estimations, forecasts, simulations, interpolations, and generalization of geographical data and information. For example, the future growth and density of population, use of land, intensity of cropping, migration pattern of workforce, industrialization, urbanization and growth of slums may be predicted with the help of models. Models are also very useful in the forecast of weather, change of climate, change in sea level, environmental pollution, soil erosion, forests depletion and evolution of landforms.
2. A model helps in describing, analyzing and simplifying a geographical system. Locational theories of industries, patterns of agricultural land use intensity, trends of population growth, patterns of migration and stages of landforms development can be easily understood and predicted with the help of models.
3. Geographical data about the physical and cultural phenomena are enormous and every passing day geographical data is becoming more and more difficult to understand. Modelling is undertaken for structuring, exploring, organizing and analyzing the obtained data through discriminating pattern and correlation.
4. Alternate models can be used as 'laboratories' for surrogate observation of systems of interest, which cannot be observed directly, and also experimenting and estimating the effects and consequences of possible changes in particular components. These are also useful for generating future scenario of evolution.

5. Models help in improving the understanding of causal mechanism, relationships between micro and macro properties of a system and the environment.
6. Models provide framework within which theoretical statements can be formally presented and their empirical validity then put under scrutiny.
7. Modelling provides linguistic economy to geographers and social scientists who understand their language.
8. Models help in the testing of hypotheses, and building of theories, and laws.

Features of a Model

The salient features of a model are as under:

1. The geographical reality of the earth's surface and man-environment relationship are quite complex. Models are the selective pictures of the world or part of it. In other words, a model does not include all the physical and cultural attributes of a macro or micro region. In fact, model is highly selective attitude to information to geographical information.
2. Models give more prominence to some features and obscure and distort some others.
3. Models contain suggestions for generalization. As stated above, predictions can be made about the real world with the help of model.
4. Models are analogies as they are different from the real world. In other words, models are different from the geographical reality or the real document of the earth.
5. Models tempt us to formulate hypothesis and help in generalization and theory building.
6. Models show only some features of the real world in a more familiar, simplified, observable, accessible, easily formulated or controllable form, from which logical inferences and conclusions can be drawn.
7. Models provide a framework wherein information may be defined, collected, tabulated, processed and arranged.
8. Models help in squeezing out the maximum amount of information from the available geographical data.
9. Models help to explain how a particular phenomenon comes into existence.
10. Models also help to compare some geographical phenomena with the more familiar ones.
11. Models cause a group of phenomena to be visualized and comprehended which otherwise could not be understood because of its magnitude and complexity.
12. Models form the stepping stones to the building of theories and laws—both the general and particular.

Types of Models

Given in the definition at the outset, it will be clear that the term 'model' has been used in a number of ways in different contexts. Owing to the great variety, it is difficult to define even the broad categories of models without ambiguity. One division is between the descriptive and normative. The *descriptive model* is concerned with some stylistic description of reality whereas the *normative model* deals with what might be expected to occur under certain stated or assumed conditions. Descriptive models may be concerned with the organization of empirical information, and also termed as data classificatory (taxonomic), or experimental design models. Contrary to this, normative models involve the use of a more familiar situation as a model for a less familiar one, either in time (historical) or a spatial (geographical) sense and have a strongly predictive connotation.

On the basis of stuff (data) from which they are made, models may be classified into *hardware, physical* or *experimental.* The physical or experimental model may be iconic (idol-shaped) in which the relevant properties of the real world are presented with the same properties with only a change in scale. For example, maps, globes and geological models are physical or experimental models. Models may be an analogue (simulation) having real world properties represented by different properties. Analogue or simulation models are concerned with symbolic assertion or a verbal or mathematical kind in logical terms.

General or Systematic Classification of Models

Given above is the fact that the complexity of geographical landscapes and geographical situations is such that models are of particular importance in the study of geography. A large number of models have been designed, adopted and applied by geographers. A more simple classification of models illustrated with examples has been given in the following pages:

Scale Models

Scale models also known as hardware models are perhaps the easiest type to appreciate as they are the direct reproduction, usually on a smaller scale of reality. Scale models may be either static, like the models of landforms, land surface of a geological model, or dynamic, like a wave tank or river flume. Dynamic models are perhaps more interesting and useful in geographical work. The great advantage that a dynamic model has over reality is that the operative processes can be controlled. This allows each variable to be studied separately. In a wave tank, the effect of material size, wave length and wave steepness on a beach slope can be measured quite accurately if two variables are held constant while the third is varied. If the resultant beach slope angle is plotted against each variable in turn the points obtained in each case may either fall in a nearly straight line indicating a significant relationship, or in a diffused scatter suggesting little or no relationship. Close relationships revealed by the model may not be apparent on a natural beach where the wave variables cannot be controlled.

There are, however, difficulties in applying the results of model studies of this type to a natural situation. One of these is the problem of scale. If wave size and material size are scaled up in the same proportion, the sand of the model would become large cobbles in nature, and these two materials do not react similarly to waves. Again, if sand in nature is scaled down to model size, it would be silt or clay which also responds differently from sand under wave action. Despite such difficulties, scale models have yielded very useful results in many fields of enquiry. The fact that engineers make a scale model before embarking upon any major project such as river improvement, dam construction, canal excavation, landslides, tidal surges, flood forecasts, droughts predictions, or harbour work scheme, demonstrate the value of this type of model.

Scale models are often used by physical geographers, especially by geomorphologists. In fact, geomorphologists have carried out fundamental research with scale models in order to investigate processes that are difficult to observe under natural conditions, such as the processes of weathering, river action, glacial movement, wind erosion, marine processes and erosion by underground water.

Maps

A map is a cartographic representation of specifically chosen spatial information. The information is transmitted through images constructed from geographical symbols. We tend to restrict the term to visual maps, but spatial information may be represented on a computer screen, through Braille, or verbally through spoken description, and these categories of spatial representation may also be described as maps.

Maps are the models that are most familiar to geographers. They are a special type of scale model which become increasingly abstract as the scale becomes smaller. At one of the spectrum is the stereo-pair vertical air photograph which provides virtually a true scale model of the real world. It is, however, static and represents only the area shown at one instance of time. A simple vertical air photograph loses the impression of height but still shows all the visible elements of the landscape virtually true to scale. A large-scale map loses much of the details of the landscape although it can show buildings, fields, forests, roads, railways, canals and other features of this size accurately. As the scale is reduced, the information becomes more symbolic and can no longer be shown true to scale; even more details must be omitted. The map can, however, give an indication of the relief by means of contours, hill-shading and hachurs; this is missing from the simple vertical air photograph. Another advantage, which maps also have over reality, is that they show a very large area simultaneously, so that mutual space relationship can be much more easily appreciated and compared than on the ground. Many maps use symbols to show specific features or distributions such as population density; these are even more abstract and further removed from reality that they represent. A new insight into a familiar area can be given by drawing a diagrammatic map where the scale is not correct for an area, but is adjusted to show population or some other climatic or economic variable to scale. Modifications in area, distance and direction are also needed in maps covering the world or large parts of it. A curved surface cannot be correctly reproduced on a plane or flat piece of paper. In fact, it is impossible to show a three-dimensional earth on a two-dimensional plane or sheet of paper. The earth may be truly represented on a globe, but globes have very little utility for geographical studies.

Simulation and Stochastic Models

Simulation means imitating the behaviour of some situation or process by means of a suitably analogous situation or apparatus, especially for the purpose of study or personal training. Stochastic means randomly determined or that which follows some random probability distribution or pattern, so that its behaviour may be analysed statistically but not predicted precisely.

Simulation and stochastic models have been developed to deal with dynamic situations rather than static state shown on a map. This type of model simulates particular processes by means of random choices, hence the term 'stochastic', one which is connected with chance, occurrence. It can be illustrated by its application to drainage development. Starting with a pattern of grid squares it is assumed that a stream source exists at the centre of certain randomly chosen squares. Random numbers are again used to determine in which of the four possible directions, each stream will flow and a line is drawn to represent its course as far as the centre of the adjacent squares. By repeating the process (with certain reservations that approximate to reality) there emerges a complete drainage pattern. Thus, a conclusion can be reached that the natural drainage pattern has some element of chance about its make up.

Simulation models can also be of use as a means of analyzing a large number of variables, which is a recurring problem in geography. For instance, the development of coastal spit can be shown to depend on a number of distinct processes of wave types. These different processes can be built into a model in such a way that each of them is allocated a specific range of random numbers. Each random number that comes up, results in the operation of the appropriate process in a random order, but in specific proportions. If the simulated spit resembles to real one, then, one can conclude that the processes probably operate in the proportion specific in the model. Once a realistic model has been found, it can then be used to predict future development of the spit provided the processes continue to operate in similar proportions.

Stochastic and simulation models have also been successfully used in the field of human geography to study the spatial diffusion of a variety of phenomena, including the spread of diseases such as malaria, smallpox, fever and AIDS or innovations such as the use of a particular piece of machinery, tractors, fertilizers, pesticides and weedicides. The simulation is made realistic by imposing barriers that can be crossed with a varying degree of difficulty. Random numbers are used to determine the direction of spread and effects of the barriers can then be assessed. The term 'Monte Carlo' is used to describe some of the stochastic models, in which chance alone determines the outcome of each move within the conditions of the model. The Monte Carlo model may be compared with Markov Chain model in which each move is partially determined by the previous move. The Markov Chain is exemplified in the random-water drainage development model described above. Both types have applied in many fields of geographical research.

Mathematical Models

Mathematical models are considered to be more reliable but difficult to construct. They obscure many of the human values, normative questions and also questions pertaining to attitudes, customs and traditions. Yet, they have symbolic assertions of a verbal of mathematical kind in logical terms. For example, suppose, I offer the following arguments:

(1) A is larger than B, and (2) B is larger than C.

(2) Now by virtue of (1) and (2) together, I offer the following theorem or conclusion:

(3) Therefore, A is larger than C.

The logical validity of this conclusion will not change with the change in time. Logically, it had to be true in 3000 B.C., 2000 B.C., 1000 A.D., and it will be true in 2025 A.D., 3000 A.D., 4000 A.D. Thus, the validity of the conclusion does not depend on specific historical period. It is ahistorical.

In the same way, the logical validity of a theory is also spatial. If a theorem is logically valid, it must be locally valid in the United States, Russia, Germany, France, China, India, Pakistan, Japan, Australia etc.

Mathematical models can be further classified according to the degree of probability associated with their prediction into deterministic and stochastic.

Mathematical models represent the equation of specific processes by means of mathematical equations which relate the operative process to the resultant situation. It is necessary, however, to have a sound knowledge of the physical processes concerned, and consequently, this type of model-building has been mainly the work of physicists. For example, a dynamic mathematical model of glacier flow has been constructed by J.F. Nye. He simplifies the basic assumptions as far as possible to make the equations sufficiently simple to solve. Thus, the glacier bed is assumed to have a rectangular cross profiled (U-shaped valley) of uniform size and specific roughness. The ice is assumed to be perfectly plastic in its response to stresses. Then, given certain stresses, the response of the ice can be calculated by means of differential equations. These can predict specific flow patterns and ice profiles for given values of the assumed conditions. The geomorphologists can play this part by measuring flow patterns and glacier dimensions in the field. The closeness with which these approximate to the calculated values is a measure of the success of the mathematical model. If the observed flow pattern agrees closely with the predicted one, then, the model can be used with some confidence to provide values for flow in parts of the glacier that cannot readily be measured in the field, but which are very important in studying the effect of glaciers on the landscape. The speed of basal flow is important in this context. Mathematical models have also advanced our knowledge on how rivers move their load and adjust their beds, and how waves operate on the coast. These models are usually in the form of differential equations largely based on known physical relationship, and it is essential to test their numerical results against observations made under natural conditions or in a scale hardware model. The models are only as successful as the assumptions and simplifications on which they are based are true and valid. They provide a very simplified situation, but one that can be expressed in precise numerical terms and hence is capable suitable mathematical manipulation. For this reason, such models are more suited to problems in physical geography.

There have, however, been somewhat different development of mathematical models in human geography. These are more in the nature of empirical relationships that can be expressed in mathematical terms. An example is the rank-size relationship. This relationship shows that within any class of occurrences there are usually a few large items and many small ones with a fairly regular distribution between them. It has been applied to towns/cities in many parts of the world. In any country or region there are a few large towns but many more small ones, and between the two, a moderate number of medium ones; the relationship is approximately linear on a double logarithmic scale.

Mathematical models have also been developed in economic geography, which is more susceptible to quantitative formulation than other branches of human geography. Such models are often not dynamic in the same way as are differential equations in physical geography, although some deal with flow of goods, etc., from one region to another.

Another mathematical model is linear programming, which is relevant to many situations in economic geography. It is a method of finding the optimum solution to a problem in which several conditions must be fulfiled. A factory will have certain requirements of labour, capital, raw materials, transport and access to markets and each of which determines the conditions that can be expressed as mathematical equations and represented graphically on straight lines. When all the equations have been plotted, they reveal the point of optimum value in terms of location. The procedure provides a definite solution based on the values assigned to the equations. If the values are accurate, then, the optimum solution will be obtained.

Analogue Models

Analogue models differ from those types of models which have already been described. In the analogue models, instead of using limitations of the original or symbols to represent it, the feature being studied is compared with some completely different features by means of an analogy. An analogy model uses a better known situation or process to study a less well- known one. Its value depends on the researcher's ability to recognize the element common to two situations. These elements constitute the positive analogy; the dissimilar or negative analogy and the irrelevant or neutral analogy are ignored.

Reasoning from analogy has long been a part of geographical study. James Hutton, in his major work published in 1795, recognized the similarity between the circulation of blood in the body and the circulation of matter in the growth and decay of landscapes. A similar circulation can also be seen in the hydrological cycle. The Davis' concept of 'normal cycle of erosion' and Ratzel's concept of 'state as a living organism' are some of the important examples in which landforms and states have been compared to living organism. Both these concepts are thus analogies.

The analogy used to further geographical knowledge must be better understood than the feature being investigated. The behaviour of metals under stress has been intensively studied, and this has allowed useful analogies to be drawn between metals and ice. Methods of dealing with one problem can often be transferred by analogy to a completely different situation. The study of kinetic waves has been applied to the movement of vehicles on crowded roads, to the movement of stones and flood waves in rivers, and to the formation of surges at a glacier snout. These very dissimilar problems have a common fact that they are one-dimensional flow phenomena and from this point of view they can be treated with the same technique.

Analogies have also proved fruitful in the study of problems in human geography; for example, those that draw on certain well-established relationships in physics. The gravity model is a good example of this type. It is based on the physical observation that the attractive force between two bodies is proportional to the product of their masses divided by the square of the distance between them. The value for the distance in the model is often squared to approximate more closely to the force of gravity as observed in physics. The attractive force may be considered in terms of transactions between two places. The number of transactions is likely to increase as the size of the places, often measured in terms of population number, increases and as the distance between them decreases. This model presupposes that there is no other force involved to limit the transaction, such as an international or language barrier. Various other physical relationships that used analogue models include the patterns of a magnetic field and the second law of thermodynamics.

Theoretical Models

Theoretical models can be divided into two categories:

(i) The conceptual model provides a theoretical view of a particular problem allowing deductions from the theory to be matched against the real situation. This can be exemplified by the theoretical consideration of the effect of a rising and falling sea level upon the coastal zone if certain specific conditions are filled. It is assumed that wave erosion is the only process operating, that waves can only erode rock to a certain depth of the order of about 13 metres (about 40 feet) and that the waves erode a wave-cut platform to certain gradient below which they cannot operate effectively. It is also assumed that the initial coastal slope is steeper than this gradient. A consideration of the prolonged action of waves eroding under these conditions, with a rising and falling sea level, leads to the conclusion that only with a slowly rising sea level, wave-cut platform of great width can be produced. The theoretical forms of the coastal zone under the various conditions specified can be established and then compared with actual coastal zones. Much more elaborate theoretical models of this conceptual type have been developed in the study of the evolution of slope profiles. These are based on the known assumed effect of different slope processes. A long series of stages of modification can be derived from this type of theoretical model, and these can again be matched with actual slopes.

(ii) The second type of theoretical models are associated with the 'theory', when this is used to denote the overall framework of a whole discipline. The framework must not be too rigid or it will cramp the growing edges of the subject, where the most exciting work is going

Figure 2.1 *Theoretical Framework of Geographical Model*

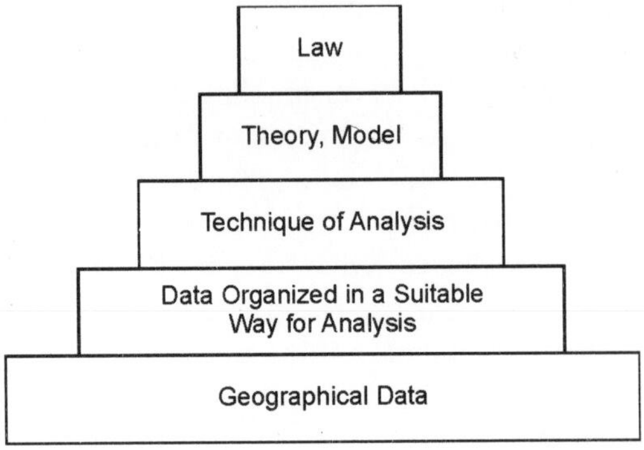

on. The ideal is a flexible framework that can contain a wide variety of geographical endeavour and yet give it coherence and purpose. Models are particularly valuable in this context as they are often common to all branches of the subject and so help to give it unity.

An analogy may help to illustrate the way in which the vast and growing amount of geographical data may be organized within a theoretical framework. Geography may be compared with a five-storied building, each storey being supported by the one below and supporting the one above (Fig. 2.1).

1. The lowest storey is the one which accommodates the data, the raw material of geographical study.
2. The data lead up to the level of model where they are organized in a suitable way of analysis.
3. The techniques of analysis, lying on the next storey, depend on the model adopted for the study.
4. Analysis leads up to the next floor, concerned with the development of theories.
5. The theories in turn lead up to formulation of tendencies and laws. These are located at the top as they are the ultimate aim of geographical methodology.

Critical Evaluation

For understanding and explaining the complex physical and cultural phenomena, models are of great significance. Modelling has, however, been criticized on many counts. Critical views on modelling vary from those which accept modelling but criticize the way in which modelling is done to those which reject modelling as a worthwhile activity in geography.

Those who agree with modelling in geography but do not agree with the way models are being prepared and hold the view that most of the models are prepared badly. The basic aim of the modeller is to present complexity by something simpler. In the exercise of modelling, the modeller may simplify the complexities of geographical realities too much or too little. Over-simplification may mislead students and generate misunderstanding, which may ultimately lead to bad prediction. Under-simplification is of little use in teaching as it does not explain the reality and gives insufficient basis for prediction.

The second objection to modelling is that the modellers may concentrate on the wrong things. Sometimes models may neglect to fulfil the basic criterion of simplifying. They go for the principal component analysis, stepwise regression and Q-analysis. These techniques often produce models more complicated than the original data. Moreover, models may incorporate some of the salient points and omit the others.

There are scholars who do not question the appropriateness of modelling as a generally applicable strategy in geography. There is a group of geographers who consider modelling as a worthwhile activity but hold the view that geographers should not be forced to apply modelling techniques to everything. According to them, modelling is not appropriate in some branches of geography, especially in human geography, regional geography, cultural geography and historical geography. In these modelling strategies have distorted the subject by putting over-emphasis on some topics and under-emphasis on others. By this strategy, generalizations have been made on the basis of few cases and many a time at the expense of specific cases.

Those who outrightly reject modelling in geography say that geography is not a pure physical science, it has a very strong component of human beings and models may not properly adjust and interpret the normative questions like belief, faith, values, emotions, attitudes, desires, aspirations, hopes and fears, and therefore, they cannot be regarded as dependable tools for explaining correctly the geographical reality.

Criticism of modelling may also be based on objections to the generalizations that modelling usually involves. It may be considered futile to construct general models to apply to geographical events, especially where idiosyncratic (regional) human actions and free will are concerned. Or, it may be that the geographer's purpose is to predict or understand specific events and situations, his or her interest may be unique (specific, regional) case for which a general model is thought irrelevant.

Many of the models in geography have also been criticized on the grounds of application of sophisticated mathematical and statistical tools and techniques. Despite the quantitative revolution, few geographers feel comfortable with mathematical symbolism and ideas, and are thus largely unconscious of the generality, clarity and elegance that mathematical modellers appreciate in a good model. Geographers apart, even students, policy makers, clients and public at large, may find mathematical models difficult to understand.

Another criticism is that no model is adequate by itself; any model must be continually subject to assessment, modification and replacement. In Feyerabend's words (1975):

> Knowledge ... is an ever-increasing ocean of mutually incompatible (and perhaps incommensurable) alternatives, each single theory, each fairy tale, each myth that is part of the collection forcing the other into greater circulation and all of them contributing, via this process of competition to the development of consciousness. Nothing is ever settled, no view can ever be omitted from a comprehensive account.

In fact, accountable growth of knowledge is not a well-regulated activity where each generation automatically builds upon the results achieved by earlier workers. It is a process of varying tension in which tranquil periods characterized by steady accretion of knowledge are separated by crisis which can lead to an upheaval within subjects, disciplines and break in continuity.

Model building also demands considerable reliable data. Such reliable data are rarely attainable in the developing and underdeveloped countries. As a matter of fact, any set of data collected in the developing countries have many pitfalls and shortcomings. Any model, theory or law developed on the basis of weak and unreliable data is bound to give only a distorted and faulty picture of the geographical reality. It has also been found that generalizations done with the help of models and structured ideas are bringing exaggerated results leading to wrong predictions.

Most of the models have been developed in the advanced countries of Europe and North America, and theories and models were constructed in these countries on the basis of data collected there. There is certainly a danger that the models developed in Europe and North America may be elevated to general truth, and given the status of universal models. In reality, we do not have universal human, cultural, agricultural, industrial and urban geography. There are different socio-cultural and agro-industrial processes, working in different parts of the world, which result into different cultural landscapes. Owing to these constraints, generalizations made on the basis of models may be misleading and faulty. Moreover, the data used by the western experts are related to a period of about one hundred years. If these models, developed on the basis of data of developed countries, are applied in the developing countries, the results and predictions may be disastrous.

References

Ackoff, R.L., Gupta, S.K. and Minas, J.S., 1962, *Scientific Methods Optimising Research Decisions,* New York.

Apostal, L., 1961, 'Towards the Formal Study of Models in the Non-Formal Sciences' in Freudenthal, H. (ed.), *The Concept and Role of Models in Mathematics, Natural and Social Sciences,* Dordtrecht, Holland, pp. 1-37.

Berry, B.J.L., 1964, Approaches to Regional Analysis: A Synthesis, *Annals of the Association of American Geographers,* 54, pp. 2-11.

—, 'A Paradigm for Modern Geography' in R.J. Chorley (ed.), *The Directions in Geography*, London, pp. 3-21.

Braithwaite, R.B., 1953, *Scientific Explanation*, Cambridge.

Caws, P., 1965, *The Philosophy of Science*, Princeton.

Chorley, R.J. (ed.), 1965, *Models in Human Geography*, London.

Coates, B.E., 'Changing Currents in Geographical Thought', *AAAG*, 26, pp. 1-37.

Chorley, R.J. and Haggett, P., 1969, *Integrated Models in Geography*, Methuen, London.

Cole, J.P. and King, C.A.M., 1968, *Quantitative Geography*, John Wiley.

Cole, J.P. and King, C.A.M., 1968, *Quantitative Geography: Techniques and Theories in Geography*, Wiley & Sons.

Davies, W.K., 1972, *Conceptual Revolution in Geography,* London.

De Geer, S., 1923, 'On the definition, Method and Classification of Geography', *Geografiska Annaler*, pp. 1-37.

Fenneman, N.M., 1919, 'The Circumference of Geography', *AAAG,* 9, pp. 3-11.

Floyd, B., 1971, 'The Quantitative and Model Building Revolution in Geography', *Geography*, Vol. LV, p.40.

Freeman, N.M., 1919, 'The Circumference of Geography', *AAAG*, 9, pp. 3-11.

Haggett, P., 1966, *Locational Analysis in Human Geography*, New York.

Gerasimov, I.P., 1962, *Soviet Geography*, Moscow.

Hartshorne, R. 1959, *Perspectives in the Nature of Geography,* London.

—, 1961, *The Nature of Geography*, Lancaster.

Harvey, D., 1969, *Explanation in Geography*, London, Edward Arnold.

Harvey, M.E., and Holly, B.P., 1983, *Themes in Geographic Thought*, London.

Hesse, M., 1966, *Models and Analogues in Science*, Norte Dame, 111.

Husain, M., 2001, *Human Geography*, Jaipur, Rawat Pub.

—, 2004, *Evolution of Geographical Thought*, Jaipur, Rawat Publications.

Johnston, R.J., 1976, *Geography and Geographers: Anglo-American Human Geography Science,* 1945, New York.

Keat, R. and Urry, J. 1975, *Social Theory and Science*, London.

Klish, J., and Valach, M. 1967, *Cybernetic Modelling*, London.

Minshull, R., 1970, *The Changing Nature of Geography*, London.

Natson, M., 1963, *Philosophy of Social Science*, New York.

Passmore, J., 1966, *A Hundred Years of Philosophy*, New York.

Wooldridge, S.W., and East, W.G., 1951, *The Spirit and Purpose of Geography*, London.

Wrighley, E.A., 1965, 'Changes in the Philosophy of Geography' in Chorley, R.J., and Haggett, P. (eds.), *Frontiers in Geographical Teaching,* London, pp. 3-20.

Yeates, M. 1968, *Introduction to Quantitative Analysis in Economic Geography*, New York.

Zvonkovo, T.V. and Sauskin, Y.G., 1968, 'Problems of Long Term Geographical Predictions', *Soviet Geography*, 9, pp.755-65.

3

System Analysis in Geography

The term 'geography' means, and has meant, different things to different people (geographers) in different times and places. There is no agreed-upon consensus on what is geography? To explain the man and environment relationship (the core of geography), geographers have adopted different methodologies—ranging from deterministic to spatial analysis based on quantitative techniques. Geographers, apart from their methodologies to explain the spatial variations and explanations of geographical phenomena, have borrowed a number of approaches and techniques from social, physical and biological sciences. The system analysis borrowed in geography from the biological sciences is an approach, which attempts to explain systematically the man and environment relationship.

What is System?

Any ordered, interrelated set of materials or items existing separate from the environment, or within a boundary; energy transformations and energy and matter storage and retrieval occur within a system.

Earth's Systems

In our daily life we use systems frequently. For example, 'hydrological system', 'weather system', 'cooling system', 'heating system', 'grading system' etc. In fact, systems surround us. System analysis began with studies of energy and temperature in the 19th century, and was developed further during the Second World War. It has moved to the forefront as a method of understanding the working of many sciences. Geographers use system methodology as an analytical tool at varying levels of complexity.

Figure 3.1 *An open system and a closed system. In either type, inputs are transformed and stored as the system operates.*

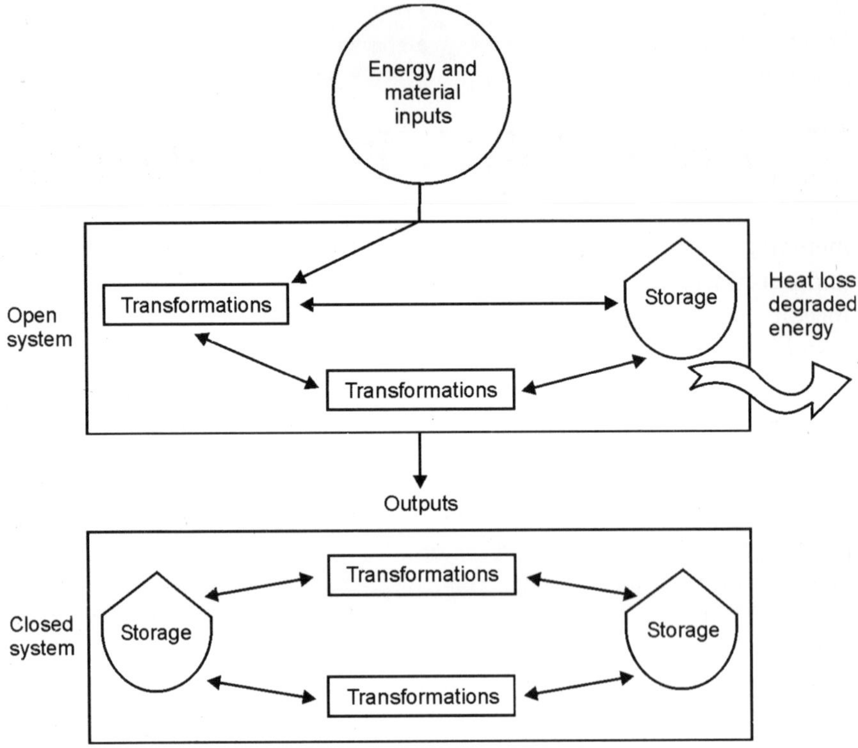

After R.W. Christopherson, 1995, p. 6.

Because the earth receives energy from an outside source—the sun—and re-radiates into space, It operates as an open system. In terms of matter, however, earth is essentially a closed system: closed to any large amounts of incoming matter (Fig. 3.1).

Earth's Energy Equilibrium: An Open System

Most earth systems are dynamic because of the tremendous infusion of radiant energy from thermonuclear reactions deep within the sun. This energy penetrates the outermost edge of the earth's atmosphere and cascades through the terrestrial systems, transforming along the way into various other forms of energy, such as kinetic energy, mechanical energy, or potential energy. Eventually, earth radiates this energy back to the cold vacuum of space in an amount essentially equal to that which entered the system.

Researchers are examining the dynamics of earth's energy equilibrium in more detail than ever before to distinguish natural changes from those produced by human intervention. Tremendous breakthrough in understanding and prediction should occur as general circulation models of earth's energy-atmosphere-water system become even more accurate. By the 1990s, at least four orbiting satellites—part of an Earth Observation System, or EOS—are expected to introduce a new era in the monitoring of earth's open energy system.

Earth's Physical Matter: A Closed System

In terms of physical matter—air, water, and material resources—earth is nearly a closed system. The only exception to this are frequent but tiny meteors, cosmic and meteoric dust, extremely rare comets and asteroids, and out-gassing of water from deep within earth's crust. Since the initial formation of the planet, no significant qualities of matter have entered the system. Just as importantly, no significant quantities have left the system. This is it, earth's physical materials are finite. No matter how numerous and daring the technological organizations of matter become, the physical base is fixed (Fig. 3.1).

Society loses track of its resources in what is essentially a once-through resource stream from virgin materials to the landfill. As earth's people come to understand the fact of finite resources, a serious effort to recycle and to make efficient use of energy and materials can begin. Recycling costs for some materials may be prohibitive, but about half the contents of and average landfill are extractable economically—glass, newspaper, aluminum, and some other metals. The fact that earth is a closed material system makes such recycling efforts inevitable.

Models of Systems

A model is a simplified, idealized representation of part of the real world. Models are designed with varying degrees of abstraction. Physical geographers construct simple system models to demonstrate complex associations in the environment, such as the hydrological cycle that models earth's entire water system. The simplicity of a model makes a system easier to comprehend. It also allows predictions, but such predictions are only as good as the assumptions and accuracy built into the model. Imposing a model too rigidly on a natural setting can lead to misinterpretation, so it is best to view a model for what it is a simplification.

We discuss many system models in this text, including the hydrological cycle, water balance, surface energy budgets, earthquakes and faulting, glacier mass budget, soil profiles, and various ecosystems. Computer-based models are in use to study most natural systems—from climatic change to earth's interior. A general circulation model of atmosphere operated by the Goddar Institute predicted the effect of Mount Pinatubo's ash on the atmosphere and lowering of global temperature.

System Analysis

This is mathematical approach to the modelling of systems which uses techniques developed in control engineering to investigate the behaviour of systems in response to external stimuli or inputs. These models consist of inputs and outputs, such as:

$$Y_{t+n} = SX_t$$

System has been defined differently by different scientists. In the words of James, a system may be defined as "a whole (a person, a state, a culture, a business) which functions as a whole because of the interdependence of its parts". If we accept this definition, then, it can fairly be said that geographers have been using forms of system concepts since the dawn of the subject. However, till the outbreak of the Second World War, no technique had been developed to enable geographers to analyze complex systems.

Geographers deal with complex relationship of living and non-living organisms in an ecosystem. System analysis provides a framework for describing the whole complex and structure of activity. It is, therefore, particularly suited to geographic analysis since geography deals with highly complex multivariate situations. It was because of this advantage that Berry and Chorley suggested system analysis and general system theory as the basic tools for geographic explanation and understanding. In the opinion of Chorley (1962), there is great significance of system analysis in geographical studies. The main advantages of system analysis are:

1. there is need to study systems rather than isolated phenomena;
2. there is need to identify the basic principles governing systems;
3. there is value in arguing from analogies with subject matter; and
4. there is need for general principles to cover various systems.

General System Theory

The general system theory was developed by biologists in the 1920s. It was Ludwig von Bertalanffy who declared that unless we studied an individual organism as a system of multifarious associated parts, we would not really understand the laws which govern the life of that organism. After sometime he realized that this idea could be applied to other non-biological systems, and that these systems had many common characteristics over a range of sciences. It was possible to develop a general system theory, which gave some analytical framework and procedure to all sciences.

A general system is a higher order generalization of a multiplicity of systems which individual sciences have recognized. This is a way of unifying the sciences. This leads to interdisciplinary approach in geographical research. In other words, the general system theory is a theory of general models.

According to Mesarevic's definition, the general system theory is concerned not merely with isomorphism and analogy in system analysis, but also with setting up some general theory for which characteristics of various systems can be deduced. It is thus concerned with the deductive unification of system analytic concept.

The general system theory provides a good framework for relating individual systems and types of systems within a unified hierarchical structure. Such a structure is useful in that it allows us to understand better the relationships that exist between various types of systems; to state categorically the conditions under which one system approximates another, and to identify types of systems that may be useful to us even though we have not yet identified real system to match them.

The general system theory can be understood in the light of a new concept of mathematics and physics known as 'cybernatics' (from the Greek *kybernete*—Helsman). Cybernatics may be defined as the study of regulating and self-regulating mechanisms in nature and technology. A regulatory system follows a programme, a prescribed course of action which produces a predetermined operation. In nature, there is a very large number of self-regulating mechanism, such as the automatic regulation of body temperature. These self-regulating mechanisms follow certain common laws and can be described mathematically in the same way. Whilst regulation is very precise in nature, in human societies it is defective.

Cybernatics places emphasis on the interactions between components rather than making sharp distinctions between cause and effect. Between two components, causal mechanism may work both ways. An impulse, which starts in one part of the system, will work its way back to its origin after being transformed through a range of partial processes in other parts of the system. This cybernetic theory enables us to understand the operation of the general system theory.

The abstract character of a system is emphasized when we realize that a system, if it is to be analyzed, must be 'closed'. An open system interacts and interconnects with the surrounding systems, and therefore, becomes difficult to analyze. All real systems (such as landscapes) are open systems. When we analyze a system, we can only consider a finite number of elements within the system and the reciprocal relations between them. The elements and connections, which are not able to consider in such an analysis, must be disregarded completely. We have to assume that they do not affect the system. In the analysis of a region, we can of course take into count individual influences and single elements, which are not geographically located within the predetermined area or region. The abstract system remains closed all the same because we enclose these elements and relationships in our conceptual model. The system is not synonymous with the model we have made for it, represented by the elements and connections we have closed to enclose or consider.

In other words, we can study only a system after we have delineated and demarcated its boundaries. This presents no mathematical problem since the boundaries draw themselves in so far as some lying outside it, although it is not all that easy to choose those elements in practical geographical research. As an example, Harvey describes a firm which functions within an economy on the basis of a particular set of economic circumstances. When we analyze the internal relations and elements within the firm as a closed system, we must regard these circumstances as unchangeable. To expand the boundaries of the system so as to include the changing social and political relationships in the society of which the firm is a part, may well alter the result of the analysis. So, even in the simple case, the drawing of boundaries creates problems.

By identifying the set of elements which we believe best describe the real system in order to model a real situation. For example, in a large industrial company engaged in several branches of activity, the head office and each of the branch offices form its constituent elements. Mathematically expressed, the system consists of:

$$A= (a_1, a_2, a_3 \ldots\ldots\ldots\ldots a_n)$$

To this expression should be added an element a_{o1} which represents the environment of the system within which the firm operated. We can then infer a new set of elements:

$$B= (a_o, a_n)$$

This includes all the elements in a system plus extra element, which represents the environment. We can then investigate the connections between these elements. Analyzing the company we can see where there are any connections between the branches, and, if so, between which branches. We can observe whether the contacts go both ways and what the contact model implies.

Structure of a System

As stated above, a system is composed of:

1. a set of elements;
2. a set of links; and
3. a set of links between the system and its environment.

Elements of a System

Elements are the basic aspects of every system. From the mathematical point of view, an 'element' is a primitive term that has no definition, like the concept of 'point' in geometry. Nevertheless, the structure of a system is the sum of the elements and the connections between them. Function concerns the flows (exchange relationships) which occupy the connections. Development presents changes in both structure and function which may take place over time.

The definition of an element depends on the scale at which we conceive of the system. For example, the international monetary system may be conceptualized as containing countries as elements; an economy may be thought of as being made up of firms and organizations; organizations themselves may be brought of as system made up of departments; a department may be viewed as a system made of individual people; each person may be regarded as a biological system; and so on. Similarly, a car may be an element in the traffic system, but may also be regarded as

Figure 3.2 *Two different views of inter-system interaction. The upper shows system A and system B interacting as units, with smaller system interactions going on within each system. The lower diagram shows systems A and B interacting at lower levels.*

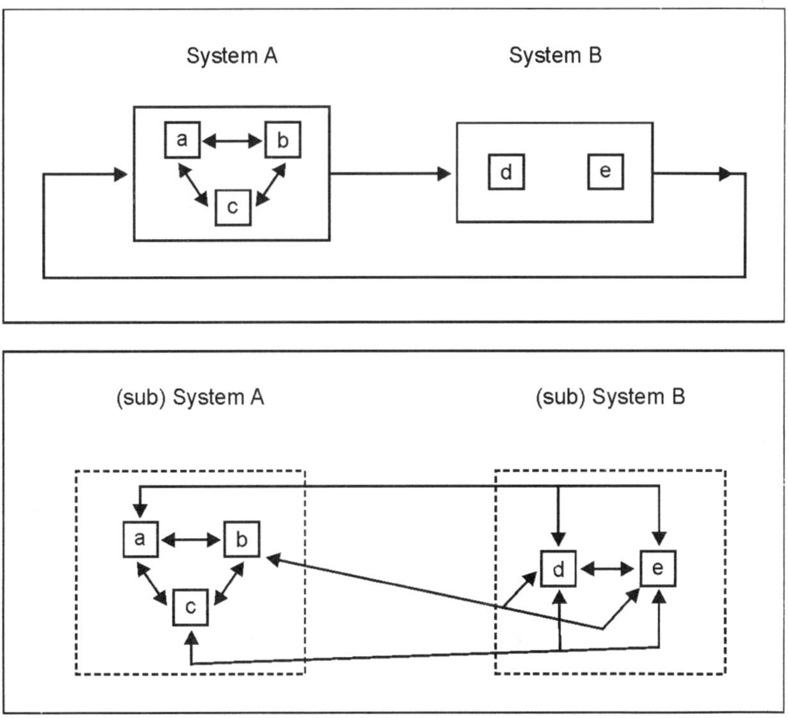

(After Blalock and Blalock, 1959)

continuing a system. Thus, it is clear from the examples that an element depends on the scale at which we conceive of the system.

The concept of element, as a component unit of a system, has been plotted by Blalock and Blalock (1959) (Fig. 3.2). The figure shows two different views of interaction. The upper diagram shows systems A and B interacting as units, with smaller system interactions going on within each system. The lower diagram shows systems A and B interacting at lower levels.

After it has been decided which scale to use, another problem in system building is how to identify the elements. Identification is particularly difficult when we are dealing with phenomena which have continuous distribution, e.g., when precipitation forms an element in a system. Identification is easiest with elements which are clearly separated, such as farms. But, from the point of view of mathematical system theory, an element is a variable. It follows, therefore, that in seeking for a translation of mathematical element in geographical context, we must construct the element as an attribute of some defined individual, rather than as the individual itself.

Links or Relationships

The second component of a system is links (relationships). The links in a system which connects the different elements in it, have been shown in Figure 3.3. These relationships (links) are as under:

(i) Series relation
(ii) Parallel relation
(iii) Feedback relation
(iv) Simple compound relation
(v) Complex compound relation

A brief account of the first three basic relationships has been given as under:

Series Relation

This is the simplest relationship. It is a characteristic of elements connected by an irreversible link. Thus, a_i–a_j (Fig. 3.3 (1)), forms a series relation and it may be observed that this is the characteristic cause and effect relationship with which the traditional science has dealt. This relationship can be explained by citing an example from India. The cultivation and productivity of rice in Punjab depend on irrigation available. Punjab is not irrigated because of the cultivation of rice there. Similarly, cotton is cultivated in the *regur* (black earth soil) of India, the soil of the Deccan plateau is not black because of the cultivation of cotton.

Figure 3.3 *Diagrams showing the kinds of relation that may exist between the elements within system.*

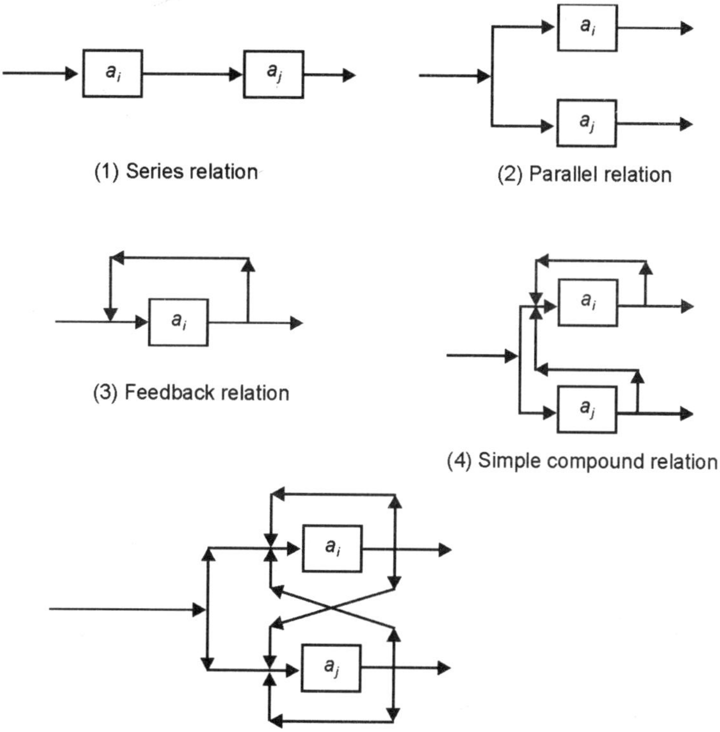

(5) Complex compound relation

(After Klish, J. and M. Valach, 1967)

Parallel Relation

This relationship occurs when two or more elements affect a third element, or inversely when one element affects two or more others. It may be noted from Figure 3.3 (2) above that a_i is affected by some other element a_j. For example the precipitation and temperature variables influence vegetation and vegetation, in turn, influences the amount of precipitation received and the general temperature conditions.

Feedback Relation

A feedback relation is a kind of link that has been newly introduced into analytic structures. It describes a situation in which one element influences itself. For example, the leguminous crops sown in a field enrich nitrogen in the soil and thus the crops get themselves favourably affected (Fig. 3.3 (3)). The feedback relationship may be direct, positive, negative or with no feedback. An interesting example of direct feedback is: A influences B which in turn influences A, or it may be indirect, with the impulse from A returning to it via a chain of other variables. With negative feedback, the system is maintained in a steady state by self-regulating processes termed as *homeostatic* or *morphostatic*. A classic example is provided by the process of competition in space, which leads to progressive reduction in excess profits until the spatial is in equilibrium. But, with feedback, the system is characterized as morphogenetic, changing its characteristics as the effect of B on C leads to further changes in B via D.

It is possible to combine these (series, parallel and feedback) relationships in a number of ways so that two elements may be connected in different ways simultaneously. The links thus form a kind of 'wiring system' connecting the elements in various ways (Fig. 3.3).

Geographical System

A system where one or more of the functionally important variables are spatial may be described as a geographical system. Geographers are primarily interested in studying systems whose most important functional variables are spatial circumstances, such as location, distance, extent, sprawl, density per area unit, etc. In the last few decades, system approach has drawn the attention of geographers. Chorley attempted to formulate structural systems in geomorphology in terms of open systems; Leopold and Langbein used entropy and steady state in the study of fluvial systems; and Berry attempted to provide a basis for the study of "cities as a system within the system of cities" by the use of two concepts of organization and information in spatial form. Recently, Wolderberg and Berry have used the system concept to analyze central place and river patterns, while Curry has attempted to analyze settlements/locations in system framework. Those geographers who focus attention upon spatial organization invariably invoke systems as Haggett's account of 'Locational Analysis in Human Geography' demonstrates.

In geography, static or adaptive systems can be easily constructed. It is difficult to make a geographical system dynamic, for that we must combine time and space in the same model. Space may be expressed in two dimensions by cartographic abstraction. We may be able to present a satisfactory explanation for such system but it is very difficult to handle and analyze it. Lund, however, has successfully analyzed these problems in the time-space model.

Some of these problems can be solved by developing geographical models which may be classified as 'controlled systems' (discussed in the preceding paras). Controlled systems are particularly useful in planning situations when the objective is known and the input in the economic geographic system has been defined. In most of the cases of closed system, one can control input but others are either impossible or too expensive to manipulate. For example, if we wish to maximize agricultural production, we may be in a position to control the input of chemical fertilizers, but control on temperature and precipitation (open system) may not be possible. Partially controlled systems are, therefore, of great interest. Our increased knowledge of environmental conditions leads us to appreciate the extent of the need for the development of planning and control system. Many of the scientists engaged in research into possible future conditions fear that the positive feedback mechanism in the form of technological development and control which have led to an exponential increase in population, industrial production, etc., will, in the long run, result in dramatic crisis of pollution, hunger and shortage of resources. One of the causes of such a crisis would be the long-term suppression of natural negative feedback mechanism.

System analysis may provide a useful systematization of our models, theories of structured ideas, but it is not necessary to refer to system analysis and its mathematical implications when we are doing practical research. For instance, a world map of iron ore production and trade may be described in systematic terms: the elements are the producing and consuming centres, the links or relations are the trading lines, the amount of iron transported along different lines depicts the function, and maps showing these situations at specific time intervals would describe the development of the system. Moreover, the system approach was technically much more demanding, and perhaps for that reason attracted fewer active researchers.

Both system analysis and general system theory have been criticized on the ground that they are intrinsically associated with positivism, i.e., these do not take into consideration the normative questions (cultural values, beliefs, attitudes, desires, aesthetic hopes and fears) and thus do not portray the real picture of geographical reality.

Population Theories

Malthusian Theory of Population

The economist Thomas Robert Malthus (1766-1834) published 'An Essay on the Principle of Population' in 1798. His motive was humanitarian as he was always thinking about the welfare of humanity. His approach to the population problem was empirical, as he based his theory of population on the experience of West European countries. It is to his ideas that subsequent thinking on the economic approach to demography may be traced. While he modified his own position, and subsequent developments of Malthusian theory are complex, his general view was that population tends to increase faster than the means of subsistence, thus absorbing the economic gains, unless controlled by what he termed 'preventive' and 'positive' checks.

Figure 3.4 *Malthusian Growth of Population and Subsistence*

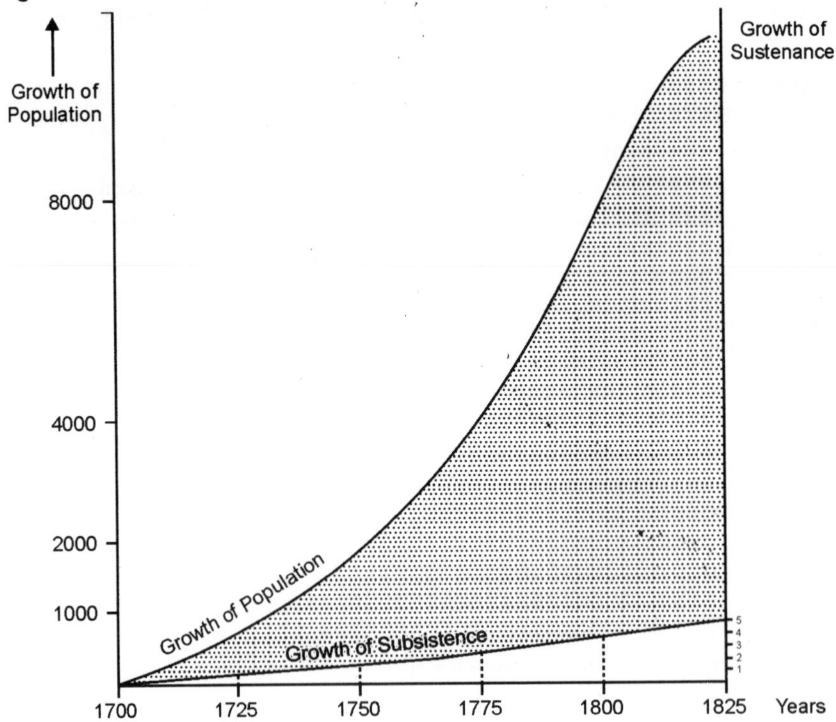

(After Majid Husain, 2006)

While assuming his theory, Malthus assumed two postulates:

1. that food is necessary to the existence of population, and
2. that the passion between the two sexes is necessary and will remain nearly in this present stage.

Having assumed these two postulates, Malthus asserted that the power of population to produce is indefinitely greater than the power in the earth to produce subsistence (Fig. 3.4).

Malthus maintained that population, if unchecked, tended to increase at a geometric rate, i.e., 1, 2, 4, 8, 16, 32, 64, 128, 256..., while subsistence (food) increases at an arithmetic rate, i.e., 1, 2, 3, 4, 5, 6, 7, 8, 9 ...

With the given geometric growth rate of population and arithmetic growth rate of subsistence, in two centuries, the population and sustenance would be in the ratio of 256 to 9; in three centuries 4,096 to 13, and in two thousand years the difference would be almost incalculable.

He also assumed that the population doubles after every twenty-five years. In case, the population growth is not checked by the people, there will be positive checks. His 'positive' checks included wars, diseases, droughts, floods, poverty, and especially lack of food; while his 'preventive' checks included principally 'moral restraint', or the postponement of marriage and 'vice', in which he included adultery, birth control and abortion.

Malthus saw the tension between population and resources as the major cause of the misery for much of the humanity. He was not, however, in favour of contraceptive methods, since their use did not generate the same drive to work hard as would a postponement of marriage. Moreover, he stressed the negative correlation between tension in life and number of children and, in order to induce in the lower classes the self-control and social responsibility he saw in the middle classes. Malthus asserted that the poor should be better paid and educated.

Malthus' views have been challenged in a great variety of ways. He certainly established the thesis that population was growing quickly (doubling after every 25 years) and that people are biological as well as social beings, depending on sexual drive and food. Yet, he has been criticized, for example, for confusing moralist and scientific approaches and for being a poor prophet of events:

1. He emphasized on preventive checks. He did not believe that the preventive checks such as delayed marriages, moral restraints, etc., would so reduce man's rate of reproduction that the positive checks would not operate more or less continuously. He was not convinced that any reforms in man's institution were likely to do much to enhance man's happiness.
2. He did not give sufficient importance to scientific solutions of the existing population problem. He was pessimistic in his conclusion about the growth of population.
3. The desire for children cannot be mixed up with passion and desire for sex. Desire for sex is biological urge while desire for children is social instinct.
4. Population has rarely doubled after every 25 years. The doubling period for population varies from country to country depending upon the stage of its economy, scientific and technological advancements.
5. Malthus believed in the supremacy of positive checks mainly because he was thinking about unnatural checks as delayed marriage.
6. Natural calamities had occurred in underpopulated areas as well. Thus, there was no causal relationship between positive checks and overpopulation.

Marx was one of the most powerful critics of Malthus, asserting that poverty is the result of the unjust social institutions of capitalism rather than of population growth. No country of Europe or North America, except possibly Ireland, conformed to the Malthusian prediction. During the 19th century, economic growth far distanced population growth in Europe, resulting in rising standards of living; and, despite this, birth rates declined first in France and Sweden and then in other parts of the continent.

It has also been argued that Malthus' reactionary views impeded the development of demography as a science.

Despite all these omissions and commissions Malthus was probably the first to appreciate the close relationship between population growth and other demographic changes, on the one hand, and socio-economic changes, on the other. It was Malthus who initiated theory building in population geography and demography, and therefore his work is pioneer and has great social relevance.

Marxian Theory of Population

Karl Heinrich Marx (1818-1883) was a creative thinker who made scientific interpretation of human history. He argued that like scientific and universal laws in physical sciences, there are also universal laws of social sciences. According to Marx, the essence of history is change in modes of production in any society; and this change is always progressive. For example, from the stage of nomads, mankind turned to the stage of settled living. Hunting, fruits and nuts gathering, and sheep rearing gave way to domestication of animals and commune/private agriculture. It was followed by urban culture and rich diversity of vocations. Consequently, population increased, private property emerged, and a stage was reached when private means of production could also be refined in a hundred ways. This led to the formation of economic classes, and there occurred a social hierarchy, codes of behaviour, rules of punishment for crime and misconduct. Subsequently, the Industrial Revolution (1779) gave birth to the capitalism and thereby emerged the classes of haves and have-nots. In Marxian thought, the division of people into classes had the effect of initiation of class struggle as the class interests were mutually hostile and irreconcilable. In brief, there was a class of employees and another of the employed. In familiar Marx terminology this could be called as 'exploiters versus exploited'.

Marx did not postulated a theory of population growth like that of Malthus. While postulating his general theory of communism and scientific interpretation of history, he have some ideas about the growth of population. As a strong believer in dialectical materialism, he considers that society, especially the feudal and capitalist society, has two major economic classes, viz., (i) the rich, and (ii) the poor. The rich were those who possess the modes of production (arable land, orchards and factories), and earned profit by exploiting the labourers and the environment (resources—land, water, vegetation and minerals). The owner of modes of production has the freedom to offer employment and the workers, being too poor, cannot resist the terms the employer offers.

Thus, the workers by contract sell their energy, skill and will to work in return for agreed wages. Under this system, the owner of means of production retains the right to terminate the contract of employment at his sole discretion and will. The employer hires and fires at will. In most of the cases the law is on his side. Where this is not so, law courts can be induced to prove helpful to this cause. Moreover, the owner is the sole judge of the utility and performance of his workers.

This has been the situation ever since man started employing other men for the production of goods in the fields, the shops, the households and the factories. The employer earns profit by exploiting the workers as they are underpaid in the form of wages. This profit was termed as 'surplus value' by Marx. Surplus value was the positive difference between the exchange value and subsistence value (wages). The profit earned by the rich led to the process of capital accumulation. In consequence to this, the poor workers (labour class) try to accumulate labour, the only commodity they possess, through rapid population growth (Marx, 1967: 218). Improvements in the technology of production and rapid growth of population among the poor workers lead to surplus labour and unemployment. This is the main cause of misery, unemployment, under-nourishment, poverty and hunger in the society. The population of the poor people grows at a faster pace as compared to that of the small proportion of population who possess the means of production (the rich and ultra-rich).

Fundamental change in history, according to Marx, is not on account of class struggle in human societies. This is also the crucial key to basic understanding of population growth. Judged in this light, not only the past pattern of growth of population can be interpreted, but even the future trends of population growth can also be reasonably projected.

Marx further advocated that it is the working population which, while affecting the accumulation of capital, also produced the means whereby itself is rendered relatively superfluous, which is turned into a relatively surplus population. This process leads to unemployment and poverty in the society. With the passage of time there occur new innovations in the methods of production, the improved technology under the growing population of poor people unemployed, which sharpen and intensify class struggle. For this reason, class struggle and poverty are seen as the main driving forces of rapid population growth among the poor workers.

Marxian theory of population was developed as a reaction to the capitalistic mode of production and governance. According to Marx, poverty and misery were not natural inevitabilities but unpleasant gifts of capitalism. Misery, poverty, unemployment and fast growth of population can disappear if capitalist form of social order is replaced by communism. He had a convinced opinion that poverty and misery are the outcomes of unemployment and underemployment arising out of inability of the capitalistic social order to provide jobs to all, regardless of the speed with which the population increased.

Marx was not in agreement with Malthus who postulated that population tends to increase faster than that of means of subsistence, thus absorbing all economic gains, unless controlled by what he termed 'preventive' and 'positive' checks.

Marx places the workers (masses) at the centre of main historic processes on which depends the growth of population and the course of history. He also asserted that on the top is, and has been, the ruling class, small in numbers but unchallenged in authority and power, and down below is that large body of masses which lives in utter misery and is content to be exploited and driven like dumb creatures.

In the opinion of Marx, there could be no universal (natural) law of population growth. The growth patterns and dynamics of population change with the change in mode of production. Mode of production is a specific set of forces of production (labour, capital, material, machinery, etc.) patterned into a specific set of relations of production developed as a result of agreement between the employer and the workers.

According to Marx, each mode of production had its own economic and demographic laws. As stated above, the demographic characteristics and growth patterns of population change with the change in mode of production. In other words, growth patterns of population of a nomadic society will be different from that of settled cultivators and the demographic attributes of an industrial urban society will be different from that of an agrarian rural society.

Thus, in the opinion of Marx, economic classes and private property are the principal twin evils, which lead to poverty, unemployment and fast growth of population. The private property hurts the general good of the community at large. In particular, property that qualifies as primary means of production such as mines, land, agricultural fields, orchards, factories, etc., cannot be permitted to be owned privately. These resources should be owned by the community at large, if the rapid growth of population is to be checked and poverty and miseries are to be eliminated from the society. Marx had full faith in the ability of communist method of production to give full employment and good living to all able-bodied workers regardless of the rate of increase in their number.

One basic postulate of Marx about the growth of population was that in the capitalistic form of society the supply of labour increases much faster than the opportunities of employment. This surplus workforce becomes an industrial reserve army of unemployed and underemployed hands. The movement of wage levels is determined by the magnitude of working population among this industrial reserve army. This critical proportion of workers among reserve army is controlled by expansion or contraction of capital. The birth and death rates as well as size of family in turn have inverse correlation with the level of wages, i.e., means of subsistence at the disposal of different categories of workers. Such a class of workers, which is more prone to become a part of reserve army or surplus population, shall have lower wage level and hence high birth and death rates—a situation found in most of the underdeveloped and developing countries of Asia, Africa and Latin America.

Marx, in his theory of population, tried to establish a relationship between capital accumulation, labour demand, surplus population (unemployed and underemployed workers), wage levels, standard of living, poverty and rates of fertility and mortality and growth of population. In his opinion, all these are closely related in the capitalistic form of society in which means of production are owned by small proportion of population and the rest are the workers who are being exploited by the employers.

Marx postulation about population exposes the weakness of the capitalistic form of social order in which individuals have the right to accumulate huge amount of wealth and in which labourers are exploited by giving them low wages, making them poor and helping them to increase population at a faster rate. The Marxian concept about the growth of population has, however, been criticized on the following grounds:

1. The unprecedented growth of population in the world is not due to the lower wages, unemployment and underemployment, but mainly due to the extension in medical amenities and health care services which reduce substantially the death rates without arresting the birth rates.
2. The increase in the population does not automatically lead to the decline in real wages as has been argued by Marx. There are numerous socio-political and economic factors which determine the wage levels and employment opportunities in a society.
3. Marx attempted to establish a positive correlation between the levels of wages and the birth rate of population growth, i.e., higher the wages, lower the birth rate. The faith and religion of the population has, however, not been taken into consideration by Marx in his postulate. There are numerous affluent ethnic and religious groups in both the developed and developing countries in which the rate of birth is significantly high.
4. Marx overemphasized and considered private property as the main cause of all evils including poverty, misery, unemployment and fast growth of population. The social norms, educational level, technological advancements and attitude towards family are all vital determinants of population growth. Thus, private property may not be blamed as the sole factor of population increase.

5. Marx theory of population growth may be relevant to the capitalistic societies and in all probability would not operate in the feudalistic, socialistic and primitive hunting and food-gathering societies.
6. The physical environment (terrain, climate, soil fertility, etc.) as the determinant of birth and death rates has been given adequate weight by Marx in the growth of population. There are scholars like Spencer, who believed that man had no control on his reproduction capacity as the forces of evolution are quite beyond his control.
7. There are demographers like Dumont who believed in the prevalence of social capillarity where the need for smaller families would be generated by the desire to better economic status.

Despite all these criticisms, it can be said that after Malthus, Karl Marx attempted to give a scientific explanation of the growth of population which was based largely on the information available from the capitalist countries. Growth of population in a region is, however, controlled by the physical, socio-cultural and economic conditions. It is because of these factors that a universal model of population growth cannot be postulated. However, the demographic transition theory, developed in the 20th century gives a more logical and scientific explanation of population growth.

Demographic Transition Theory

The demographic transition theory is one of the most important population theories, which is the best documented by the data and statistics of recent demographic history. In its original form, it was put forward by W.S. Thompson (1929) and Frank Notestein (1945). These demographers based their statements and arguments on the trends in fertility and mortality, being experienced in Europe, Anglo-America and Australia.

The theory postulates a particular pattern of demographic change from a high fertility and high mortality to low fertility and low mortality when a society progresses from a largely rural agrarian and illiterate society to a dominantly urban, industrial, literate and modern society. The three very clearly stated hypotheses involved in the process are:

(i) that the decline in mortality comes before the decline in fertility;
(ii) that the fertility eventually declines to match mortality; and
(iii) that socio-economic transformation of a society takes place simultaneously with its demographic transformation.

In the present-day world, as would be true of any point in time, different countries of the world are at different stages of the demographic transition. In the opinion of Trewartha, this is largely due to the dual nature of man. According to him, biologically, man is same everywhere and is engaged in the process of reproduction but culturally man differs from one part of the world to another. It is the cultural diversity of man

Figure 3.5 *Demographic Transition*

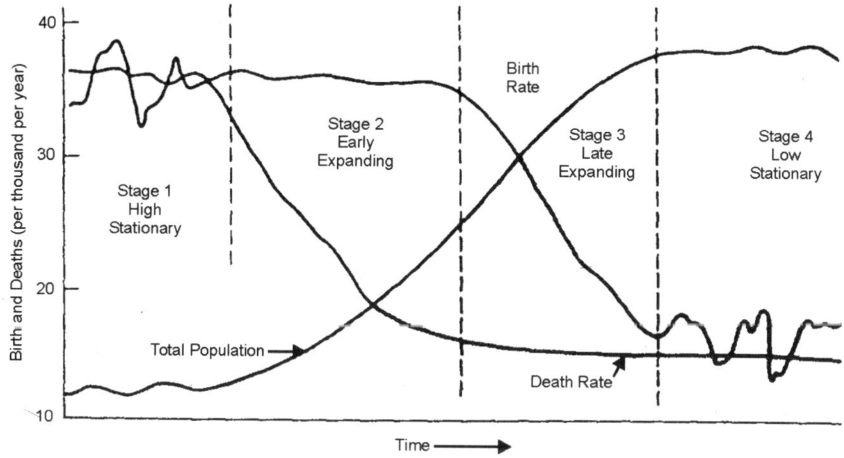

(After Haggett, 1975)

that gives rise to varying fertility patterns in different areas resulting in different stages of demographic transition.

The demographic transition theory is characterized by conspicuous transition stages. The transition from high birth and high death rates to low rates can be divided into following five stages (Figs. 3.5 & 3.6):

Stage I High and fluctuating birth and death rates, and slow population growth.

Stage II High birth rates and declining death rates and rapid population growth.

Figure 3.6 *A Model of Demographic Evolution*

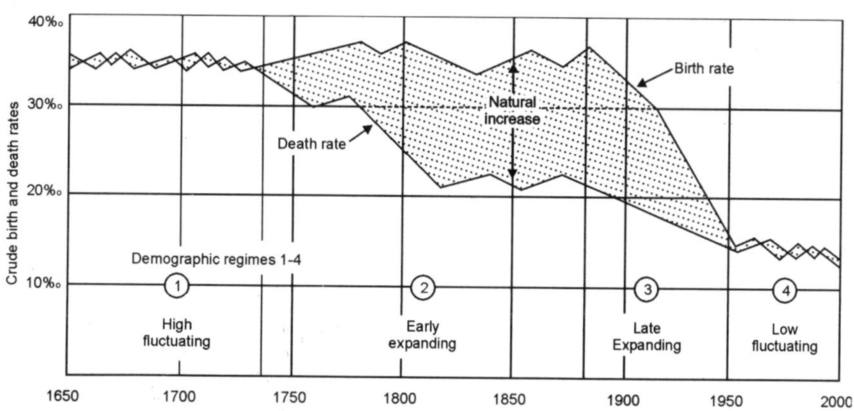

(After R. Knowles, p. 93)

Stage III Declining birth rates and low death rates, and declining rate of population growth.
Stage IV Low birth and death rates, and slow population growth.
Stage V Birth and death rates approximately equal, which in time will result in zero population growth.

In the first stage, the fertility is over 35 per thousand and the mortality rate is also high being more than 35 per thousand. The behaviour of mortality is, however, erratic due to epidemics and variable food supply. This stage, thus, postulates a stable and slowly growing population where the people are engaged in wasteful process of production. This stage mainly occurs in agrarian societies where the population densities are low or moderate, generally productivity level is low, large-sized families are an asset, life expectancy is low, the development of agricultural sector is at its infancy stage, masses are illiterate, technological know-how is lacking and urban development is limited. About two hundred years ago, all the countries of the world were at this stage of demographic transition. At present, it may be difficult to ascertain whether any country in the world still be at this initial stage of demographic transition because the data pertaining to fertility and mortality for such areas would either be lacking or would not be reliable. Moreover, the diffusion of modern technology has also been so fast, particularly in the field of medicine, that it is very difficult to find a solitary example of the country which may still be unaffected by the mortality declines, taking place all over the world. It is in this context that the first stage has been called as the pre-industrial and pre-modern stage.

The second stage of demographic transition is characterized by a high and gradual declining fertility of over 30 per thousand and a sharply reduced mortality rate of over 15 per thousand. In this expanding stage of demographic transition, while the improvements in health and sanitation conditions result in sharp decline in the mortality rates, the fertility maintains a high level, at least in the early second stage. As the second stage prolongs, the fertility also shows signs of gradual decline (Figs. 3.5 & 3.6).

A distinction has often been made between the early second stage with high fertility and declining mortality and the late second stage with slowly declining fertility and sharply declining mortality.

In the second stage, as a whole, the population expands, firstly, at a gradual increasing rate and afterwards at a gradual subsiding rate. In the wake of population explosion associated with the widening gap between the two vital rates (birth rate and death rate), the population of resource mobilization becomes significant. The life expectancy starts improving. The processes of industrialization, urbanization and modernization become prominent. The large families are no longer an asset. Consequently, the fertility undergoes a gradual decline leading to a gradual

Figure 3.7 *World Birth and Death Rates*

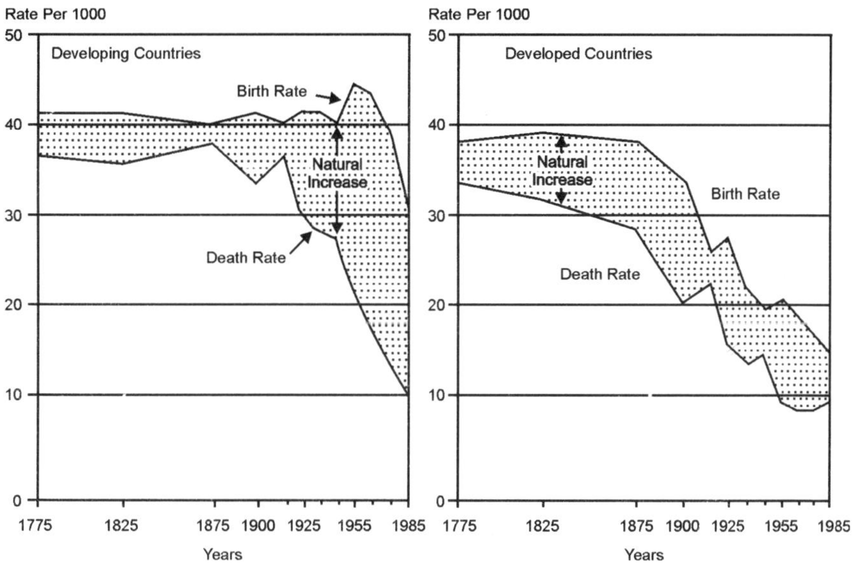

squeeze of rate of natural increase at the tail end of the second stage. Most of developing countries of the world are passing through this explosive stage of demographic transition because of widespread penetration of modern medicines and sanitation measures have drastically reduced their rates whereas their fertility rates are high. The countries like India, Pakistan, Bangladesh, Nepal, Bhutan, Iran, Yemen, Kenya, Indonesia, Malaysia, etc., are in the category of second stage of demographic transition where fertility rates have started declining but since the decline in mortality rates has been sharper, there is tremendous growth of population. The birth and death rates of the developing and the developed countries have been shown in Figure 3.7.

The third, or the late expanding phase, is characterized by a slowing in the growth rate as the death rate stabilized at a slow level and the birth rate declines. This decline is associated with the growth of an urban/ industrial society.

In the last stage of demographic transition, both birth and death rates decline appreciably. The population is either stable or grows only slowly. In this stage, the population is highly industrialized and urbanized. The technical know-how is abundant, the deliberate controls on family size are common, the literacy and education levels are high, and the degree of labour specialization is also very high. Countries of Anglo-America, West European, Australia, New Zealand, Singapore, Hong Kong and Japan are supposed to have reached this stage of demographic transition.

Although the theory of demographic transition has been appreciated widely by the demographers, it has been criticized on many counts. Some of the important criticisms of the theory are as under:

1. It is based upon empirical observations in Europe, Anglo-America and Australia. The socio-cultural and economic conditions of these countries are different from those of the developing countries and therefore what is true in the case of developed societies may not be correct in developing societies.
2. Loschky and Wildcose asserted that the demographic transition theory is neither predictive nor its stages are sequential and inevitable. For example, China has entered the third stage of demographic transition owing to strict one child policy adopted by the Chinese government in the eighties of the last century.
3. The role of man's technical innovations cannot be underrated, particularly in the field of medicine, which can arrest the rate of mortality.

Despite all these omissions and commissions, the demographic transition theory does provide an effective portrayal of the world's demographic history at macro level of generalization. From such a model one should not expect that there would be typical period of time for transition, a typical sequential of fertility and mortality patterns and typical socio-economic fabric of each stage. As an empirical generalization developed on the basis of observing the demographic trend in the West, the transition model does help in understanding the transition process for any country provided the situation context is properly understood. It would be academically unsound to expect all the countries of the world to follow the same sequential pattern as experienced by the European and Anglo-American countries, because the present context is a significant deviation from the 19th century European context.

Theory of Optimum Population

The distribution and density of population in the world are highly uneven and same is the case with natural resources. The pressure of population on the resources and the available technology determine whether the region is overpopulated or underpopulated. In other words, the extent to which resources are utilized and the way in which they are used determine whether an area is underpopulated or overpopulated. A country is said to have an optimum population when the number of people is in balance with the available resources. Optimum conditions can only be maintained if the exploitation of new resources or the development of other forms of employment keeps pace with increases in population. If the population becomes too large, the law of diminishing return begins to operate. Up to certain point an increase in number of people working on the land leads to a marked increase in production. Once the optimum population has been reached, however, a further increase may increase production but at decreasing rate, so that output per capita declines. As more people become dependent on the same resource base, each individual will become poorer.

Contrary to this, if there are not enough people to develop all the available resources, the standard of living of the population may remain lower than it could be, were its full potential realized. Thus, Brazil, with about 14 persons per sq km, may be considered underdeveloped at present. Yet, before it was colonized by the Europeans, it may not have been underpopulated even though there were fewer people, because the range of resources utilized by the indigenous population was also far smaller than the resource has exploited today. Similarly, in terms of present-day technology, Central Asian Republics are considered to be underpopulated. There are vast mineral resources which could support industrial development in Kazakhstan, Uzbekistan, Turkmenistan, and Russian Siberia. But, in the past, Central Asia was inhabited by mainly by pastoralists who knew nothing of modern technology. The resources which they were capable of exploiting were often overstrained. Consequently, during the medieval period, waves of Central Asian people emigrated and invaded the neighbouring and distant regions in search of food, fodder, water and land, and got spread as far as Eastern Europe, South-West Asia, South Asia, and China.

Underpopulation or overpopulation, therefore, must be considered mainly in terms of technology and economic stage of development of the country/region concerned, and the standard by which this is measured in that of industrial countries. An advanced country can be considered as one where agriculture, industry, communications, trade and commerce, and social services are all well-developed and the resources of the country are fully and judiciously utilized.

Central Place Theory

The central place theory, postulated by Walther Christaller (1933, trans. into English in 1966), concerned with the way the settlements evolve and are spaced out. It is a theoretical account of the size and distribution of settlements within an urban system in which marketing (service sector) is the predominant urban function. Christaller based his theory on the data obtained from Bavaria (South Germany).

The central place theory envisages a hierarchical structure of settlements with seven or more levels in a meso-region. This settlement system has a metropolitan city in its apex. In this respect, it is comparable to both the primate city and rank-size settlement systems. The number of settlements at a level increases inversely with its status in the hierarchy. Thus, the central place theory, like the rank-size rule, envisages an inverse relationship between settlement number and their size. Moreover, it is essentially a normative theory, providing a statement of what might appear in certain idealized circumstances.

Christaller based his theory on two concepts:

(i) the range of a good (the maximum distance that a consumer will travel to purchase that good alone); and

(ii) the threshold for a good (the minimum volume of business necessary for an establishment (shop) selling that good alone to be commercially viable.

On the assumption of constant utilities for both consumers and shopkeepers, Christaller then assumed that different goods would have different ranges and thresholds, which would determine both the number of establishments in an area with a given population and their distribution.

Fundamental Concepts

The fundamental concepts on which the central place theory is based are:

(a) the principle of centralization, and

(b) the principle of hierarchy.

These principles govern the structure of all phenomena.

Principle of Centralization

All matter in the universe, whether organic or inorganic, has a core or focal point and a periphery. For example, the sun is located at the centre of the solar system, the neutron at the centre of atom; the human cell has a nucleus with protoplasm around it. Similarly, regions have core and peripheral areas. The village has its focal point in the temple, mosque, church or the *chaupal* (meeting place in a village). Further, the smaller settlements are organized around a larger settlement which serves as a focal point for all the settlements around it. In other words, small towns serve as points for a number of villages and hamlets. It is not, however, necessary that the focal point has to be at the geometric centre of the settlements. Centralization is thus not a geometrical concept; it is a concept concerning the relationship between a central settlement and peripheral settlements.

Principle of Hierarchy

Most phenomena are arranged in a hierarchical fashion. For example, the sun, the planet and the moon form a three-tier hierarchy in the solar system. Humans, apes and monkeys form a hierarchy among the primates.

Man-made hierarchies change through time, according to change in the way of life. Nevertheless, most socio-economic phenomena have a hierarchical organization. Thus, we have everywhere an administrative hierarchy, a military hierarchy, and every institution has its concomitant hierarchical order. Hierarchy manifests itself in terms of territorial division and places. Thus, the state, district and tehsil form an administrative territorial hierarchy, which is complemented by the place hierarchy of state capital, district headquarters and tehsil headquarters.

Thus, the principle of centralization and principle of hierarchy have universal validity.

There are six major derivative concepts ingrained in the central place theory: (i) concept of central place, (ii) concept of complementary area, (iii) concept of central goods and services, (iv) concept of range of goods, (v) concept of threshold, and (vi) concept of centrality.

(i) Concept of Central Place

A central place is a settlement which serves as a focal point for a number of other settlements which are dependent on it. A place where a festival is held once in a year, however, is not a central place, though it may influence a number of settlements. Nor are weekly markets in rural areas central places. These markets come to life once a week, but do not necessarily involve a permanent settlement or establishment such as shops. Moreover, all urban places are not necessarily urban places. A mining town or a manufacturing town is not a central place, unless it has some tertiary activities catering to the needs of the surrounding villages (Fig. 3.8).

Figure 3.8 *Central Place*

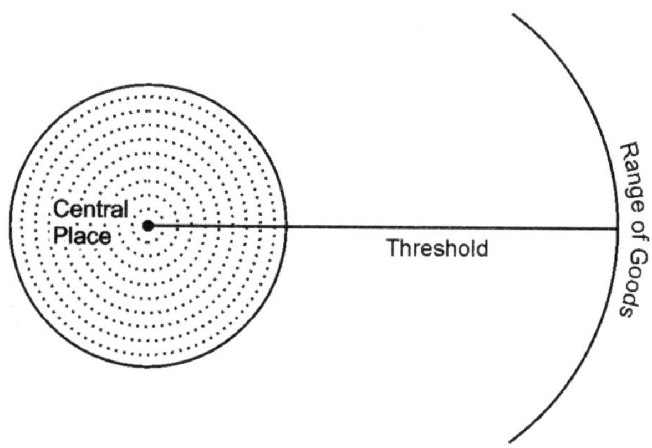

(ii) The Concept of Complementary Area

The complementary area is an adjunct to the concept of central place. It would be larger for bigger and more important central places. In a theoretical sense, it is a collection of settlements of lesser importance.

(iii) Concept of Central Goods and Services

The offering of goods and services in a place constitutes the primary function of a central place. Central place functions are essentially tertiary activities. The goods offered by central places may range from ordinary consumption (cloth, footwear, stationery items, electrical goods and grocery) to items of daily use to expensive equipments, clothing and jewellery. Large central places offer, in addition, goods and services that are less frequently required and which cost more, such as expensive clothing, jewellery, radio, television sets, motor-bikes, cars, tractors, higher educational institutions and costly medical facilities.

(iv) Concept of Range of Goods

While goods and services are offered at central places, the consumer in the complementary area has to travel a certain distance to the central place to obtain these. For basic necessities one would not ordinarily walk/travel more than a kilometre, but for luxury items one might be willing to travel a hundred kilometres or more to a big city (Fig. 3.9)

Figure 3.9 *Range of Goods and Threshold*

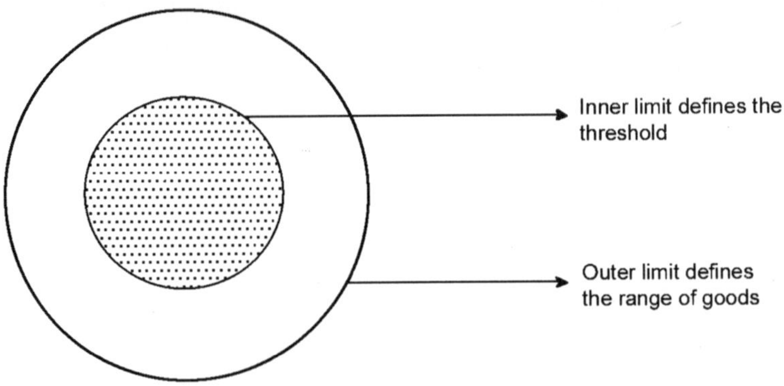

(v) Concept of Threshold

In order that a good be offered in central place, the establishment offering it must have a sufficient number of customers to make it profitable. This minimum number of customers required is the threshold for the goods or service (Figs. 3.10 & 3.11).

(vi) Concept of Centrality

The concept of a central place implies in itself the importance of a centre as a focal point. This importance is generated by offering of goods and services, the size of the complementary area and the number of people coming to the central place.

Figure 3.10 *Distance Decay and Range of Goods-I*

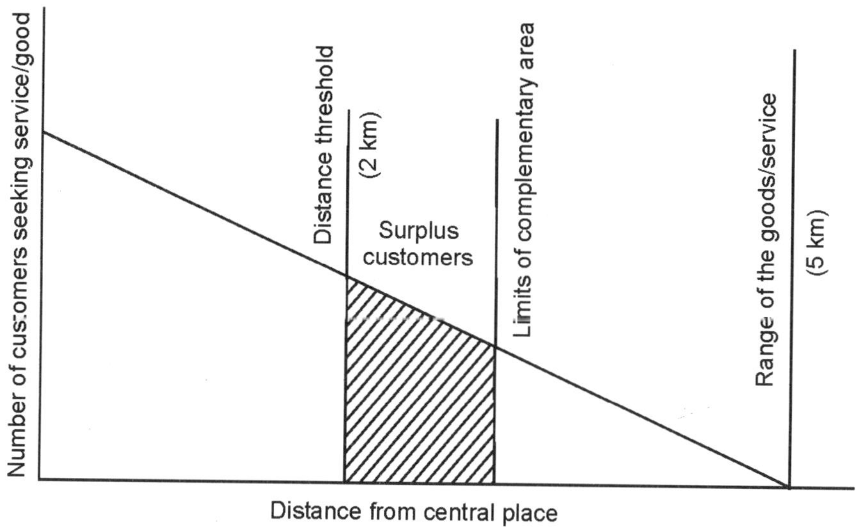

(After R. Ramachandran, 2001: 210)

Figure 3.11 *Distance Decay and Range of Goods-II*

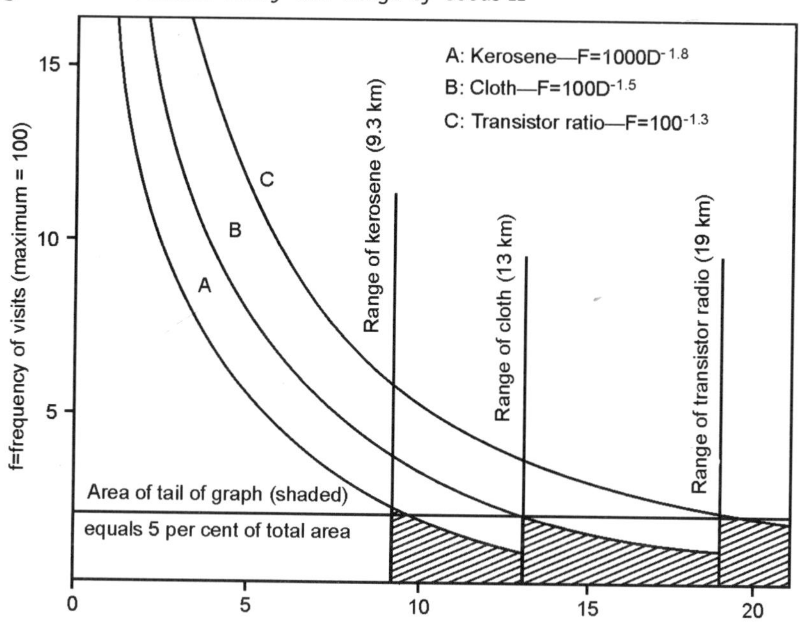

(After R. Ramachandran, 2001: 242)

Assumptions of the Central Place Theory

Christaller assumed that the central place theory will be applicable only if the following geo-ecological and socio-cultural conditions exist:

1. The landscape is an isotropic surface (uniformity in terrain and climatic conditions).
2. The settlements distribution is uniform with respect to resources.
3. The distribution of population and their income level are almost the same.
4. People have almost the same income level.
5. Both the purchaser and seller (shopkeeper) are rational or economic persons who try to optimize their profit.
6. The entire landscape of settlements is to be fully served by a set of uniformally spaced central places, wherein no part of the area is left unserved by a centre and no area is served by more than one centre. This is possible in a hexagonal arrangement of central places. Because the circular complementary areas would either result in leaving several areas unserved by any centre, or they would overlap, thus generating wasteful competition between centres. The different packing patterns of settlements with varying levels of efficiency have been shown in Figure 3.12.

Figure 3.12 *Basic hamlet grid and the development of hexagonal service areas: (a) hamlets spaced 4 km apart on an equilateral triangle grid; (0 pattern of hamlet and village service areas governed by a k=3 hierarchy.*

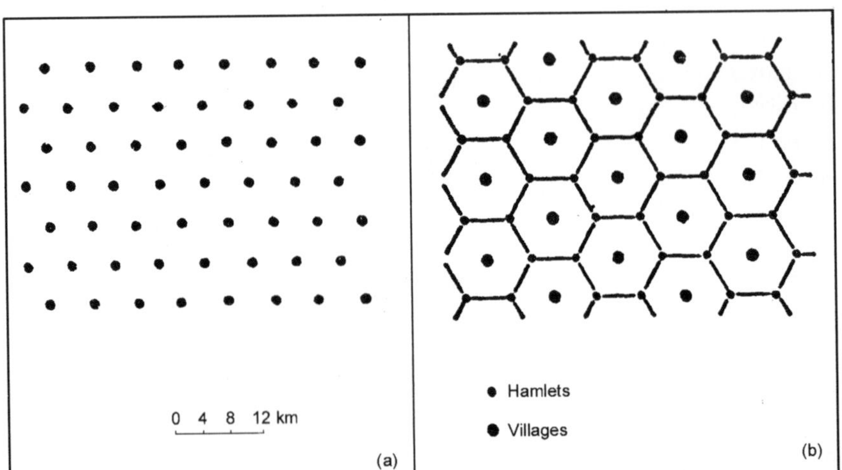

(After R. Knowles, et al. 2002: 227)

The Pattern of Nesting

For each hierarchical level of central places, there is a corresponding network of hexagonal complementary areas. The lower order hexagonal networks nest within the higher order network, since they all cover the same territory. The geometrical pattern of nesting is again determined by certain optimally considerations.

Christaller suggested three ways in which hierarchical spatial structure would be organized (Fig. 3.13). These are as under:

1. marketing principle (k=3)
2. transportation principle (k=4), and
3. administrative principle (k=7).

Figure 3.13 *Different Packing Patterns of Settlements with Varying Levels of Efficiency*

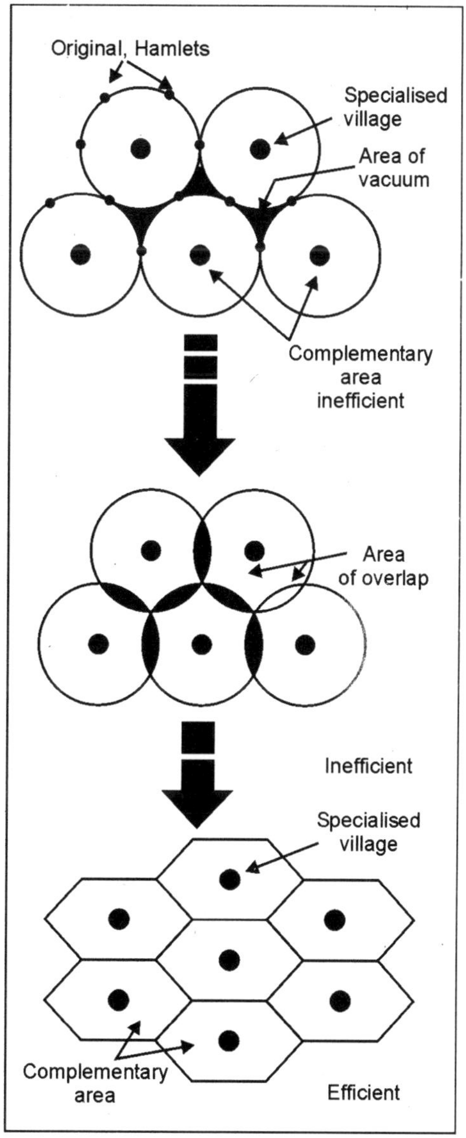

1. Marketing Principle (k=3 Network)

In his k=3 (or marketing principle), the number of settlements at each level of the hierarchy, below the second, is three times the number at the next highest. Thus, with one at the highest, seventh, order, the numbers are 1, 2, 6, 18, 54, 162 and 486. Figure 3.14-A illustrates this arrangement with a three-level hierarchy only.

2. Transport Principle (k=4 Network)

In Christaller's k=4 model, the criterion is to minimize the length of roads to join all adjacent pairs of central places. As shown in Figure 3.14-B, this

Figure 3.14 *Patterns of Nesting of Complementary Areas*

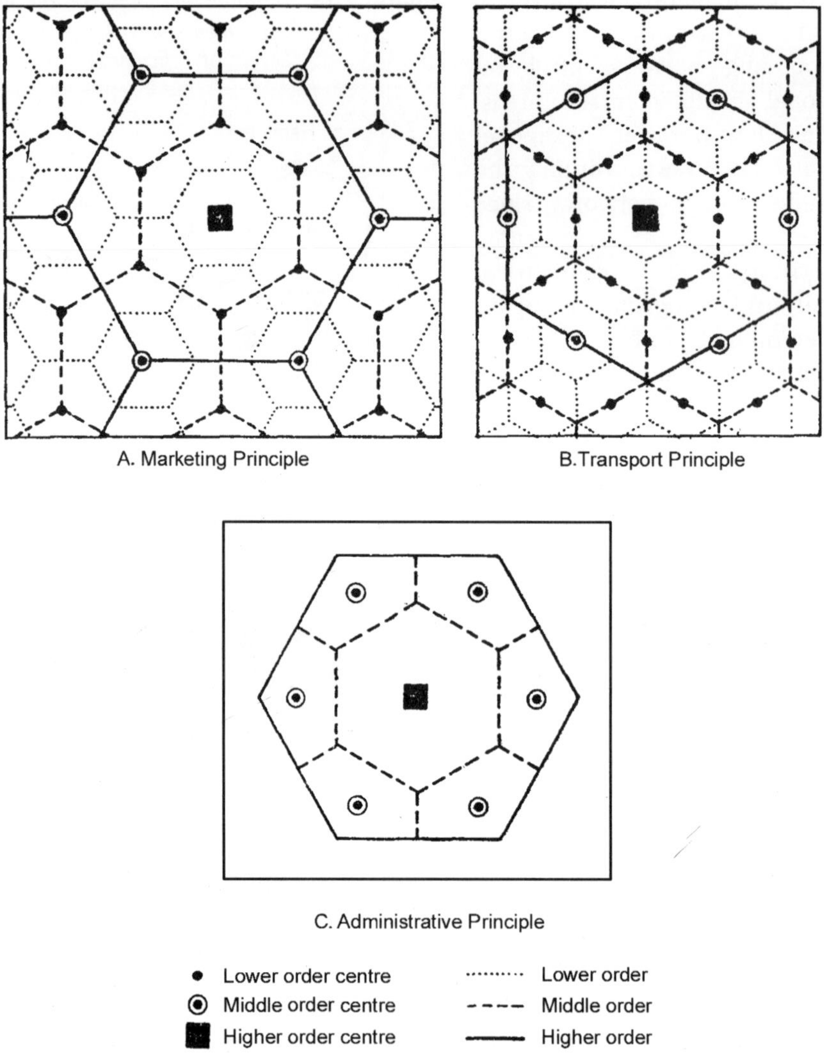

(After R. Ramachandran, 2001: 215)

involves each settlement being placed centrally on each side of a hexagon, and each is on the boundary of two rather than three hinterlands. The number of settlements is thus greater than in k=3 model (in the ratio 1, 2, 8, 32, 128, 512 and 2,048).

3. Administrative Principle (k=7 Network)

The administrative efficiency and necessity demands that complementary area boundary of a lower order centre lies entirely within the

Figure 3.15 *Central Place Theory. The size and spacing of central places, plus their hinterlands (left) and routes (right), according to three variants of Christaller's model; (a) the market principle, which minimizes the number of centres; (b) the transport principle, which minimizes the road length; and (c) the administrative principle, in which hinterlands are nested hierarchically.*

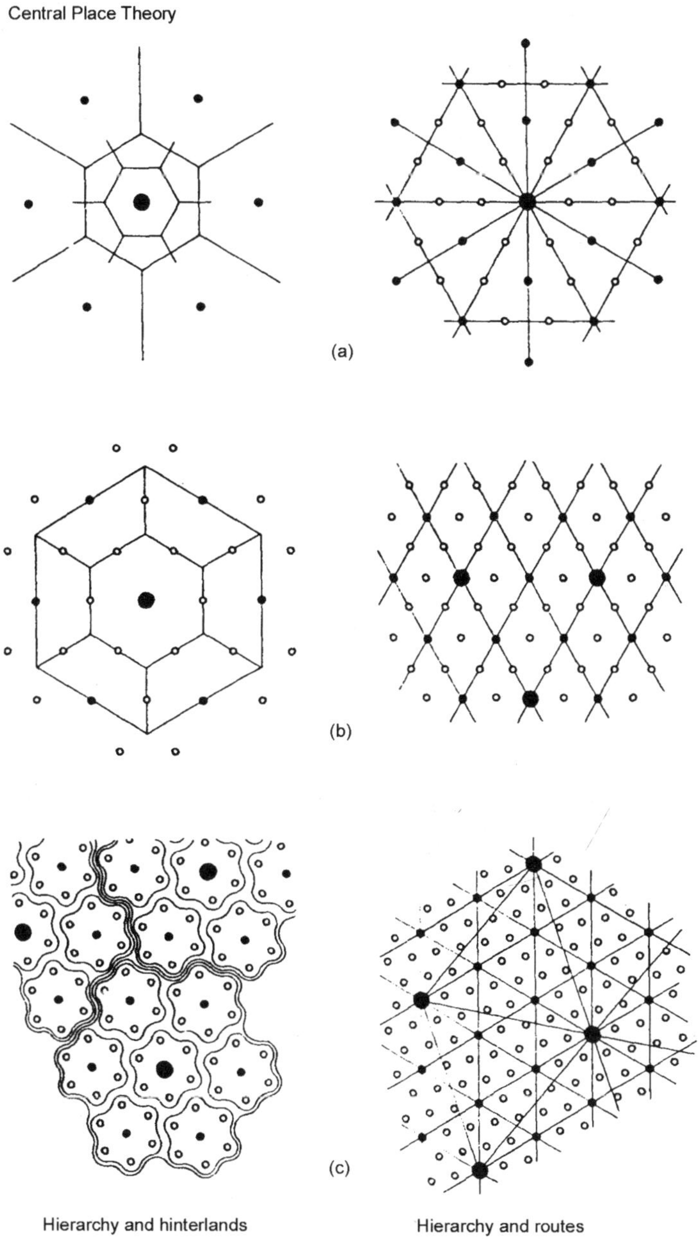

(After R.J. Johnston, 1991, *Dictionary of Human Geography*, p. 58)

complementary area boundary of a higher order. For example, in India, all district boundaries. In this network, each lower order hinterland nested exclusively within the hinterland of one higher order central place only, producing a sequence of 1, 6, 42, 294, 2,058, 14,406, and 1,00,842 settlements (Fig. 3.14-C). In the case of districts in India k=7 hierarchy is evident, although the geometric fit is not exact and needs modification (Fig. 3.15).

Central Place System in India

India has a six-level hierarchy of settlements from the point of view of administration. This hierarchy of places and areas has practical relevance and meaning in the everyday life of people in India. Thus, we have national capital, state capitals, district headquarters, tehsil towns, and finally the block development centres and gram panchayat centres (a gram panchayat may include one or more revenue villages).

Criticism

The central place theory has been criticized on the following grounds:

1. An isotropic surface is ideal, rarely found in the real world.
2. The behaviour of consumers and sellers (establishment owners) is not always rational. Many a time man's behaviour is satisfying.
3. The hexagonal pattern of central places is rarely found in real life.
4. The theory is suitable primarily for agricultural regions, because modern factors, such as the impact of transportation, communications, internet shopping, mall shopping and mobile communication have reduced the distance between the consumers and the sellers. In fact, the world has been reduced to the level of a village.
5. The fixed value of 'k' shows poor approximation with reality.
6. The model does not take into consideration the location of industries which invariably modify the centrality of settlements and threshold of goods. No real world settlement system can be expected to conform to all the assumptions of the Christaller's model.
7. Christaller's theory, in particular, was used many times as the basis for searches for hierarchically structured, hexagonal arrangements of central places, both in rural areas and within cities, and for the distance-minimizing patterns of consumer shopping choice.

Losch Location Model

The first attempt to develop a general theory of location with the major emphasis on demand was made by A. Losch in 1940. It was translated into English in 1954. This sought to explain the size and shape of market areas within which a location would command the largest revenue. Losch's study was based on the urban places of entire Germany in which he took into consideration 150 commodities and services, including the manufacturing and industries.

Losch simplified the world to a flat uniform plain, held supply constant, and assumed that demand for produce decreased with an increase in price. If this price increase was the result of an increase in transport costs, then, the demand would decrease with distance from a production centre. Losch treated each function having a separate range, threshold and hexagonal hinterland. Wherever feasible, establishments in these functions were clustered in the same settlements, but in his system all central places with a function with a particular threshold need not contain examples of all functions with smaller thresholds. The result is a more complicated pattern of central places than in Christaller's simpler representation—and one which can incorporate other urban functions (such as manufacturing and industries).

In Figure 3.16 demand decreases as price increases. Assuming that the price increase is the result of transport costs, by rotating the demand curve around the production point P, the shape of the market area is seen to be circular.

The competition increases as other producers develop on the plain, the market areas become hexagonal to avoid overlap and the exclusion of some areas. They also become smaller as large profits are competed away. Each product will have a different market area (threshold) depending upon the relative importance of transport costs in its price, and different patterns of production centre, some of these patterns are rotated round a

Figure 3.16 *Comparative Figures of Christaller and Losch*

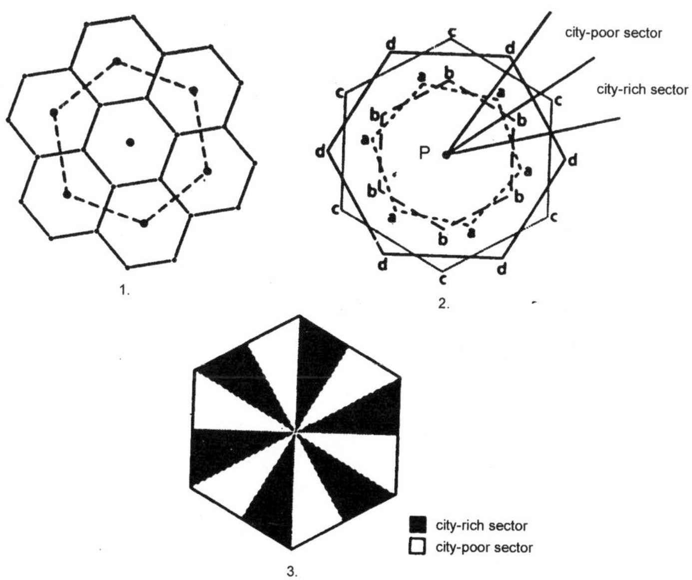

common production centre, some of these patterns will coincide, forming points of maximum demand which should develop as concentrations of industry (Fig. 3.16).

Losch's model is less restrictive than Christaller's, because rather than bundle functions into orders he treated each as having a separate range, threshold and hexagonal hinterland. Wherever feasible, establishments in these functions were clustered in the same settlements, but in this system all central places with a function of a particular threshold need not contain examples of all functions with smaller thresholds. The result is more complicated pattern of central places than in Christaller's theory predicted a stepped hierarchical form to the city-size distribution (i.e., all places in an order had the same population), while Losch's theory is consistent with a continuous distribution of population sizes.

The salient features of Losch's model are as under:

1. Christaller's model is based on the study of Boveria province of South Germany, while Losch sampling is from all over Germany.
2. Christaller's concept is based on the consumers and seller's behaviour, transport and administrative services, while Losch took into consideration 150 commodities and services and developed over 40 networks.
3. Losch's model is less restrictive than that of Christaller's. Christaller treated each function as having a separate range, threshold and hexagonal hinterland. Losch also used hexagonal service areas, but allowed various hexagonal systems and levels to coexist and overlap, thus overlapping the k=3, k=4, and k=7 hierarchies (Fig. 3.17). Whatever feasible, establishments in these functions are clustered in the same settlements.

Figure 3.17 *Losch's Model of Region and Settlements*

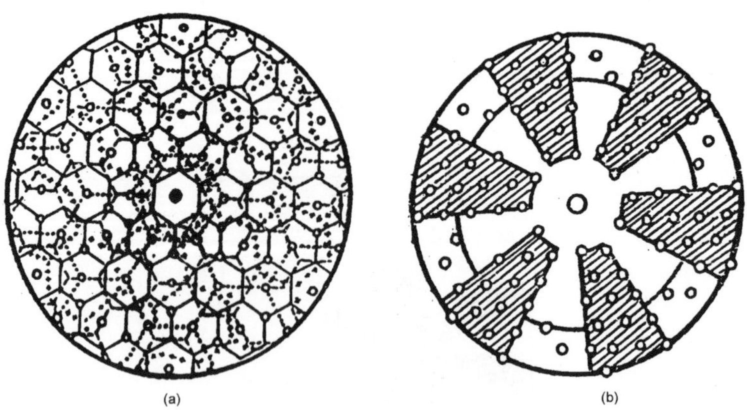

(From August Losch, *The Economics of Location*, New Haven: Yale University Press, 1967, p. 125, Figs. 28 and 29)

4. In Christaller's model the higher order central places have all the functions of the lower order, while in the case of Losch it is not necessary.
5. Losch's model incorporates other urban functions also such as manufacturing and industries.
6. Christaller's theory predicted a stepped hierarchical form of city-size distribution, i.e., all places in an order had the same population, while Losch's model is more closely in line with the theoretical results of the rank-size rule.
7. Losch divided the thresholds into the sectors of the rich and the poor.
8. Both the models stress on the distance minimizing pattern of consumer's shopping choice.
9. The theories of Christaller and Losch are central to much of the early work.
10. Undertaken during geography's Quantitative Revolution in the 1960.

Both the models have been of great importance in the regional planning both in the developed and the developing countries.

Losch's theory has been criticized for its abstract nature and its failure to take into consideration the problems arising from the locational interdependence of manufacturing factories and plants. Losch is also criticized for overemphasizing demand. Moreover, this model presents a very complicated picture of the services and manufacturing thresholds. The overlapping of networks creates the problems of cartographic representation and often difficult to understand.

von Thunen Model of Agricultural Land Use

This is a model for analyzing agricultural location patterns. It is a pioneer work developed by a Prussian (German), Johann Heinrich von Thunen. He developed a crop theory and crop intensity theory in 1826. In the construction of his model, von Thunen used the data of his agricultural estate—Macklenburg, near Rostock in Germany. He operated and managed this estate for forty years until his death. Most of the data used in explaining his theory was obtained by him through practical experience, including detailed cost accounting of his estate. He attempted to construct a theoretical model of land use pattern, giving a particular arrangement of towns and villages in a situation, experienced in Mecklenburg. The main objective of von Thunen's analysis was to show how and why agricultural land use varies with the distance from a market. He had two basic postulates:

1. The intensity of production of a particular crop declines with the distance from the market. Intensity of production in a measure of the amount of inputs per unit area of land; for example, the greater the amount of money, labour and fertilizers, etc., that are used, the greater the intensity of agricultural production.

2. The type of land use will vary with distance from the market.

Assumptions

The model of von Thunen is applicable only if the following conditions exist:

1. An 'isolate estate' (no links with the rest of the world) with one city at the centre of an agricultural area.
2. The city is the sole market for the surplus production from agricultural area, and the agricultural area is the sole supplier of commodities to the city.
3. In the market of the city, all farmers receive the same price for a particular crop.
4. This agricultural area is an isotropic surface, having homogeneity in terrain, topography and climatic conditions.
5. The farmers are rational who behave as economic men and conduct agriculture to optimize their profit.
6. The farmers have a full knowledge of the needs of the market.
7. There is only one form of transport (in those days horse-carts and boats). The transport network in the region—both roads and navigable canal—was poor and the cost of transport increased at a constant rate.
8. The town existed in the centre of the agricultural land which had no counter magnets (market) in the vicinity.

von Thunen's model examines the location of several crops in relation to the market. The model is constructed around the concept of *rent*, and assumes that all farmers will produce on their land that commodity which yields the 'highest' rent, and so will maximize their profit. This net profit is termed as land rent (L), and its value is controlled by the production costs per unit of the commodity (a) its market price (p), the yield per unit of land (E), and the distance from the market to the production point (k). These are combined to give (Dunn, 1954):

$$L = E(p - a) - Ek.$$

In this context, land rent differs from economic rent in that it takes no account of opportunity costs and so ignores what might be earned from alternative uses of the land.

In the simplest applications of the model, which have attracted geographers for several decades, transport costs are taken as the only variable (i.e., prices, production costs and yields are held constant for each commodity): rates per unit of distance vary, being highest, for bulky and/or perishable items such as timber and dairy products. Thus, land rent can be shown to decline away from the market point according to the 'Distance Decay Model', but the rate of decline differs by commodity. The

Figure 3.18 *Two Crops Types Locational Rent and Distance from Market*

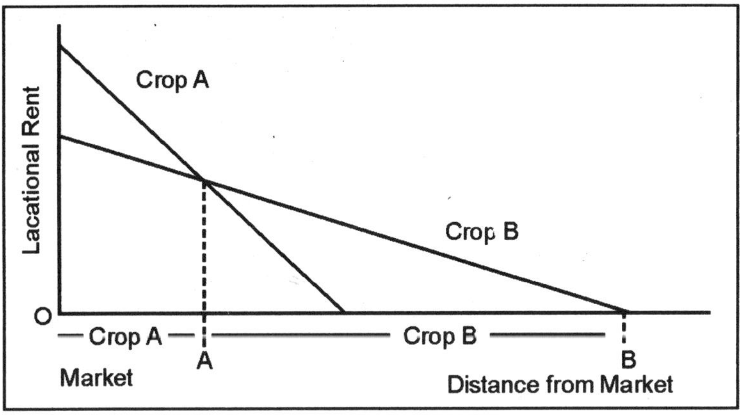

Figure 3.19 *Economic rent for two crops: wheat and potatoes. Taking into account transport costs, potatoes are most profitably grown between 0 and A, and wheat between A and B.*

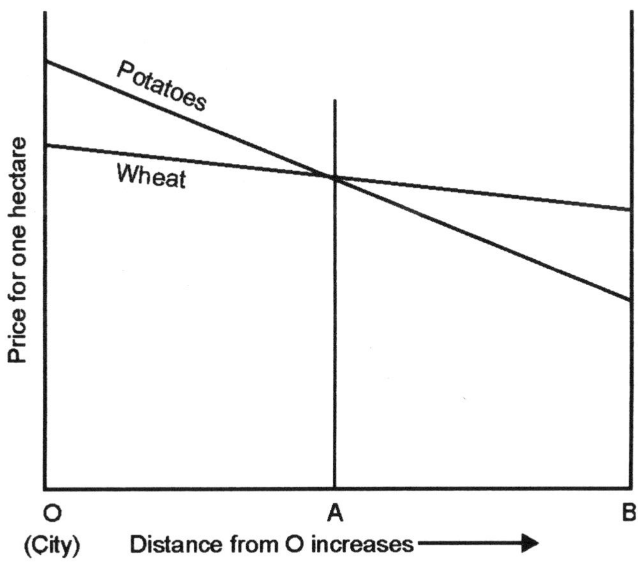

(After R. Knowles, et al., 2002: 141)

market price for each commodity determines the maximum rent, at the market point, and the result is differing land rent slopes, as shown for three commodities in the Figure 3.18 and 3.19. If those distance decay curves are translated from one- to two-dimensional space, they suggest a zonal pattern of land use organization around the market point.

The economic rent considering three crops (horticulture, forest products, and intensive arable cereals) has been plotted in Figure 3.20.

This figure shows a simplified model of concentric zones. It may be observed that zone I in which the economic rent is high is devoted to horticulture (fruits and vegetables), while zone II is devoted to forest products (fuelwood, etc.) as the transportation cost of fuelwood is high. The zone III is that of intensive arable land devoted to cereal crops.

Figure 3.20 *Concentric Land Use Zones*

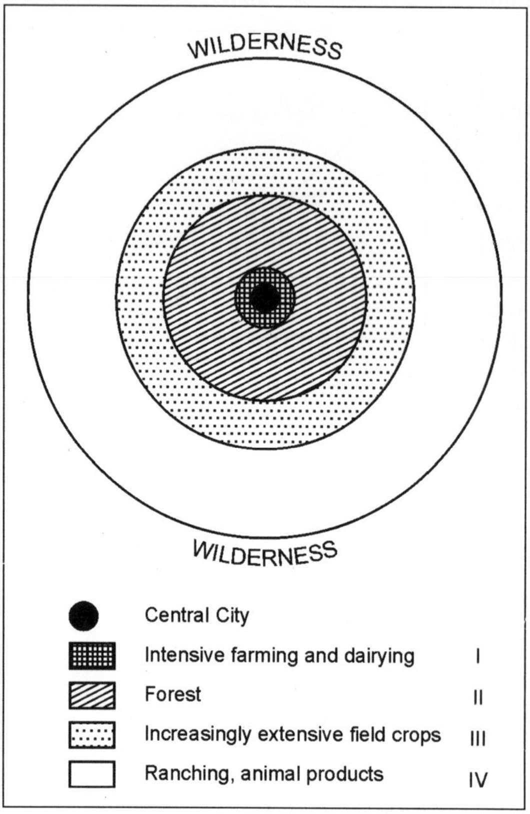

(After H.J. de lij, et al., 2000)

On the basis of these assumptions, von Thunen constructed a land use model, having a number of concentric zones around each town (market). The perishable, bulky and/or heavy products, according to this model, would be produced in the belts nearer the town. The more distant belts would specialize in products which were less in weight and volume but fetched higher price in market as they would afford to bear relatively higher transportation costs. The final model was conceived having specialized agricultural enterprises and crop-livestock combination. Each belt, according to von Thunen, specializes in the production of those agricultural commodities to which it was best suited (Fig. 3.21).

The concentric zones of von Thunen's model are:

Zone I Market Gardening and Milk Production
Zone II Firewood and Lumber Production
Zone III Crop Farming without Fallow
Zone V Crop Farming, Fallow and Pastures
Zone VI Three Field System
Zone VI Livestock Farming

Zone I: The land use model, propounded by von Thunen, suggests that the production of fresh milk (in the context of Europe), vegetables, fruits and flowers was concentrated in the zone I, nearest to the city (market) because of the perishable nature of these products. In this zone the fertility of land was maintained by means of manuring and, if necessary, additional manure was brought from the city and transported to the fields located at short distances from the market (Fig. 3.21).

Figure 3.21 *von Thunen's Agricultural Zones*

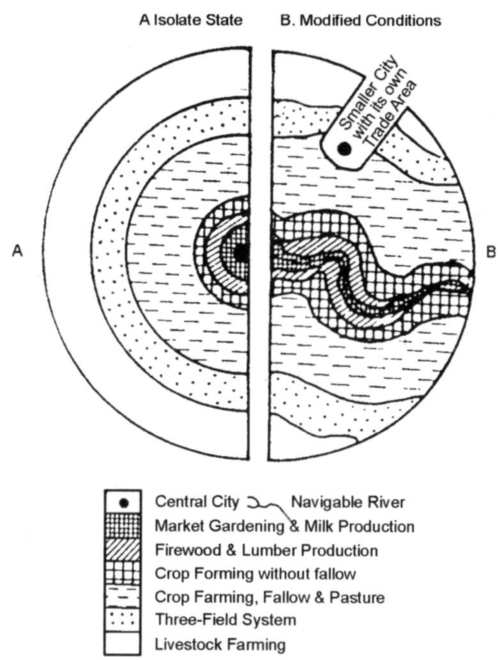

Source: A Grotewold, *Economic Geography*, Vol 35, 1959

Zone II: This zone was used for production of wood, a bulky product in great demand in the city as a fuel in the early part of the 19th century. He showed, on the basis of his empirical data, that forestry yielded a higher location rent, since its bulkiness meant relatively higher transport cost.

Zone III: Beyond the forest belt were three zones where rye was an important market product. The difference between the zones was in the intensity of cultivation. As the distance from the market increased, so the intensity of rye production decreased with a consequent reduction in yields. There was no fallowing and manuring to maintain soil fertility.

Zone IV: The land being less productive, agriculture in the fourth zone is less intensive. This zone is situated away from the market. Farmers in this zone used a seven-year crop rotation. In the crop rotation, there was one year of rye, one year of barley, one year of oats, three years of pastures, and one year fallow. The products sent to the market from this zone were rye, butter, cheese, and occasionally, live animals to be slaughtered in the city. These products do not perish so quickly as fresh milk, fruits and vegetables, and could, therefore, be produced at a considerably greater distance from the market.

Zone V: In the fifth zone, rye is an important cereal crop. In this zone, the farmers followed a three-field system, whereby one-third of the land was used for field crops (rye), another one-third for pastures, and the rest left fallow.

Zone VI: The farthest of all zones was devoted to livestock farming (ranching). Because of distance to market, rye did not produce so high a rent as the production of butter, cheese or live animals (ranching). Consequently, only cattle were kept for slaughtering in the outer most zone.

Finally, von Thunen incorporated two examples of modifying factors in his classic model (Fig. 3.21). The effect can clearly be seen of a navigable river where transport was more speedy and cost only one-tenth as much as on land, together with the effect of a smaller city acting as a competing market centre. Even, the inclusion of only two modifications produces as a much more complex land use pattern. When all the simplifying assumptions are relaxed, as in reality, a complex land use pattern would be expected.

A more complicated picture under commercial agriculture has been presented in Figure 3.22. In the modified model of von Thunen, the influence of fertility, subsidiary town, information, etc., has been incorporated. The isorent lines at an interval of 200 have also been plotted. It will be further seen from Figure 3.22 that the concentric zones of von Thunen get modified under the impact of various physical, socio-economic and cultural factors. The influence of availability of information also substantially modifies the concentric zones of agricultural land use.

Figure 3.22 *The von Thunen Zones: Under Several Complicating Conditions*

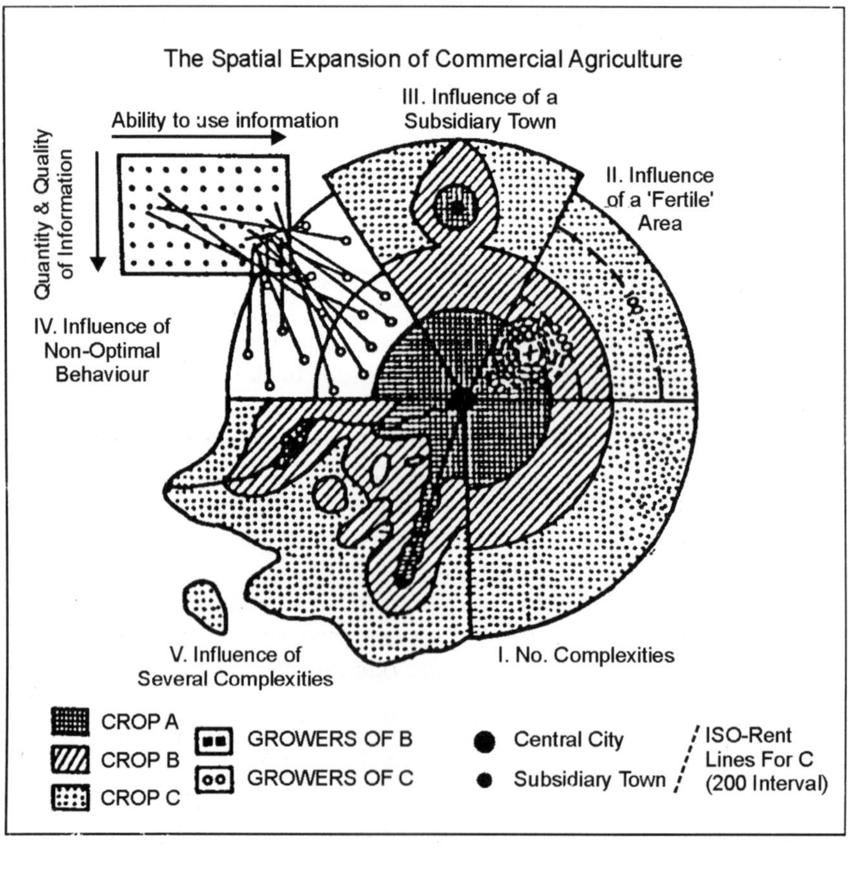

Europe: Intensity of Agriculture

It is a great tribute to von Thunen that his work still commands the attention of geographers. The economic-geographic landscape of Europe has changed enormously since his time, but geographers still compare present-day patterns of economic activity to the Thunen model. Such a comparison was made by Samuel van Valkenburg and Colbert Held, whose map of Europe's agricultural intensity reveals a striking ring-like concentricity (Fig. 3.23). The overriding spatial change since von Thunen's days is the improvement in transportation technology, which permitted the isolated state to expand from micro to macro scale. Thus the model is no longer centred on a single city but on the vast urbanized area lining the southern coast of North Sea, which now commands a continent-wide Thunen agricultural system.

The map of Europe produced by Valkenburg shows the productivity of land per acre on the basis of six very widely raised crops, viz., wheat, barley, oats, potatoes, sugar-beet, and hay. For each crop, the average yield per acre for Europe as a whole is taken as an index of 100 and the specific yield in each country is calculated accordingly. The outstanding fact is that the Netherlands and Belgium lead in agricultural intensity, with Denmark, North Germany, Switzerland, France and England closely approaching them. If one regards the major markets of Europe for

Figure 3.23 *Europe: Intensity of Agriculture*

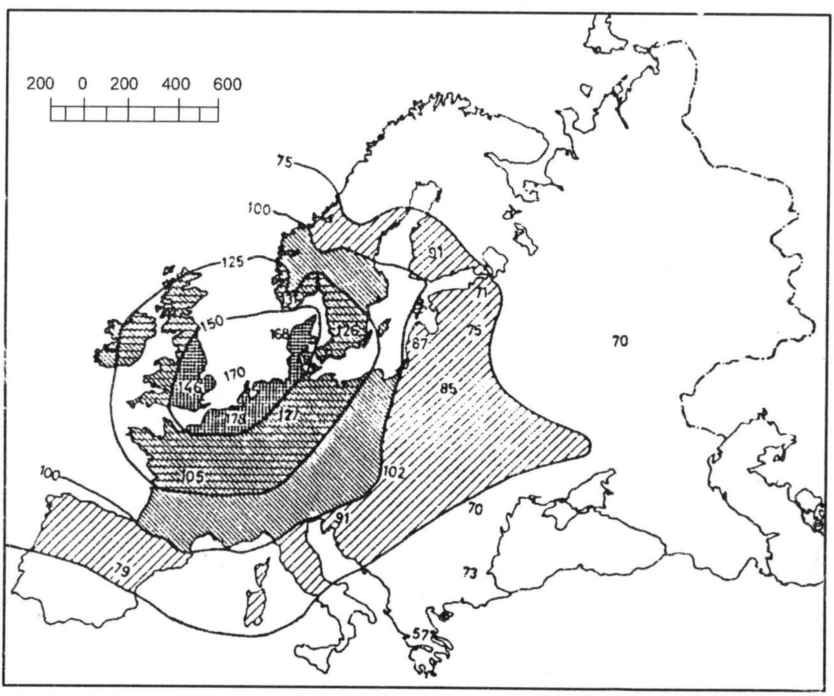

agricultural products as north-east France, the Netherlands, Belgium, south-east England, North Germany and Denmark, the model seems to be applicable on continental scale. In fact, the intensity of agriculture and the yield per acre decrease going away from north-west Europe. However, it is not as simple as this, since to mention just one factor, the periphery of Europe has much less useful agricultural land (the Alps, Pyrenees, and Appennines). As a consequence of modern means of transportation and communication, the concentric agricultural zones of von Thunen have been modified along the roads and railways.

Agricultural Regions of the United States

Agricultural is still an important element in America's human geography, because it is the most space-consuming economic activity. In fact, vast expanses of US (and Canada) landscapes are clothed with fields of grains. In addition, great herds of livestock are sustained by pastures and fodder crops. The growing application of high technology mechanization to farming has steadily increased both fruit and vegetable zones (1), and a surrounding belt of dairying; the forestry ring was displaced to the outermost limits of the regional system because railroads could now transport wood quite cheaply; (2) the field crop ring sub-divided into an inner, mixed crop and livestock ring to produce meat (the corn belt) and the outer zone that specialized in the mass production of wheat grains (Wheat Belt) and (3) the ranching area remains in the outermost position, a grazing zone that applied young animals to be fattened in the meat-

Figure 3.24 *Agricultural Regions of the United States*

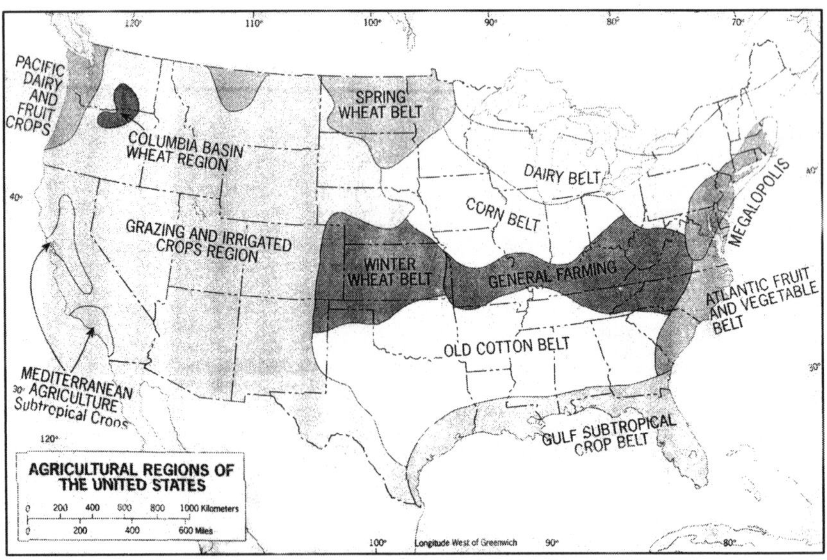

(After H.J. de Blij, et al., 2000: 172)

producing Corn Belt as well as supporting an indigenous sheep-raising industry. The 'supercity' anchoring this macro-Thunian system was the north-eastern Megalopolis, well on its way towards coalesce and already the dominant food market and transport focus of the entire country (Fig. 3.24).

Although the circular rings of the model are not apparent in Figure 3.24, many spatial regularities can be observed (remember that von Thunen, too, applied his model to reality and thereby distorted the theoretically ideal pattern). Most significant is the sequence of farming regions as distance from the national market increased, especially westward from Megalopolis towards central California, which was the main directional thrust of the historic inland penetration of the United States. The Atlantic Fruit and Vegetable Belt, Dairy Belt, Corn Belt, Wheat Belts, and Grazing Regions are indeed consistent with the model's logical structure, each zone successively farther inland astride the main trans-continental route-way.

Relevance of the Thunen Model to Indian Conditions

In many of the underdeveloped and developing countries of the world, in the villages, cropping belts are found. For example, in the villages of the Great Plains of India, concentric patterns of land use can be observed around the rural settlements. The highly fertile and heavily manured lands around the village settlements are devoted to the perishable commodities like vegetables, potatoes and orchards, while the land lying in the middle belt is given to cereal crops like rice, wheat, barley, maize, pulses and sugarcane. The land lying in the fringe zone is generally devoted to inferior cereals like bajra, guar and millets.

Prof. M. Shafi tested the model in the Koil tehsil of the Aligarh district (U.P.). In his opinion, the land use intensity decreases away from the intensively tubewell irrigated areas. Prof. Jhujar Singh and Prof. Dhayan Kaur also tested the von Thunen model in the different parts of Punjab and they concluded that the model cannot be applied in its original form as the extension of irrigation facilities and dense network of means of transportation and communications have affected the cropping patterns and land use intensity significantly.

After the introduction of tubewell irrigation in the Great Plains of India, this concentric pattern of crop land use has, however, been largely modified as the farmers with better economic status are able to produce perishable crops even in the distant fields from the market and settlement. The consolidation of holdings in India has also modified the crop intensity rings as each of the farmers is interested in growing the commodities for his family consumption as well as some marketable crops for earning cash to clear his arrears of land revenue and irrigation charges. Despite these changes and modifications, the land use intensity decreases with the increase in distance from the settlements, provided the physical environment and living standard of the farmers are the same. The diffusion of high yielding varieties (HYVs) has made the application of this model almost irrelevant both at the micro and meso levels. The fast development of means of communications and transportation has made it possible to transport perishable commodities at long distances. The model, thus, no longer is operative in India in its original form.

Criticism

The main criticism of the von Thunen's model is that it is based on unrealistic assumptions. Some of the important criticisms are as under:

1. The existence of isolated estate and only one market centre for the disposal of agricultural commodities do not apply in the present time. The modern transport system deviates from the principle of constant tonne-mile (km.) cost. In the modern system of transport, for example, larger the distance, lower is the transportation cost per tonne/kilometre.
2. The demand factor was not incorporated by von Thunen in his model. In fact, the demand of an agricultural commodity has a significant impact on the cropping pattern and crop land use intensity.
3. At present, no estate (agricultural holding) has an isolated entity in the fast changing world in which distances have been reduced appreciably by the swift means of transportation and communications. Consequently, the world markets are shrinking at a faster pace. The farmers have international markets to dispose of their commodities. The transport agencies in many regions of specialized agriculture transport the produce at a cheaper rate to the distant markets, which modifies the locational rent and the crop land use patterns.
4. Man is not a rational or economic person always as assumed by von Thunen. In many of the East European countries, Russia and China, the cooperative and collective farming has been developed in which the decision-making process lies with the political authority.
5. The concentration on spatial factors has led to less recognition of the role of environmental factors as determinants of land use patterns.
6. As with many economic models, von Thunen's simplifies the 'real world' in order to understand it, in ways that some find unacceptable. The process of decision-making by farmers is considerably oversimplified. For example, the element of risk and uncertainty are ignored (Game Theory and Wolpert, 1964).
7. At present, the cargo ships are fitted with refrigeration facilities and milk, butter, cheese, and meat are sent to markets in Britain from New Zealand and Australia, a distance of over 17000 km and now it is not necessary that perishable commodities be produced near the market.

8. The advantages of physical factors should not be ignored. The fertility of the soil, slope of the land, climate, drainage, and other infrastructural facilities will continue to play their part and cannot be taken for granted.
9. The spatial component of von Thunen model is thus no more than an ideal type. Its analytical framework, however, has much wider applicability.

Despite all these limitations, the validity of this model has been evaluated with considerable success at a range of spatial scales from the global (Peet, 1969) to the individual village and farm holding (Blaike, 1971; Chisholm, 1979). The model also stimulated the modelling of zonal patterns of intra-urban land use in the same way (Alonso Model).

While criticizing the model, it should be kept in mind that it was developed in 1826 when the means of transportation and communication were not so well developed and isolated estates were found in greater parts of the world.

However, the work of von Thunen is useful in two ways. First, it focuses attention on economic factors, particularly transport costs, and distance to market, contrary to the work of earlier geographers, who were subordinated by the factors of physical environment when attempting to land use patterns. Secondly, it attempted the concept of locational rent theory. This concept has great significance in both the rural and urban land use studies. Moreover, the assumptions of von Thunen led to more researches in the field of locational rent theories, crop intensity and cropping patterns. It, therefore, can be concluded that von Thunen's work with all its limitations, still provides a useful framework for organizing agricultural activities at the micro, meso and macro levels.

Weber's Theory of Industrial Location

Alfred Weber, a German regional economist propounded the theory of industrial location in 1909. The theory of industrial location, also known as the 'Least Cost Location Theory', was published in his classic work entitled as *Uber den Standort der Industrien*, and issued in its English translation 20 years later in 1929. It has probably had more influence on industrial location theory than any other single contribution. This model is a device for analyzing the location of industry, originated by Weber and elaborated as the *Locational Triangle.* The Weber's model provided the foundation for variable cost analysis, which dominated the study of industrial location for many decades.

Objectives

The main objectives of Weber's theory were as under:

1. To ascertain the minimum cost location of an industry.

2. To establish that transport cost plays a vital role in the selection of industrial location.
3. To prove that irrespective of socio-economic and political conditions, the location of industries depends on transport cost. Transportation cost in the location of industries is universal.

Assumptions

Weber's theory of industrial location is applicable only when certain optimal conditions are available:

(i) The area/region under consideration has a self-supporting economy.
(ii) There is an isotropic surface. In other words, there is uniformity in landform, weather, climate, distribution of population, labour, and human resource.
(iii) There is perfect competition in the market.
(iv) The labour is static and the wages are uniform in the region.
(v) There is uniformity and stability in the socio-economic and political environment in the region.
(vi) The industrialists and the labourers are rational and economic persons who try to optimize their profits and wages respectively.
(vii) Transportation cost increases uniformally and proportionately according to weight in all directions.
(viii) There is a uniform demand for a product at all stations, resulting in a uniform price, and therefore the plant (factory) located at the point of least costs would get the highest profits.

Weber, on the basis of quality and purity, divided the raw materials into two categories:

(i) ubiquitous raw materials (found everywhere, e.g., water, air, sunshine, land, soil); and
(ii) fixed raw materials confined to particular places (iron, copper, manganese ores, etc.).

The raw materials have also been classified as (i) pure raw materials, and (ii) impure raw materials. If the weight of a raw material remains the same even after processing and manufacturing, it is called a pure raw material. Contrary to this, if after manufacturing, the raw material is reduced in weight, it is called an impure or weight-losing raw material.

Postulations

The factors that control the location of industries are as under:

(a) Influence of transport (general factor).
(b) Influence of labour cost (general factor).
(c) Influence of industrial agglomeration (local or special factor).

Possible Locations of Industries

1. *Linear location*: when industry is located between the source of one raw material and the market.
2. *Non-linear location*: when industry is located between the market and the source more than one raw material.

Linear Location of Industry

In the case linear location, the following are the possible locations (Fig. 3.25):

(i) at the market,
(ii) at the source of raw material, and
(iii) at any intermediate location.

The selection of location of industry, in this case, entirely depends on the nature of raw material and the degree of weight-loss during manufacturing. The cotton textile and leather goods industry generally have a linear location.

Figure 3.25 *Linear Location of Industry*

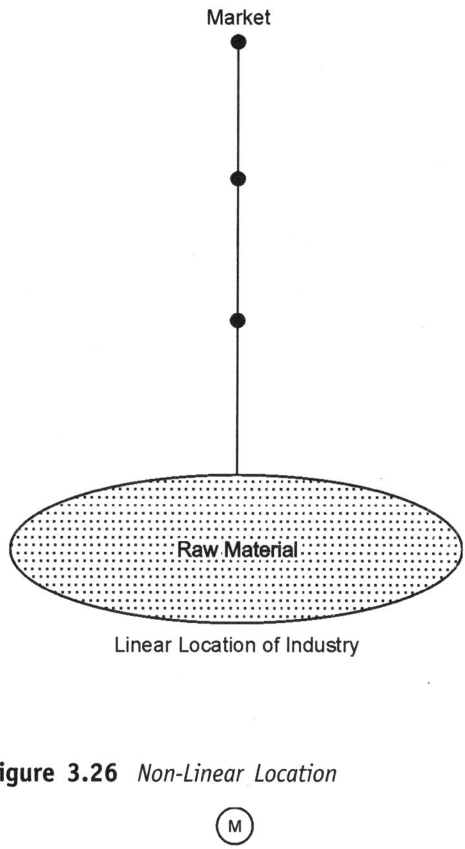

Linear Location of Industry

Non-Linear Location

When two raw materials are used, 'the influence area' for the location should be a triangle. The possible locations of the industry would be (Figs. 3.26 & 3.27):

(a) at the market,
(b) at the source of raw material M_1,
(c) at the source of raw material M_2, and
(d) at any intermediate point between the above three.

Figure 3.26 *Non-Linear Location*

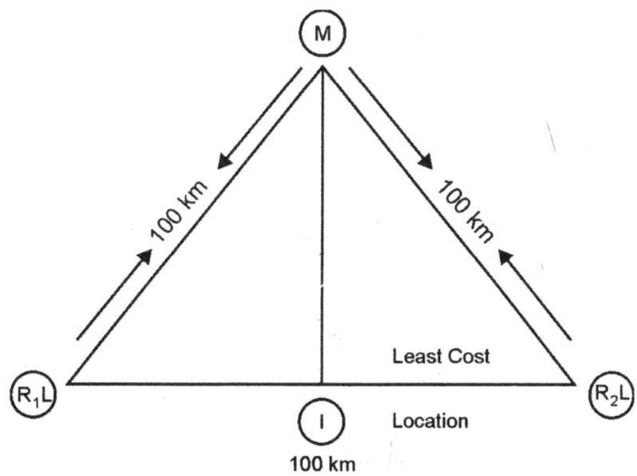

Figure 3.27 *Weber's Locational Triangle*

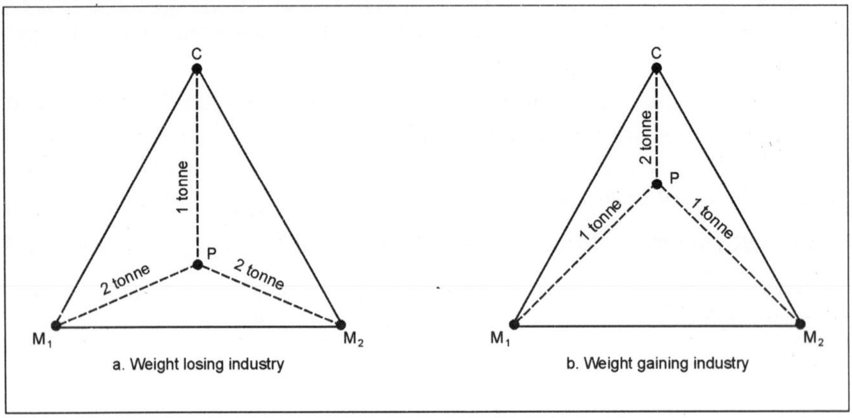

In this triangular area, the least cost location may emerge through transport cost analysis.

The location of an industry in triangular area is closely influenced by the nature of raw material (whether pure or impure). The material index of each raw material and distance of the market from the raw material source decides the least cost location.

Weber's attempt to derive 'pure' rules of location involved a number of simplifying assumptions. These were materials, markets, and cheap labour locations at fixed points, and movement was possible in any direction at the same cost per unit of distance. From this idealized world Weber abstracted sources of two materials (M_1 and M_2) and a single market point (C) which form a locational triangle. In this triangle area, a least cost location may emerge through transport cost analysis. To find the point of least cost, Weber first sought the least transport cost location, which he considered the most important influence, using a 'locational triangle' (Fig. 3.27).

Reality is simplified to two raw materials, M_l and M_2, and one consumption point (market) C. The least transport cost point P, is the point at which the total cost of moving raw materials and finished products is least. These transport costs are calculated by multiplying the weight of material or product by the distance carried, resulting in a 'pull' being exerted on the production point by each of the corners of the triangle.

(a) If one raw material M_l is ubiquitous and the other raw material M_2 localized and impure, the industry will be located at the localized raw material source (M_2). In the case of two pure raw materials, the least cost location will be on market.

(b) A complex situation may arise if both the required raw materials are localized impure or weight-losing and the raw materials and the market are at an equilateral triangle (suppose 100 km in each direction). Suppose, to produce one tonne of finished product, the raw material required from each source is two tonnes, and the rate of transportation of the raw material is Re. 1 per tonne per kilometre, the cost structure of the four possible places will be as under:

If the industry is to be located at R_1L (Fig. 3.26), the total transportation cost will be:

(2 tonnes $\times$ 100 km $\times$ Re 1.00) +100=Rs. 300

If the industry be located at R_2L, the total transportation cost of the product to the market will be:

$(2 \times 100) + 100 -$ Rs. 300

If the industry is to be located at the market M_l, the total transportation cost of the raw material and the finished product will be:

$(2 \times 100) + (2 \times 100) =$ Rs. 400

If the industry is to be located at 'I' or intermediate point between the two raw materials, the transportation cost of the raw material and the finished product will be:

$(2 \times 50) = (2 \times 50) + 86.6 \times 1 =$ Rs. 286

So, 'I' or intermediate location will be the least cost location (Fig. 3.26).

Influence of Labour Cost

Weber next examined the effects of labour costs on location since he considered that industries would be located away from the point of least transport costs to the point of least labour costs if savings in labour costs were greater than any additional transport costs involved in such a move.

The influence of labour cost may be explained with the help of Figure 3.28 showing the 'isotimes' (the lines joining the points of additional equal transport costs of raw material to the cheap labour centre). All along the lines or isotimes the transportation cost is the same per unit distance. Suppose, if the least transportation cost Rs. 400 is at R_2L and the total labour cost per tonne of production is Rs. 500, the total cost of production (transportation cost + labour charges) at R_2L would be Rs. 400+Rs. 500 = Rs. 900.

In case the labour cost at 'O' situated at 12th isotime is only Rs. 250 and the wages of the workers is Rs. 250. In this case the total cost of production will be Rs. 400+Rs. 250 + Rs. 120 (additional transportation cost) = Rs. 770.

Figure 3.28 *Isotimes: The lines joining the points of equal transport costs*

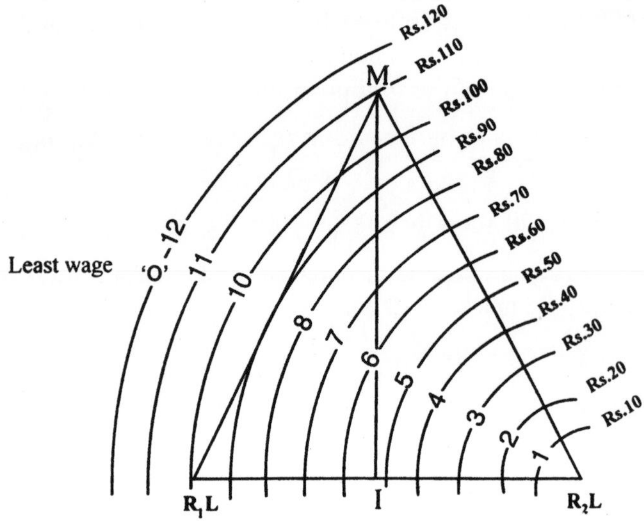

In case of cheap labour, the factory will be located at 'O' as the cost of production per tonne will be cheaper, being only Rs. 770 as against Rs. 900 least transport cost location of raw material (R_2L).

Weber also used the concept of 'isodapane' (cost contours) is his model. An isodapane is a line joining the points of equal additional transportation cost of the two materials and delivering the product to the market (Figs. 3.29 and 3.30).

Weber, with the help of isodapane, identifies the areas in which the cost of production and supply of the finished product would be the same. In the given Figure 3.29 the dotted area shows the sites of equal total transport costs.

Having combined the effects of transport and labour costs, Weber thirdly examined the effect of industry's tendency to agglomerate. In Figure 3.31 A, B, C, D and E are least cost locations, but the firms located there could cut their production costs by Re. 1 per unit of production if at least three of them operated in the same location. However, they must not incur increased transport costs of over Re. 1 per unit of production. In Figure 3.31, the critical isodapane of Re. 1 has been drawn round each producer and it is clear that firms C, D and E could reduce their total costs by locating in the shaded area.

Criticism

Weber's attempt to find the least transport cost location and then to examine how this would be modified by other considerations has obvious

Figure 3.29 *Isodapane: The total transportation cost*

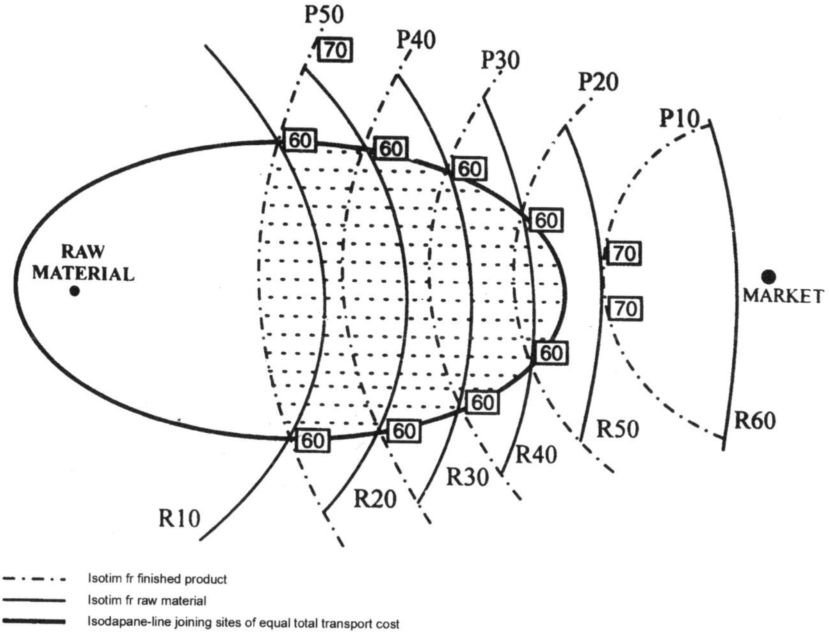

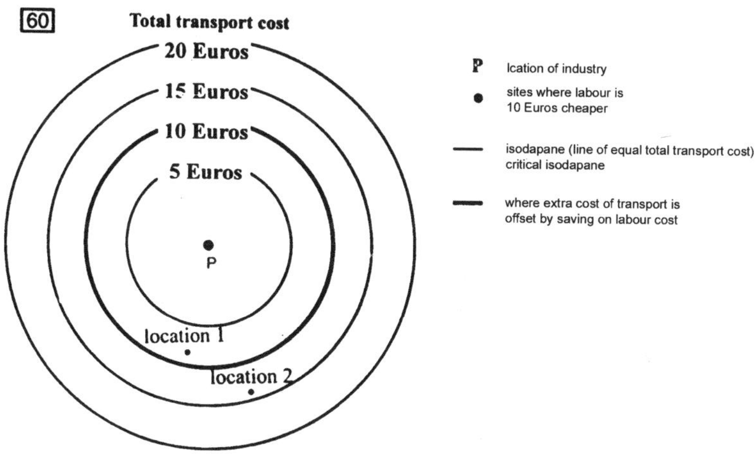

limitations, and has been criticized for being too abstract. Some of the criticisms of Weber's model are as under:

(1) He gave more emphasis to supply, while the role of demand in the location of an industry has almost been completely ignored. Weber overemphasized the role of transport in the establishment of an industry.

Figure 3.30 *Critical isodapane: The effect of labour and transport costs on location*

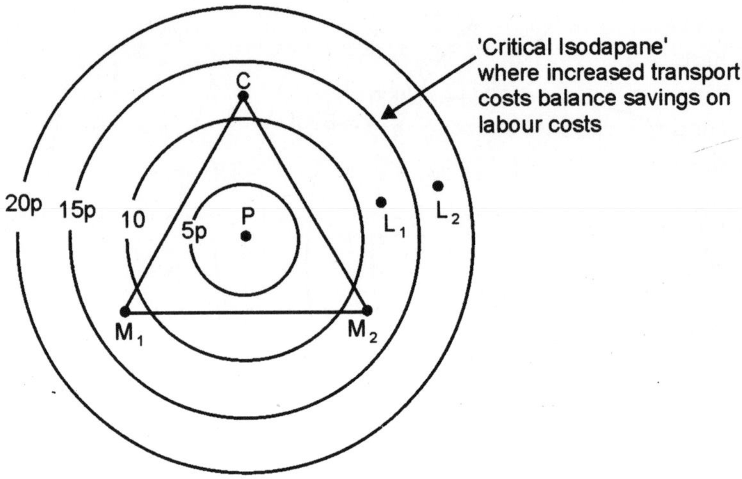

Figure 3.31 *Effect of Agglomeration on Location*

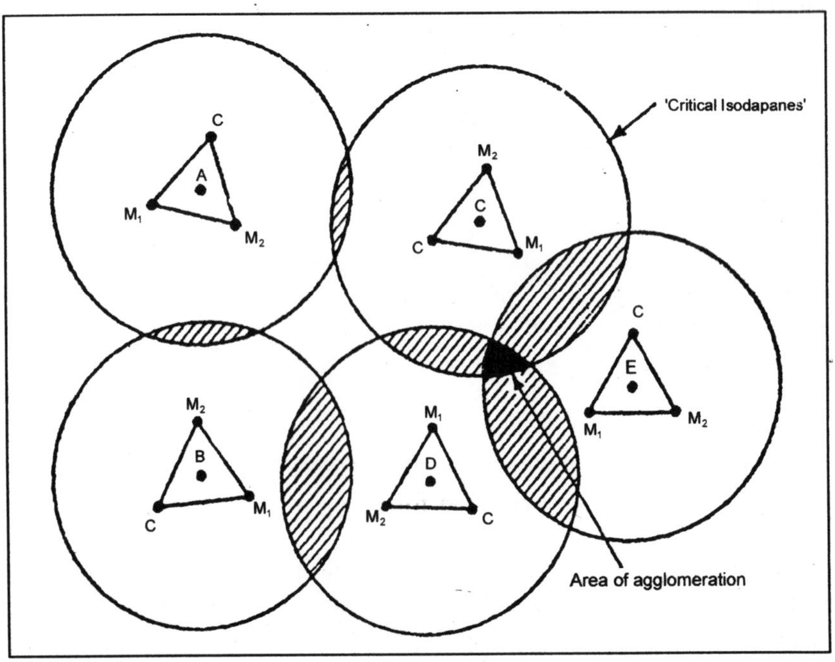

(2) Transportation costs of raw materials are cheaper than that of finished product. Moreover, transportation rate is generally cheaper at long distances.
(3) Weber, in his agglomeration concept, ignored the problem of space, high cost of land and high rent in the industrial areas.
(4) The assumption of perfect competition of Weber is only an ideal condition. In the long run, it is very difficult to sustain perfect competition in the real world.
(5) The historical factors, which are so important in the location of an industry, have not been considered.
(6) His assumptions about transport cost and effects of agglomeration have also been questioned.

Despite all these limitations, the theory is important because of its pioneering nature and its effects on later researchers. The real test of theory is that it should accord with reality, and empirical studies as W. Isard's work on US steel industry and W. Smith's work on weight-losing industries in Britain have, however, shown the validity of many of Weber's conclusions.

Rostow's Stages of Growth

This is a five-stage sequence of economic and social development postulated by the American economic historian W. W. Rostow (1971). Rostow saw economic growth as occurring in five stages. In this model, Rostow argued that all societies may pass through the five stages to achieve socio-economic development (Fig. 3.32).

The stages present an attempt to generalize "the sweep of modern history". Their elaboration in book form is described by its author as a 'non-communist manifesto', written in deliberate opposition to what were perceived to be Karl Marx's views on the relationships between economic and non-economic behaviour. Marx emphasizes on the material basis of society, and looks the historical development of social relations to comprehend societal change. Historical materialist analysis assumes the importance of ideas and argues that "life is not determined by consciousness, but consciousness by life".

As social beings, men and women develop "their material production and their material intercourse", thereby alter "their history and products of their thinking" (Marx and Engles, 1970: 47). In geographical research, historical materialism rose to prominence in the 1970s; it attempts to explain patterns and processes of spatial and environmental change as the result of the specific social relations of capitalism or other modes of production.

The model of economic development, derived from the stages in both the teleological (a theory that events can only be accounted for as stages in

Figure 3.32 *Stages of growth: Rostow's stages theory of economic development*

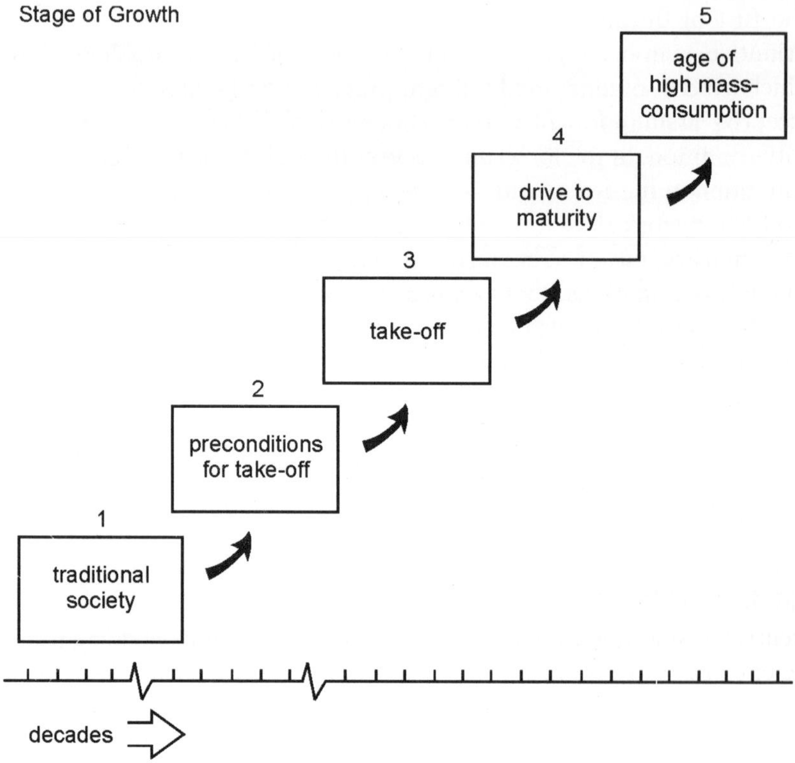

(After Keeble, 1967: 250)

the movement towards a pre-ordained end: the end may be determined by those involved in the event, as with various forms of planning, or it may be extremely defined, as in many religions) and mechanical in the sense that the end result (stage 5) is known at the outset (stage 1), and mechanical is that, despite the claim that the stages have an inner logic "rooted in a dynamic theory of production", the underlying motor of change is not explained. As a result, the stages become little more than a satisfactory system. The model itself is based on data for 15 countries only, plus outline data for others.

The stages of growth of Rostow for socio-economic development of a primitive society to the modern developed society are as under:

1. Traditional society
2. Precondition for take off
3. Take off
4. Drive to maturity
5. Age of high mass consumption

Stage I: Traditional Society

The first of the five stages—the traditional society—is characterized by 'primitive' technology, hierarchical social structures (the precise nature of which are not specified) and behaviour conditioned more by custom and accepted practice than by what Rostow takes to the 'rational' criteria. Initially, the technology is primitive and social structures are rigid and tradition-bound. Acceptance of new ideas and innovations are minimum. Production per capita is low and change and transformation are rare. The raw materials are exported and the finished goods are imported. People remain busy in the fulfilment of basic needs (food, clothing and shelter) and there is minimum use of the comfortable and luxury goods. This is the traditional society. These characteristics combine to place a ceiling on production possibilities (Fig. 3.32).

Stage II: Preconditions for Take Off

The second stage is a transitional stage, i.e., the 'preconditions for take off'. Here, possibly because of outside stimuli, investments rise, the infrastructure begins to be developed, and there is growth in the agricultural and industrial sectors. It is upon these bases that the next stage 'take off' occurs. Outside stimuli to change (including, for example, colonialism and the expansion of capitalism) are admitted in the transitional second stage (the preconditions for take off). Colonialism is the establishment and maintenance of rule, for an extended period of time, by a sovereign power over a subordinate and alien people that is separate from the ruling power, e.g., the ancient Greek colonies, or British settlers in India. Characteristic features of the colonial situation include political and legal domination over an alien society, relations of economic and political dependence and exploitation between imperial power and colony, and racial and cultural inequality.

The second stage emphasizes a rise in the rate of productive investment, a start on the provision of social and economic infrastructure, the emergence of a new, economically based elite and an effective centralized national state. Again, no specification of social relations is given. However, the opportunities for profitable investment presented by the preconditions for take off are likely to be ignored by capital and they pave the way for the third stage.

Stage III: Take Off into Sustained Growth

The stage of take off is a short period of time during which the economy and society are transformed. Investments and savings rise, and new industries grow in the primary and manufacturing sectors. This is described by Rostow as "the great watershed in the life of modern societies". It is a period of around 10-30 years, during which growth dominates society, the economy and the political agenda (although the social relations which facilitate this dominance are not described) and investment rises, especially in the leading sectors of manufacturing industry. This leads to urbanization and industrialization. The literacy rate, longevity and per capita income also rise.

Stage IV: Drive to Maturity

The self-sustaining growth results in the 'drive to maturity' (stage IV). This stage is characterized by diversification as most sectors grow, imports fall and productive investment ranges between 10 and 20 per cent of national income. Industrial development in this stage diversifies, imports fall, and investment is still high. The increasing importance of consumer goods and services and the establishment of the welfare state indicate that the final stage of the 'age of high mass consumption' has reached.

Stage V: Age of High Mass Consumption

This is the stage in which every citizen of the country has access to all the available goods of comforts and luxury. The sources of recreation are enormous, the leisure time is, however, reduced as people have to work extra time to obtain the latest goods and services for consumption. The literacy rate is almost 100 per cent, the longevity is high and the per capita income is also high. In the religious terms, this is the stage of paradise in which one gets everything that he desires.

The Rostow's model assumes that capitalism is the underlying structure of the society. It does not seek to explain the sequence of changes. In addition, the terminology is somewhat vague.

The insistence within the model upon placing growth in a wider historical and social context and upon a disaggregated approach, which reflects the uneven nature of development, marks a substantial advance upon abstracted and formal theories of economic growth. But, at the same time, these characteristics expose its socially universal and ahistoric features.

The stages of growth are unrealistic, as they could apply to anywhere at any time—to China as well as Brazil, to the former USSR as well as to the USA. However, the strategic implications are clear: following Rostow's logic, capitalism is a necessary consequence of development. All societies that are not currently capitalist in form will become so: "there is no alternative". Socialist societies are simply in an arrested stage preceding the breakthrough to capitalism. Such underlying implications are not made explicit. By concealing the specific social relations of production of the stages—and most especially of the first and second stages—capitalist societies may be reduced and extended by apparently universal processes of growth. This is the true meaning of Rostow's concept of stages of growth. If it were to read 'a capitalist manifesto', its ideological objective would be revealed and their achievement limited or subverted.

Heartland Theory of Mackinder

A Scot by birth, Mackinder was the first Professor of Geography in Britain at the London University, Director of the London School of Economics and member of the British Parliament. Mackinder interpreted history as essentially a struggle between land and sea powers. This geopolitical concept, first coined by the British geographer Sir Halford Mackinder in 1904, was presented before the Royal Geographical Society of London. Mackinder presented his paper 'The Geographical Pivot of History' in order to elucidate his formula of 'geographical causation in world history'.

On the basis of the physical map of the world, Mackinder concluded that on the global scale, the earth surface consisted of a huge and continuous landmass (Europe, Asia and Africa), and some isolated islands, e.g., North and South America, Australia, Great Britain and Japan. He named this huge united landmass of Europe, Asia and Africa as the 'World Island' (Fig. 3.33), which accounted for the two-third of the total land area of the earth, being 11 million square kilometres, and also seven-eighth or about 88 per cent of the total population of the world. The rest of the continents supported only one-eighth of mankind. The concept of 'Heartland' was later used in cold war discourse too to denote an area of Eurasia. Mackinder suggested that the Columbian era of sea power, which had

Figure 3.33 *Mackinder's Heartland Model*

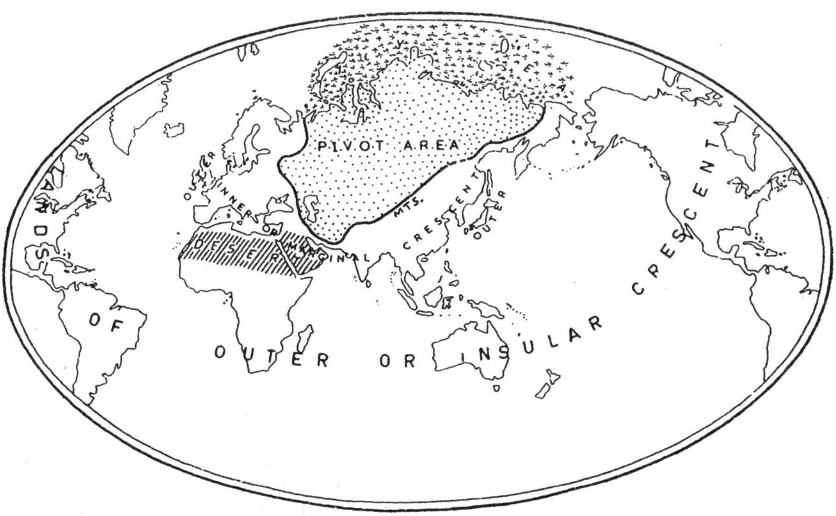

(After Mackinder, 1904)

Figure 3.34 *The East European Corridor: Mackinder's 'Pivot Area'*

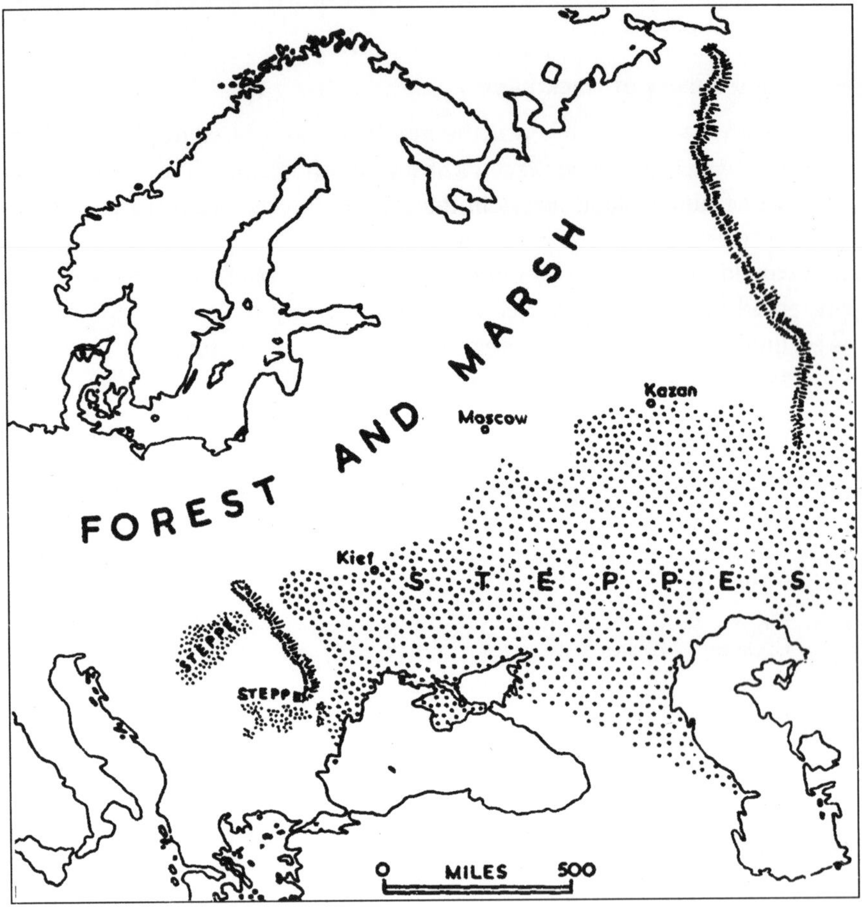

(After Mackinder, 1904)

given Europe its pivotal role for the past four centuries, was coming to a close and was being eclipsed by the ascendancy of land-based powers and in particular with a new 'geopolitical pivot of history', namely, the 'Heartland' of Europe-Asia (Fig. 3.34).

Mackinder's Heartland Theory regards political history as a continuous struggle between land power and sea power with the ultimate victory going to the continental power. He envisioned the landmass of the world arranged in three tiers:

1. *The Heartland*: In the first tier, there is a huge area of interior and Arctic drainage located in inner Eurasia. This is surrounded by mountains on three sides and by the ice-bound Arctic on the north. This was his so-called 'pivot area' that he later named as the 'heartland'. It stretched from the Volga river in the west to eastern Siberia

in the east, and from the folded mountains like Kolyma, Stanovoy, Yablonoy, Sayan, Altai, Tien-Shan, Pamir-knot, Himalyas, Hindukush, Kopat-Dagh, Elburz, and Caucasus and the related mountains and rugged plateaus in the south to the Arctic in the north-east. The distinguishing feature of Mackinder's Heartland area was that it was not accessible to sea powers, and was, therefore, strategically secure like a fortress. The Heartland was, however, vulnerable to land forces in the south-west, through eastern Europe, between Ural mountains and the Caspian Sea, where a narrow corridor, covered by steppe grasslands, offered an easy entry (Fig. 3.34).

2. *Inner Crescent*: The 'pivot area' was pictured by Mackinder as surrounded by an 'inner' or 'marginal' crescent, consisting of an arch of coastland, and characterized by drainage into navigable seas. This Inner or Marginal Crescent considered of the whole of Eurasia outside the Heartland. It included Europe west of the Ural mountains, South-West Asia outside the Iranian upland, South-East Asia, and most of China.
3. *Outer or Insular Crescent*: Outside the Inner or Marginal Crescent lay Mackinder's third tier, which he named as the Outer or Insular Crescent. Besides the continents of North and South America, Australia, and Africa south of the Sahara, the island regions of Great Britain and Japan were also designated as 'outer' because of their complete insularity from the mainland of Eurasia.

For Mackinder, whoever could gain control of the 'World Island' would be in an almost unstoppable position to dominate the entire globe. In his opinion, the Heartland having industrial and agricultural resources, could conquer Europe, the Middle East, and Far East (China, Japan etc.). The other landmasses, e.g., Inner or Marginal Crescent (rest of Europe, India, South-East Asia, and much of China), Outlying Islands (England and Japan), Outer or Insular Crescent (North and South America and Australia and New Zealand) would follow later. Mackinder (1919: 113) conceptualized his theory on the following premises:

Who rules East Europe commands the Heartland?
Who rules Heartland commands the World Island?
Who rules the World Island commands the world?

The key to control the Heartland, Mackinder later argued, lay in Eastern Europe, reflecting a powerful strand of pre- and post-Versailles geopolitical thinking concerning the need to separate the two land powers of Russia and Germany through the creation of 'buffer states'. Concluding his 1904 essay, Mackinder wrote that a sea power could, in the long run, be stronger than the land base that sustained it (its home base).

Modified Concept of 'Strategic Heartland', 1919

In 1919, Mackinder published a somewhat laboured and extended version of his 1904 paper in a book form entitled, *The Democratic Ideals and Reality*. In this book he renamed the 'pivot region' as the Heartland, and redefined the Heartland as the region to which, under modern conditions, sea power can be refused access, though the western part of it lies without the region of Arctic and Continental drainage. His modified 'Strategic Heartland' included the Baltic Sea, the navigable middle and lower Danube, the Black Sea, Asia Minor, Armenia, Persia, Tibet, and Mongolia (Fig. 3.35). In his original presentation of 1904, Mackinder had talked only in terms of Old World including Europe Asia and Africa north of Sahara. But in his 1919 presentation, his 'World Island' included the whole of Africa.

Figure 3.35 *Mackinder's 'Strategic Heartland' of 1919*

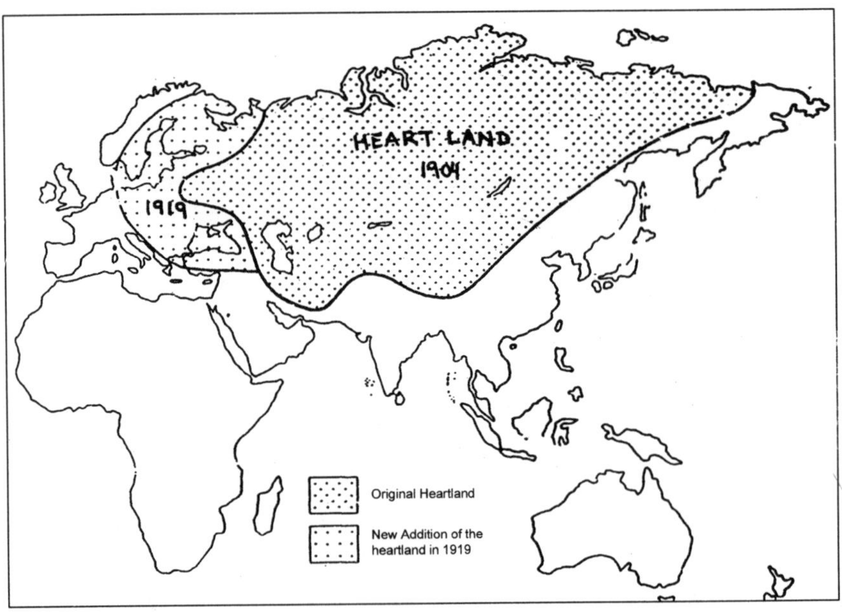

The trritorial waters of India extend into th sea to a distance of twelv nautical miles measured from the appropriate base line

The strategic argument is simple. The 'Heartland' is the strongest fortress on the earth, commanding resources of huge trans-continental area. Any power that could organize it effectively was bound to emerge as a great power in the world politics and such a power can establish a world-wide empire.

Modified Concept of the Midland Basin, 1943

Until 1919 Mackinder had confined his attention to Eurasia and the Old World. By the mid-1920s, it began to dawn upon him the western Europe and North America constituted for all practical purposes, a single community of nations. He wrote: "In the United States, the most abundant rainfall and the most productive coalfields are to be found in the east, but in Europe they are in the west. Thus, the west of the Europe and the east of North America are physical complements of one another, which are rapidly becoming balanced halves of a single community" (Mackinder, 1924: 252). This changed emphasis, in his perception of the politico-geographic reality, was to mirror in yet another major modification of his global strategic view. This paper entitled, 'The Round World and the Winning of the Peace' appeared in the *Foreign Affairs* in 1943. In this paper Mackinder clearly recognized (what he had failed to do in 1919) that the two sides of the North Atlantic were bound together but mutually supporting links of sea and air communications, and that the combined industrial and military strength of the United States and western Europe was a force to be reckoned with. He described this geostrategic area comprising the North Atlantic Ocean, the eastern United States, and western Europe, as the 'Midland Basin' (Fig. 3.36). He regarded this area as an effective counterbalance to the emerging political power potential of the Eurasian Heartland. In 1943 he excluded from the Heartland, that part of the Soviet Union lying to the east of the Yenisei river. He named it as Lenaland. Owing to its rugged topography and forest cover, this region was of little economic value.

During the Second World War, Mackinder's theory was put to the test. The Heartland (or pivot area) could have become the focus of power if either Russia had united with Germany or Russia had been overthrown by China and Japan. In his opinion, the Heartland is the greatest natural fortress on earth. For the first time in history, it is manned by a garrison sufficient both in number and quality (Mackinder, 1943: 595-605). What he meant to say was that the shattering defeat of Germany had turned the Heartland region from power vacuum to power centre.

Mackinder now visualized the emergence of two great centres of power in the world—the Midland Basin and the Heartland Basin—separated by the German realm of east and central Europe. He envisioned the two centres of power to be surrounded and insulated by a girdle of deserts—Saharan, Arabian, Irania, Tibetan, and Mongolian—extending through the rugged and desolate Lenaland to Alaska, the Canadian Arctic, and the deserts of western America. Mackinder thought that this empty land and deserts will prevent the spread of conflict to the outside world (Fig. 3.37).

Critical Appraisal of the Theory of Heartland

Though the pioneer geo-strategic model of Mackinder influenced the global strategic thinking in the modern times, he has nevertheless been

Figure 3.36 *Heartland and Midland Basin*

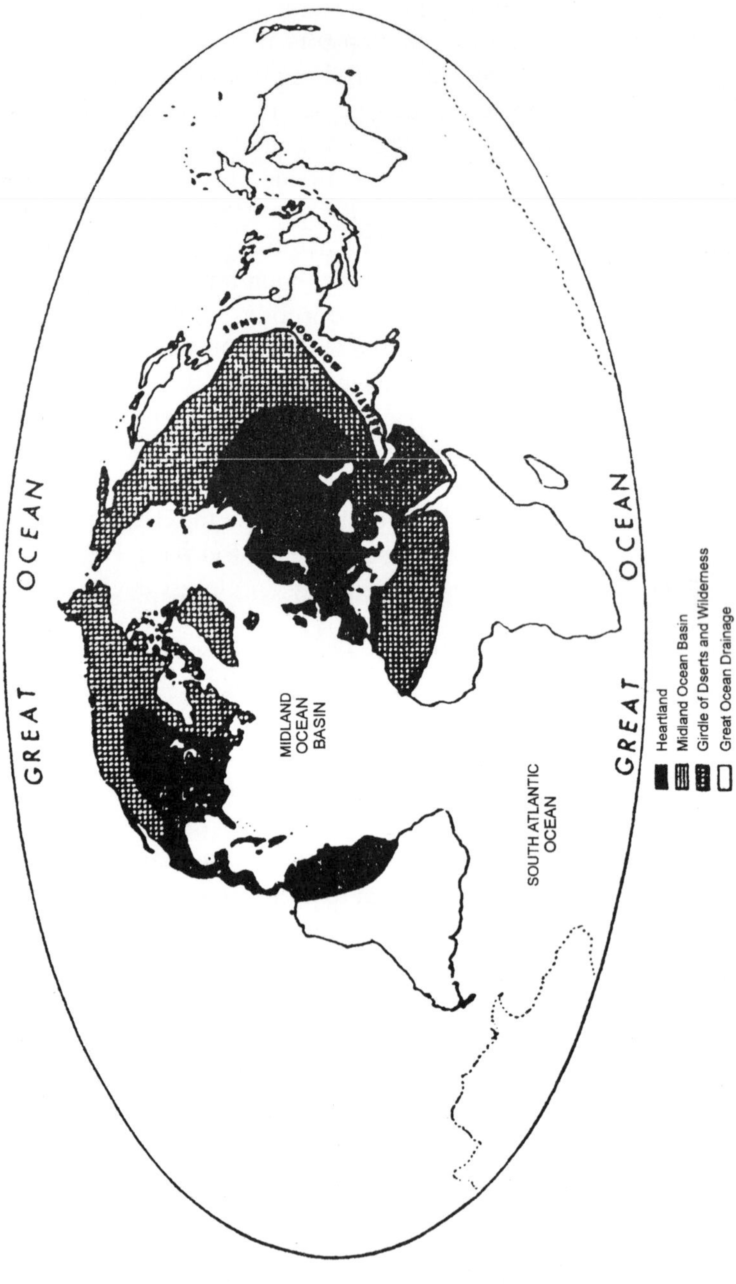

(After Mackinder, 1943)

Figure 3.37 *The Arctic Mediterranean*

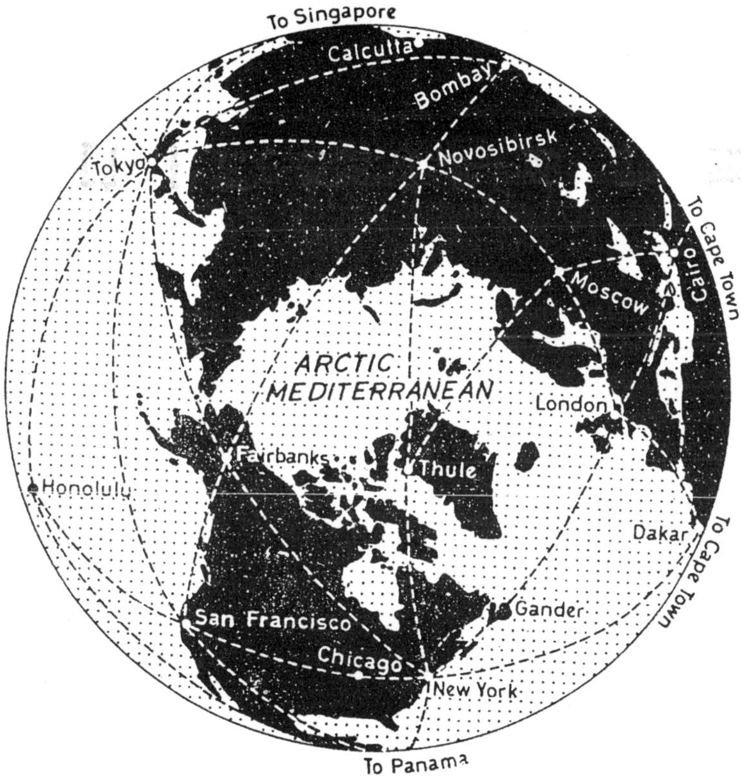

criticized on several counts. Some of the criticisms of the Mackinder's Heartland Theory are as under:

1. He simplified history in a deterministic fashion as a struggle between land and sea powers, which in fact was far from truth. History is influenced by the physical, socio-economic and cultural factors.
2. Mackinder wrongly equated power potential with sheer geographical area—a factor that made him to overinflate the resource and power potential of the inner Asian Heartland.
3. Mackinder overlooked the fact that the Heartland was a region of permanent difficulties because of its interior location and, owing to that, the extremes of climate. It is too cold in winter, parts of it hot and dry in summer. Owing to these locational-climatic factors, the Heartland is at a relatively disadvantage as compared to certain other larger and well-endowed areas of the world.
4. He look a rather simplistic view of possibilities of the development of railways in the sparsely populated Heartland.

5. The Heartland is not as resourceful as it is made out to be; the greater part of the Heartland is wasteland incapable of supporting large population.
6. One of the greatest shortcomings of his thesis lay in that he tried to explain the present, and prognosticate the future, in terms of technology that had already become outmoded. He wrote at a time when, in a general sense, the railway age was ending and the air age starting.
7. Heartland is not capable of supporting a large population from the produce of land. Moreover, despite the increased mobility created by the railways, automobiles and the aircrafts in the centre of the Eurasian land (Heartland), it could not be ignored that this area is ringed to the north, east, south and south-west by some of the greatest obstacles to transportation in the world. Ice and freezing temperatures for a large part of the year, and towering mountains pierced only by a few difficult passes from its borders.
8. With the development of war technology and fast moving fighters and missiles the fortress now lay open to air raids from all sides, as much as any other place.
9. Another serious consequence of Mackinder's thinking about the present in terms of the past technology was that at the beginning of the 20th century he was still explaining global strategy in terms of Mercator-projection map, which created a false picture regarding the unlimited expanse of Arctic ice separating North America from Eurasia. The reality is that North America and the Heartland lie face to face across the narrow expanse of Arctic. Since the shortest air routes connecting the Soviet Union and the United States lie across this narrow sea, the Arctic had got transformed as the Mediterranean of the modern times.
10. With the rapid march of science and technology, surface configuration of earth (global distribution of landmasses and oceans) has become very much less significant politically than it was at the time when Mackinder first presented his 'pivot area' thesis. Indeed, "when ballistic missiles armed with thermonuclear warheads can be fired from nearly any point upon the earth surface, we ... have reached the end of the line of geopolitical concepts and theories which purport to explain and to forecast the overall design of international politics by reference to the configuration of lands and seas".
11. The real world was not flat but spherical, and that the flat earth map used by Mackinder to utilize the pivot area (Heartland) thesis had led to a false sense of world geography and, hence, of global relationship between the major centres of power. The Mercator projection map has fostered the impression that geography did not favour a central role for the United States in the world affairs.

12. It is a fact, after the Second World War, "geopolitics has migrated from Germany to North America".

Spykman's Rimland Concept

Spykman propounded the Rimland Theory and published it in his book entitled *The Geography of Peace* (1944). He developed the Rimland Theory in opposition to the Heartland Theory of Mackinder. Working from the same premises as Mackinder, Spykman gave a very different interpretation of the relative importance of Heartland vis-à-vis the surrounding tier, the Inner/Marginal Crescent, partly continental and partly oceanic, which Spykman renamed as the 'Rimland' (Fig. 3.38).

Spykman considered geographical features (topography, terrain and climate) as important determinants in the foreign policy because of its emphasis on spatial variations. The theory says that the national power is the final determinant of the security of a state, while lasting peace is possible only through a collective security arrangement of either an armed league of nations or an international balance of power arrangement. The theory approached the problem from the point of view of the USA. Spykman saw the USA surrounded by Eurasia and Africa. Japan and England were seen as centres of military and economic power and separated by the Pacific and the Atlantic Oceans. According to him, the US strategy in case of war should be to avoid unification of the Old World

Figure 3.38 *Spykman's Concept of the Rimland*

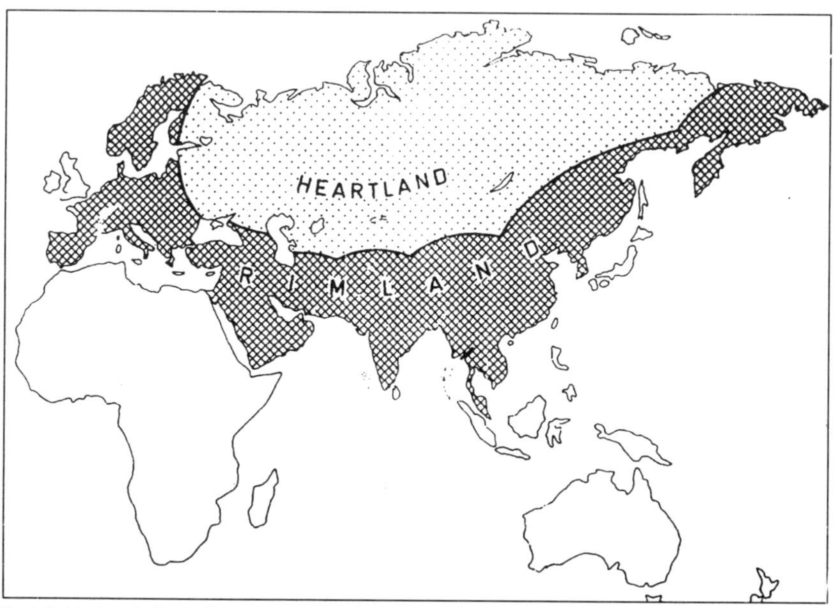

The territorial waters of India extend into thee sea to a distance of twelve nautical miles measured from the appropriate base line

Spykman's concept of the Rimland

powers against it. He saw the Soviet Union as the strongest land power and a unified Rimland as a menace to both Russia and USA.

Spykman gave emphasis on maritime mobility as the basis of new type of geopolitical structure, the overseas empire, in contrast to the former pattern of great land powers based on the control of contiguous landmasses. Spykman emphasized that it is sea power which has made it possible to conceive of the Eurasian continent as a unit and it is sea power which governs the relationships between the Old and the New Worlds. To him, sea power appeared as the key factor in global strategy. He wrote that if at all, there is need for a slogan for global power politics, it must be:

> Who controls the Rimland, rules Eurasia.
> Who rules Eurasia control the destinies of the world.

To Spykman, the Heartland appeared "less important than the Rimland" and he was convinced that it is a combination of land and sea powers controlling the Rimland that would in all probability control "the essential power relations of the world"; inevitably so, because the combination would emerge as the owner of far greater resources, manpower, and mobility, than the Heartland, or any other combination.

Spykman opined, the actual facts of Russian economy and geography offer no basis for the assumption that the 'Heartland' is or will be in the very near future a world centre of communication, mobility and power potential. In his opinion, there were many reasons to support such view. In the first place, the distribution of world climate made it certain that unless revolutionary changes in agricultural technology took place, the centre of agrarian productivity shall continue to be located in western Russia rather than in central Siberia. In this region the actual extent of arable land was only a very small part of total area. The essential bases for industrialization, at that time, also were mostly found to the west of Ural mountains. Consequently, the Heartland is not capable of supporting a large population from the produce of the land.

Writing at a time when the Second World War was still on, Spykman advocated that the allied powers should base their future policy on preventing any consolidation of Rimland and the enemy. With the defeat of Germany in the Second World War and the emergence of the USSR as the sole master of the Heartland, Spykman's prescription became the basis of the American policy of containment of communism. Looking at the geo-political and geo-strategic importance of the Rimland, the United States had consistently tried to build a tier of defence against the Soviet Union. The North Atlantic Treaty Organization (NATO), Baghdad Pact subsequently known as the Central Territorial Organization (CENTO), and the South-East Asian Territorial Organization were made by USA to keep an eye on the defence of the Rimland and to prevent the Sovient influence in the warm waters of the Rimland. The containment of communism, in the opinion of Domino, was necessary because the fall of any single country in the Rimland to the Soviets would have inevitably led to defeat of American interests in the other adjacent countries.

The Rimland Theory has been criticized on the ground of advancement of war technology and nuclear deterent. It has also been criticized on the ground that Spykman underestimated the role of world community and that of the UNO. Moreover, at present, world is like a village and the international law does not permit any territorial expansion. In fact, state is no longer a living organism in the sense of Ratzel's *Lebrensraum*. It is the time of economic imperialism and not that of political colonization.

Boundaries and Frontiers

Boundaries and frontiers have different meanings in geographical literature. Boundaries are lines demarcating the outer limits of territory under the sovereign jurisdiction of a nation-state. Frontiers, on the other hand, are zones of varying width, separating the ecumenes (fully developed and politically and economically integrated parts) of a given pair of states. These may consist of uninhabited or sparsely populated areas of marginal utility at the current level of technology, so that the states on either side of the frontier may not feel the need to define the precise real limit of their political jurisdiction (Fig. 3.39).

Figure 3.39 *Boundary and Frontier between Chili and Argentina*

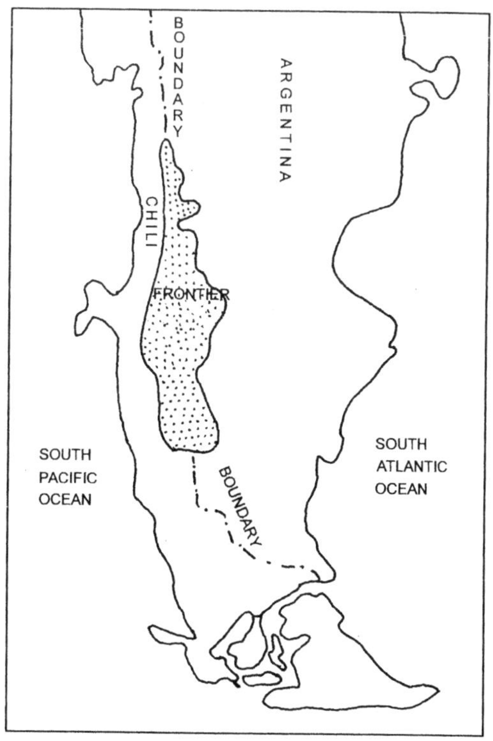

Boundary

As stated above, boundary is a line making the limits of a unit of land or demarcating the outer limits of territory under the sovereign jurisdiction of a state, often a geographical region, but also of economies or societies, such as a ghetto. Boundaries may be physical or man-made. Physical boundaries follow natural features such as rivers (e.g., the Rio Grande between Mexico and USA), and geometric boundaries follow lines of latitude (e.g., the 49th parallel between the USA and Canada), and longitude (as in large parts of the boundary between Botswana and Namibia).

A distinction is made between *antecedent boundaries* (which demarcate territories before they are settled, like the 49th parallel, or before they have been colonized, as in the case of many African boundaries established by the colonial powers at the Congress of Berlin 1884) and *subsequent boundaries* (which evolve together with the society they encompass).

Frontier

Frontier is that part of boundary which lies on the limit of the settled area. It differs from the boundary because the term 'frontier' indicates outward expansion into an area previously unsettled by a particular state. Some frontiers have occurred where two nations advance from different directions, leading to boundary disputes. A *settlement frontier* marks the farthest advance of settlement within a state while the *political frontier* is where the limit of the state coincides with the limit of settlement.

Distinction between Frontiers and Boundaries

According to Kristoff (1958), the main distinctions between boundaries and frontiers are as under:

1. Frontiers are 'outer-oriented', whereas boundaries are 'inner-oriented'. In the case of a frontier, the main attention is directed towards the outlying areas which are both a source of danger and a coveted prize. The hinterland—the motherland—is seldom the directing force behind the pulsations of frontier life. The boundary, on the contrary, is inner-oriented. It is created and maintained at the will of the central government.
2. The frontier is a manifestation of 'centrifugal forces', whereas boundary is that of 'centripetal' ones. This distinction derives respectively from their 'outer' and 'inner' orientation in relation to the ecumene (the inhabited areas of the world as opposed to the non-ecumene which is sparsely populated or not inhabited at all).
3. A frontier is an 'integrating factor' between states on either side; a boundary, on the contrary, is a 'separating factor'. Being a zone of transition between the sphere of one way of life and another, the frontier represents forces which are neither fully assimilated into, nor satisfied with, either state. It provides an excellent opportunity for

mutual inter-penetration and sway between frontier communities of the neighbouring states. The frontier, therefore, is an integrating factor. In contrast, although physical, geographical, cultural and political factors may, at times, tend to make it inconspicuous, the boundary must remain essentially a barrier, impeding integration across the two peripheral borderlines.

4. Frontiers are transitional between geographical regions, rather than between states. They are, therefore, 'geographical' rather than 'political' in nature. Boundaries, in contrast, are purely 'political' in origin and function.
5. Frontiers are 'areal' and boundaries are 'linear' in character. The former may be described as 'natural' in so far as they are parts of the earth's surface. In some cases, frontiers fall in the category of 'geographical' regions, in as much as they possess the quality of individuality used on their function as transition zones. Boundaries are artificial, since they are selected, defined and demarcated by man.
6. Frontiers are a phenomenon of the past, whereas boundaries belong to the present. This is because unlike the case of the frontier, the linear boundary is inseparable to the functioning of the modern state.
7. Finally, a frontier, whether physical, linguistic, religious, or ethnic, cannot be moved. It may change its character, and lose much of its frontier function, but it must remain *in situ.* In contrast, boundaries are by no means immovable. Until the Second World War almost every shift in the balance of power, between neighbouring pairs of states, used to be reflected in a shift in the location of the boundary line.

Classification of International Boundaries

There are two important systems of classification of international boundaries: (i) the physical, also called the *genetic classification.* It is based on the nature of relationship between the boundary line evolution of the cultural landscape of the state whose sovereignty it defines, delimits and separates; and (ii) boundaries may be classified in accordance with their form, that is, their demarcation and fixation on the ground. A boundary may often be drawn to follow some conspicuous physical feature, such as mountain range, a river, or a lake. These are physical or physiographic boundaries. Boundaries may also be drawn to follow a geometric line (geometrical boundaries) or to separate certain ethnic communities (ethnic or anthropo-geographic boundaries). In most boundaries, however, more than one criterion of delimitation may be involved. Most boundaries are, therefore, complex in nature. This classification is called the *morphological classification*.

Genetic or Functional Boundaries

The genetic classification of boundaries is based on the relationship that a boundary line had shared with the surrounding cultural landscape at the time of its demarcation.

(i) Antecedent Boundaries

Such boundaries predate the evolution of natural landscape. They are the most common type to come across in the New World. Here, international boundaries were agreed upon at the conference table even before the concerned territory was fully explored, and colonized.

(ii) Subsequent Boundaries

Subsequent boundaries are those whose definition and demarcation had followed the evolution of the cultural landscape. Such boundaries often conform to ethnic-cultural division of the landscape, specially, the divisions of language and religion. Most boundaries in eastern Europe, and those between India and Pakistan, and India and Bangladesh belong to this type.

(iii) Superimposed Boundaries

These are subsequent boundaries of a special type. Superimposed boundaries were also drawn after the cultural landscape had fully evolved. The difference lies in that the former type conforms to the cultural division between neighbouring communities, and were decided upon through mutual agreement. The latter, in contrast, do not conform to the socio-cultural divisions. They were imposed upon the concerned communities, either by outside powers or the overbearing unit between the two. Most colonial boundaries in Africa are of this type. In many cases, single communities were divided into two or more states. The boundaries of Ghana, Nigeria, Togoland, Dahomey and Somalian Republic are some of the examples of superimposed boundaries.

(iv) Relic or Relict Boundaries

These represent boundaries, which have lost political function, but which may still be discernible in the cultural landscape. Such boundary lines result when a smaller state is absorbed by a larger one, or when former boundaries between states are abandoned and redrawn. The boundaries between Poland and Germany, between Russia and Germany, between East and West Germany, and the former boundary between Spanish America and Anglo-America (USA and Canada).

Morphological Classification

(i) Physiographic Boundaries

Physiographic boundaries are those that were drawn to follow some conspicuous feature of the physical landscape. Since these boundaries follow some natural features of physical landscape, they are sometimes wrongly referred to as natural boundaries, as contrasted to boundaries drawn to follow certain geometrical lines or divisions of language or religion. The latter are sometimes referred to as artificial boundaries. This distinction is not correct. All boundaries are man-made, hence, are artificial. Some of the examples of physiographic boundaries are mountain boundaries, river boundaries, boundaries in lakes and straits, and forests, swamps and deserts.

(ii) Mountain Boundaries

Mountain boundaries have been the most favoured type since they have traditionally served as natural barriers. Being firmly fixed on the ground, mountains were considered to be highly stable. However, revolutionary changes in transport and communication, and the opening of the sky as a highway, have greatly reduced their function as protective barriers. As the Chinese invasion on India across the Himalayas in 1962 proved, even the loftiest of the mountains are no longer impregnable.

The location of boundary line along a mountain range often poses difficult problems since, most mountain ranges do not possess a well-defined crest line. Even where the crest lines exist, they are often divided by transverse valleys. Besides, most mountain ranges consist of several semi-parallel ranges, each with its separate crest line. Therefore, contrary to popular belief, mountain boundary, coincide between the crest line and the water divide is seldom found. Both the Indus and the Brahmaputra originate from near the Manasarovar lakes, and thus drain both Himalayas and trans-Himalayan ranges to the sea by the combined river system of the Ganga and the Brahmaputra.

The boundary between Chile and Argentina along the Andes mountain is an interesting example which posed difficulties by the selection of mountains as boundaries. The boundary was decided upon. The boundary was, however, always obscure and, as Holdich put it, "enveloped in a cloud of conjecture". As such, the boundary treaty of 1881 'failed' in as much as it admitted of diverse interpretations when the terms of it were applied to the existing physical features of the 'Patagonian Andes' (Fig. 3.40). In the treaty, it had been agreed that the boundary should follow "the highest crest which may divide the waters". But, the fact that the highest crest and the water parting do not coincide for a

Figure 3.40 *Boundary of Argentina and Chile in Andes. Complications in the choice of international boundary along a mountain range: example of the Chilean-Argentinian boundary along the Andes.*

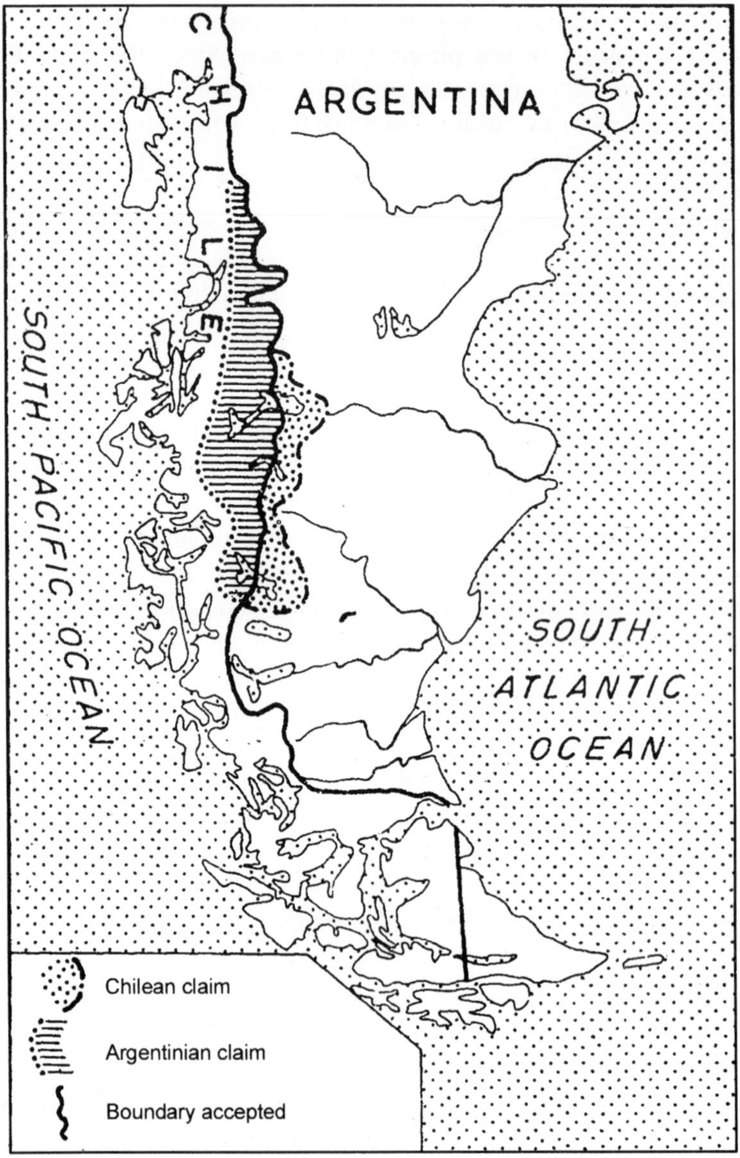

(After R.D. Dikshit)

distance of many hundreds of kilometres was unknown at the time of agreement. How diverse interpretation were permitted by 1881 treaty is evident from a map of the boundary. The dispute along Indo-China border in the Himalayas is another example of complications inherent in the choice of mountains as international boundaries. The MacMahon Line, as the boundary between Tibet and India in the eastern sector is called, was delimited in 1914 by the treaty signed by Great Britain (for India), Tibet and China, though it was never actually demarcated on the ground.

(iii) Rivers as International Boundaries

Many of the international boundaries are based on river streams. The advantages of selecting a river as boundary are as under:

(i) It is a clearly marked feature on the map.

(ii) It is a more narrowly defined (almost linear) feature than mountains and hills.

(iii) Wide, unfoldable streams offered a barrier to communications, and, as such, were thought to possess some military value by providing a line of defense against an advancing army.

Despite these advantages, the river boundaries have some disadvantages too. First, drainage basins generally tend to exert a unifying rather than a separating influence. The India-Pakistan boundary through the Indus basin, and the Indo-Bangladesh boundary are two important cases in point.

Generally, the mid-stream is defined as the international boundary. After each flood the mid-stream may change its course transferring the territory of one country to another. The problem of administration, and crime control are also serious in case of river boundaries. The Rio-Grande is an interesting example of river boundary between USA and Mexico (Fig. 3.41). In the flood plain section, this river adopts a frequently zigzag course. Following each change in the main channel, some areas along the river, forming part of one country, become included in the territory of other.

Figure 3.41 *Rio-Grande: Boundary between USA and Mexico*

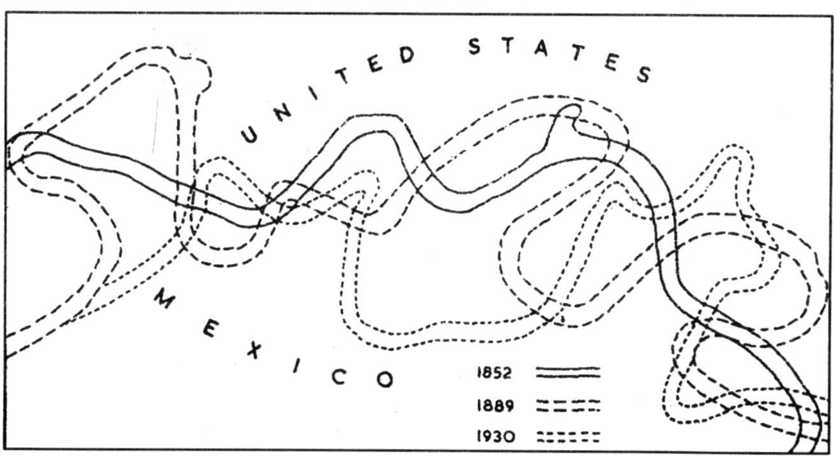

Figure 3.42 *Rivers as Boundary between Uttar. Pradesh and Bihar*

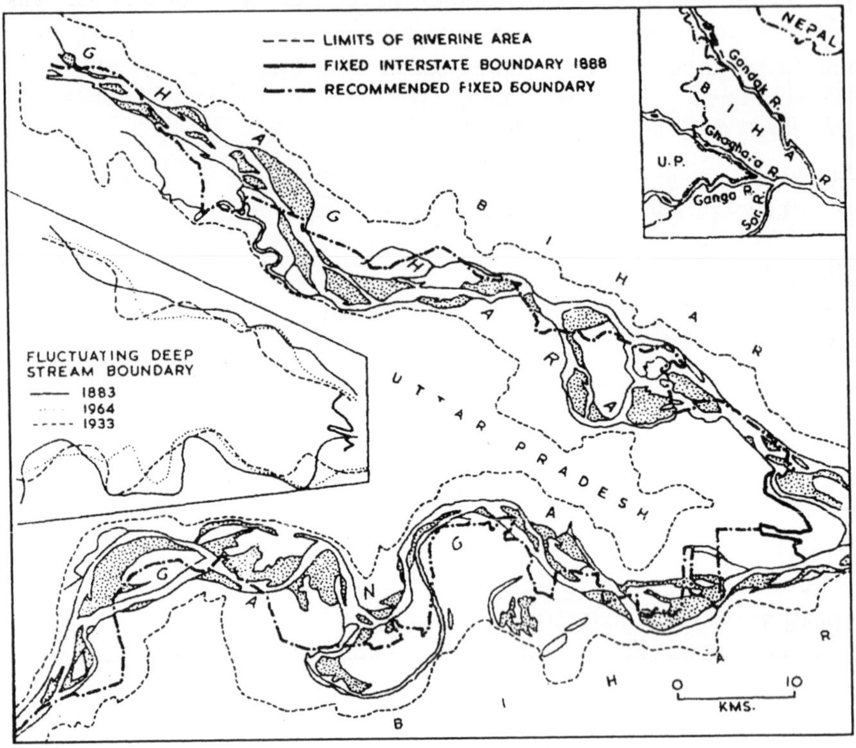

(After R.D. Dikshit)

On the border of Bihar and Uttar Pradesh flow the Ghagra and the Ganga. In this section, these rivers frequently change their course which become a cause of dispute between the two states (Fig. 3.42).

(iv) Boundaries in Lakes and Straits

According to the international law, the boundaries in lakes and straits are drawn at equal distance from the base of the respective countries from which the territorial sea is measured (Figs. 3.43 and 3.44).

(v) Forest, Swamp and Deserts

By their nature, the forests, swamps and deserts act as cultural divides. The Sahara desert, throughout the history, separated the European-dominated culture of the Mediterranean basins from the distinctive African culture. On the borders of Finland, Russia, Poland and Lithuania, forests form the boundary. The forest boundaries are much less effective as cultural barriers. Marshes have also served as a basis for drawing boundaries between Belarus and Poland during the two world wars.

Figure 3.43 *International Boundaries in Lakes*

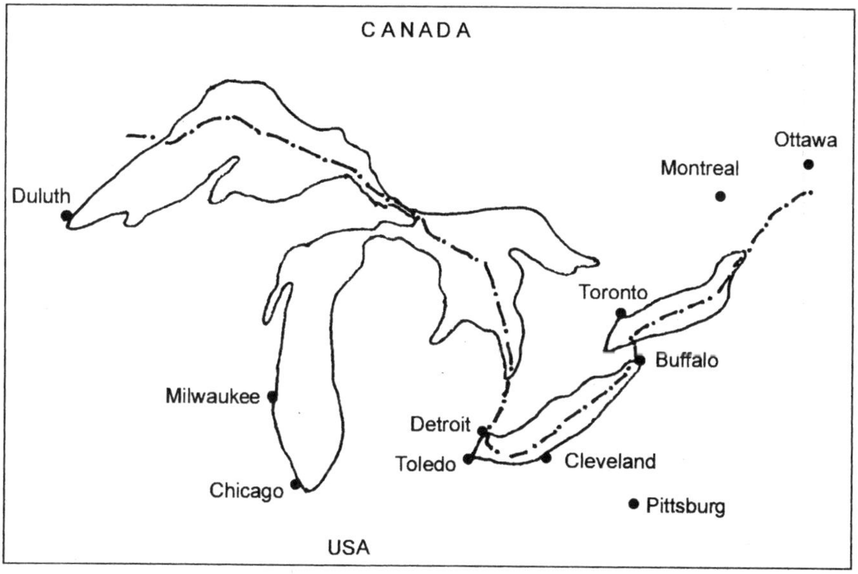

Figure 3.44 *International Boundaries in Straits*

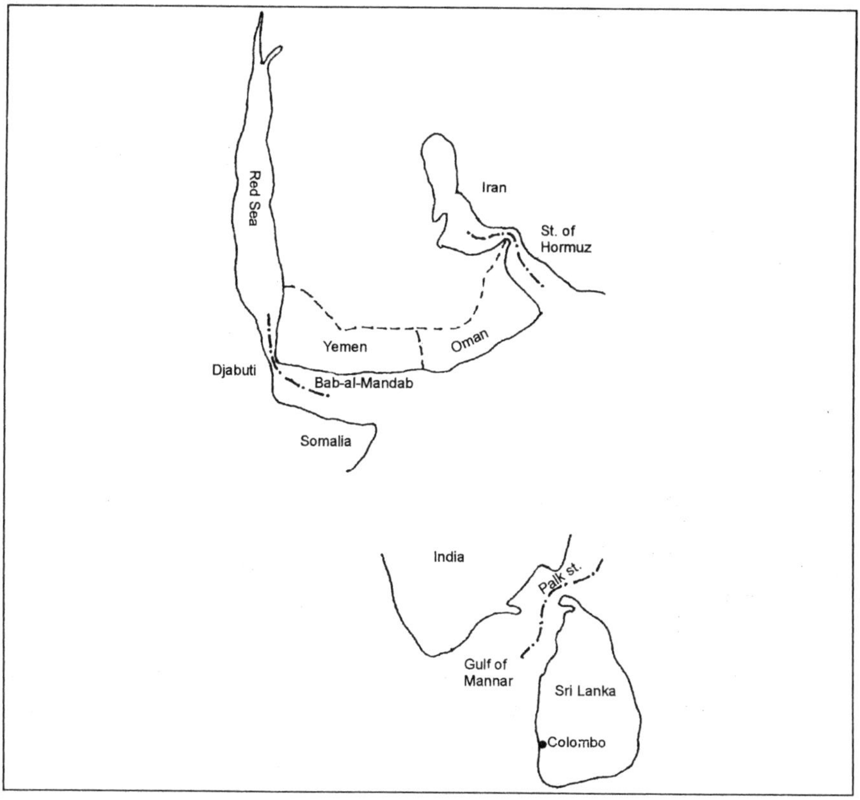

Figure 3.45 *Geometrical Boundaries in US and North Africa*

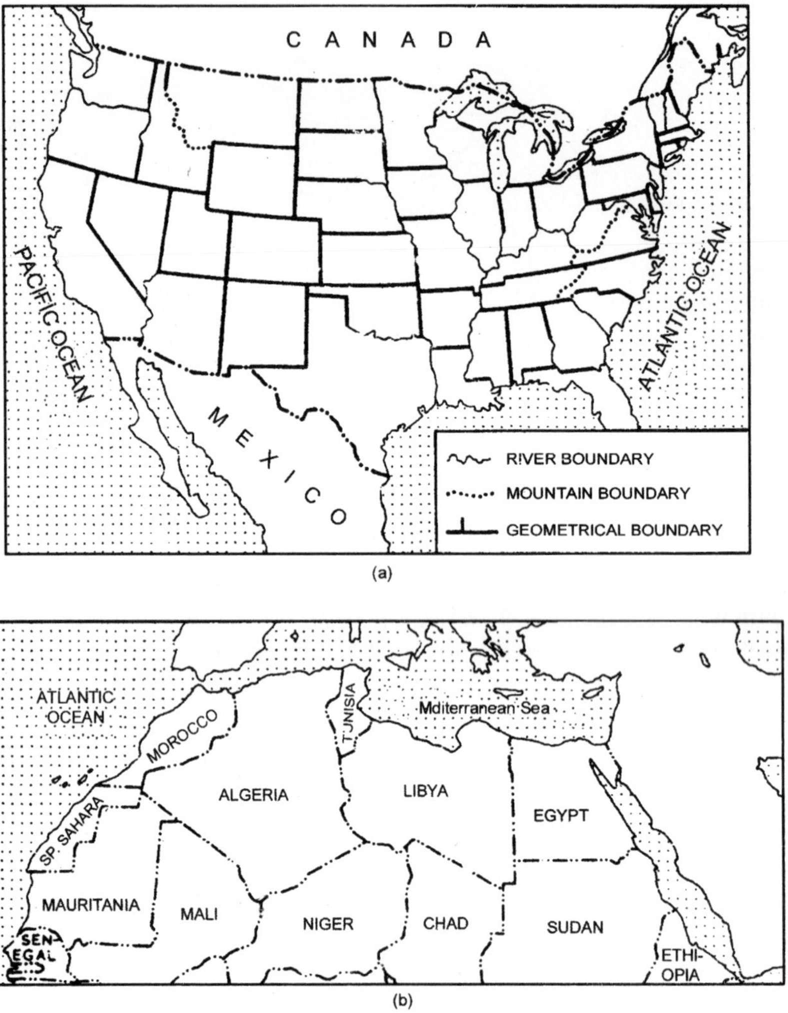

Geometrical Boundaries

There are straightline boundaries following lines of latitude or longitude. The state of Gambia in West Africa has its boundaries fixed by arcs drawn from the centre of river Gambia. The straightline (geometric) boundaries are common in North and South America and Australia (Fig. 3.45).

Anthropogeographic Boundaries

These boundaries were drawn to separate political communities, speaking different languages or following different religions or divided by some other cultural elements creating differences in nationalism. These boundaries are the features of the post-First World War period, when boundaries in central Europe and eastern Europe were readjusted with a view to separate different nationalities on the principle of allowing each nationality the right of sovereign political status. Since people speaking different languages, following different religions, and sharing common traditions do not, in reality, live in neatly defined areas, adoption of the geographical distribution of a certain ethnic element as the basis of political sovereignty may often necessitate large-scale transfer of population, resulting in great hardship to the dislocated people. The sad plight of refugees on both sides of the border in India and Pakistan, in 1947, is the most glaring example in recent history. The Indo-Pak example also highlights how much boundary lines disrupt regional economic unity, creating serious developmental problems.

Maritime International Boundaries

Prior to 1958 most of the law of the sea derived from customary law, but subsequently three United Nations conferences on the Law of the Sea have resulted in a series of conventions which have had some success in bringing the oceans within a single body of international law. The most recent is UNCLOS III, agreed on 17 December 1982 at Montego Bay, Jamaica, and signed by the majority of maritime states. Although it has still not formally come into force, its provisions are now being

Figure 3.46 *Law of Sea: Three-dimensional nature of sea divisions*

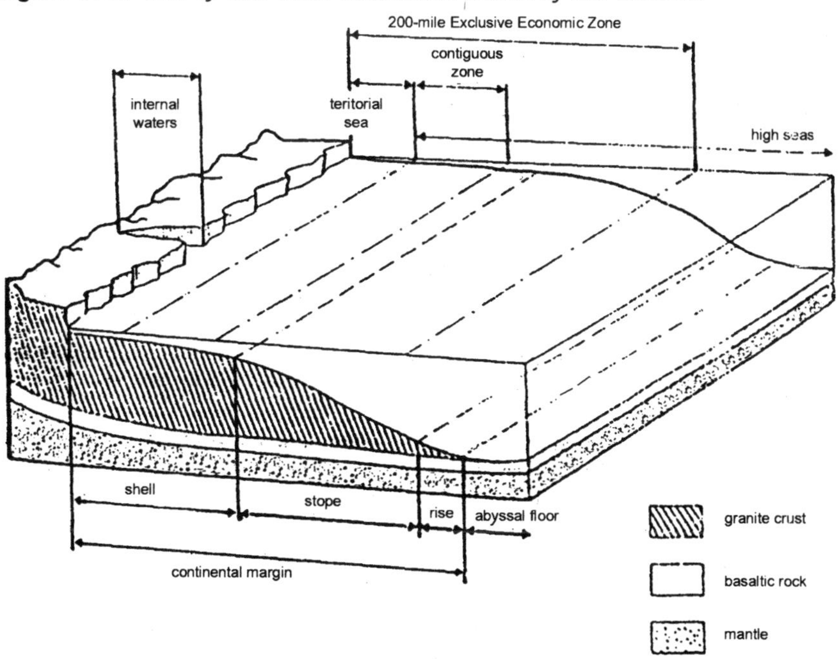

(After Couper, 1978)

implemented by most of the countries. The convention defines seven maritime jurisdictional zones (Fig 3.46):

(i) *Internal Water*: All waters landward of the baseline from which the territorial sea is measured, such as rivers, lakes, bays, ports and any waters of the low-tide line.

(ii) *Territorial Sea*: States exercise total sovereignty over these waters, except for rights of innocent passage. They extend out to sea for 12 nautical miles from the baseline, unless this impinges on the territorial seas of a neighbouring state, when compromise has to be agreed.

(iii) *Contiguous Zone*: An area 12 nautical miles beyond the limit of the territorial sea, within which states are free to apply customs and other national regulations.

(iv) *The Continental Shelf*: An area extending 200 nautical miles from the baseline, within which states may claim virtually exclusive rights to the seabed resources.

(v) *Fishing*: Most states now claim and exercise exclusive fishing rights up to 200 nautical miles out to sea from their coasts.

(vi) *Exclusive Economic Zone*: Synonymous with the redefined continental shelf describe in (iv) above.

(vii) *The High Seas*: These include all waters other than those defined in (i) to (vi) above, and there is complete freedom of movement within them. However, proposals are still under negotiation for the exploitation of seabed resources to be subject to UN control—a controversial proposal for many states.

Finally, all habitable islands are subject to the same regulations as coastal states, but it is not permissible to claim an Exclusive Economic Zone around small, uninhabitable rocks.

Limits to Growth Model

A phrase that gained currency as the title of the report issued in 1972 by the Club of Rome on global resources and pollution trend (Meadows et al., 1972). The classic work in the broader history of thought on limits to growth remains Thomas Robert Malthus' 'Essay on the Principle of Population', first published in 1798, which depicted food supply as a fundamental check to demographic and economic growth; the modern neo-Malthusianism represented by the Club of Rome report as extended the argument to other "natural resources and to the earth's capacity to absorb pollution".

The Limits to Growth reported the results of computer modelling of the world system carried out by the authors and by J.W. Forrester (1971) of Massachusetts Institute of Technology (MIT). D.H. Meadows, along with, J.W. Forrester, had developed the new concept 'System Dynamics' in 1972 to find out the extent and nature of global resource depletion and how it will affect the growth and prosperity of human civilization in future. The entire project was sponsored by the Club of Rome—a non-government scientific forum—and MIT in USA. A computer model developed by Forrester was used and the data fed in for the period 1900-1970. The processed data and the findings were published in 1972 under the title 'Limits to Growth'.

The growing human consumption pattern, alarming population growth, vulnerability of environment and scarcity of resource were the focal themes of the report.

Meadows, in his report considered the interrelationship between:

(i) population growth,
(ii) economic development including agriculture and industry,
(iii) environmental degradation, and
(iv) impact of renewable resource depletion.

The rate of population growth, economic development and anomaly in the rate of resource extraction have been revealed by this report. It has been stressed that the exponential growth rate of population will jeopardize economic development and enhanced consumption rate will accelerate resource extraction. In turn, the world will witness unprecedented shortage of both renewable and non-renewable resources.

The calculations of Meadows revealed a complete gloomy picture of the future world. According to his estimate, agricultural development would not be able to keep pace with rising food demand; most of the prime minerals like iron ore, copper, manganese, tin, and fossil fuel (coal, petroleum and natural gas) will be exhausted or depleted within coming 50 years, technological development will be stagnated and mankind will fail to develop substitute resources.

Mahatma Gandhi voiced the same apprehension: "Nature has enough for our needs, but not enough for our greed."

Forrester-Meadows Model

As stated at the outset, the special model on 'Limits to Growth' was developed by Prof. Jay Forrester of the Massachusetts Institute of Technology, whereas the specific analysis to 'Limits to Growth' was undertaken by his colleagues Dennis Meadows, Donnella Meadows, Jergen Ronders and William W. Behrens. The model utilizes a dynamic system approach to suggest what was likely to happen if current economic, demographic and resource utilization trends were to continue in the future. The model looks at the dynamic interaction of five major variables, namely, (i) population, (ii) pollution, (iii) natural resources, (iv) industrial output per capita, and (v) food per capita.

Assumptions

The model is based on six major assumptions:

1. A finite stock of exploitable non-renewable resources.
2. A finite amount of land that can be used to grow food, i.e., a finite amount of arable land.
3. A finite capacity of the environment to absorb pollutants.
4. A finite yield of food that can be obtained from each unit of land.
5. Exponential growth of population, population and industrial output as long as resource supplies and interactions with the variables in the system permit.
6. Forms of technological change (such as birth control, agricultural yield and capital productivity) are built into the model provided there is money for them and that environmental technology (such as resource recycling and pollution control) is also developed.

Projections

The 'Limits to Growth' model is a projection, not prediction. It projects consequences of continued growth in world population, food supplies, industrial output, resource depletion and pollution over a long duration to the year 2010 with dramatic results. They extrapolated growth in each factor as it had occurred between 1900 and 1970 showing how it would grow if it continued at this rate until the year 2100. They, then, altered, in turn, assumptions about how each of the variables would grow, feeding the projections into their computers and seeing what the outcome of each scenario would be.

They found that the factors they analyzed were highly interdependent and that changes to single factors often merely pushed problems on to others. Technical innovations to modify or deal with the consequences of growth in each of the factors, they concluded, could slow but not halt crisis, which could only be prevented by actual halts in growth.

The model projected different scenarios or at least seven different permutations, each of which assume alternations to growth in different combinations of the factors considered. In each run, they assumed a solution to the problem that precipitated overshoots and collapse in the previous run.

The First Run of the Model: Standard Run

The first assumption is the standard run in which the model assumes continued growth on 1900-70 trends in all of the factors. The exponential growth in all the variables leads to 'overshoot and collapse'. High levels of industrialization lead to resource depletion. Depletion and the diversion of capital from investment to the search for resources leads to the collapse of industry and of service and agricultural sectors dependent on it. Lack of food (malnutrition and undernourishment) and health services leads to population decline. Halts in growth happen in all these areas (Fig. 3.47).

Figure 3.47 *Limits to Growth: The First Run of the Model: Standard Run*

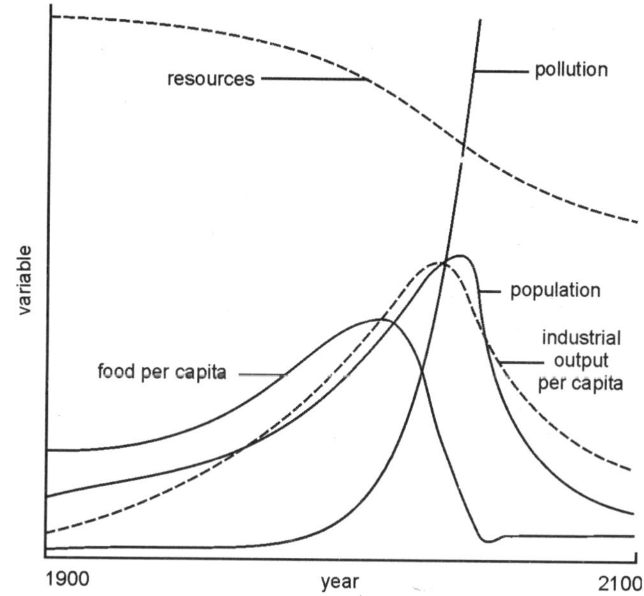

The Second Run

Since resource depletion is a primary problem and has been a key factor in collapse on the first run, it is assumed in the second run that technological development will double resource availability. This time, high industrial

Figure 3.48 *The Second Run*

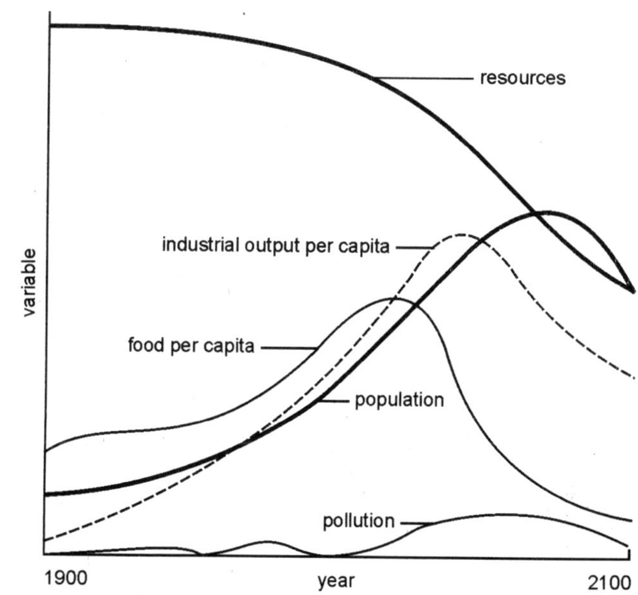

output leads to unabsorbable pollution levels. Population growth and lack of food lead to an increased death rate. Industrial growth, elongated by extra resources, still eventually leads to resource depletion (Fig. 3.48).

The Third Run

The third run assumes that technological developments double resource availability (e.g., nuclear power, new techniques for sea-bed mining and the use of low-grade ore) and cut resource use to a quarter of present levels (e.g., through new techniques of recycling and reclamation). However, unlimited resource does not prevent the same key problem of the previous run—unsustainable pollution levels. The lives of industrial output, food production and service industry are elongated but collapsed eventually and this hits pôpulation again. In the first run depletion of resources was the main problem. The second and third runs resumed solution to this but led to collapse caused by pollution (Fig. 3.49).

Figure 3.49 *The Third Run*

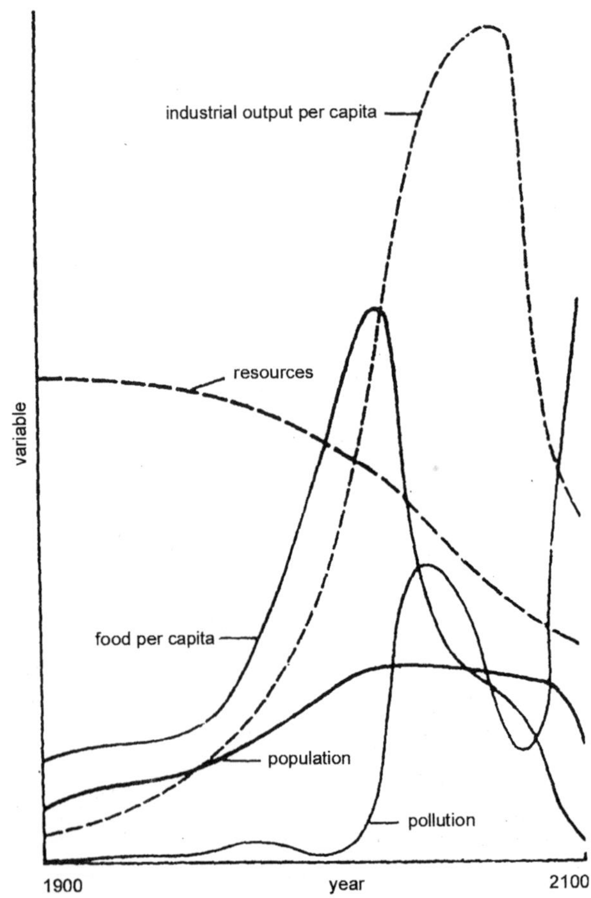

The Fourth Run

The fourth run assumes technical solutions to pollution: nuclear power and recycling (which cut down on some polluting emissions) and pollution control, for example. These measures will add a four-fold reduction in pollution to extra resources availability of the previous run. Resource depletion and pollution crisis are averted and population and industrial output rise. But, a rising population and industrial output cause exhaustion of the arable land through overexploitation and appropriation by industry. This leads to food crisis. If capital is diverted to agriculture to solve the problem, it leads to collapse of industrial output. Population falls as a result of food shortages. The next two runs assume technical solutions to food production and population levels to check the food crisis of the last run (Fig. 3.50).

Figure 3.50 *The Fourth Run*

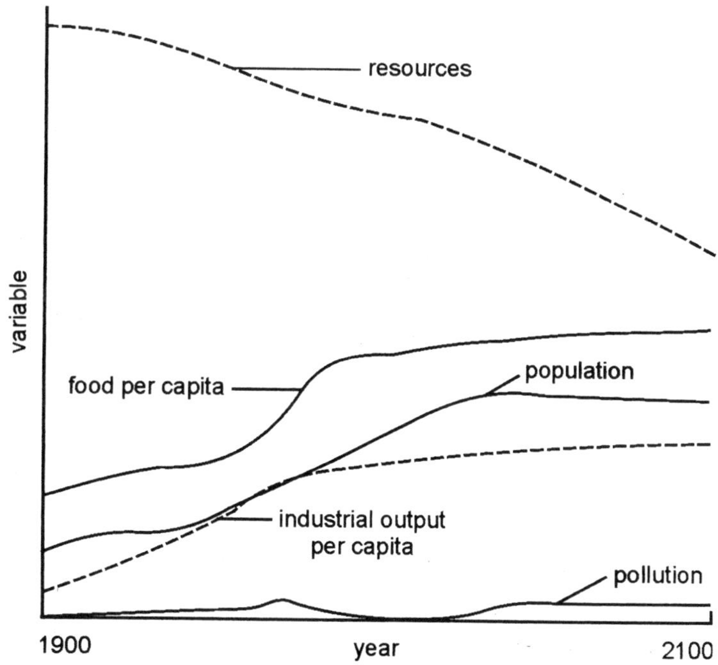

The Fifth Run

The fifth run, besides the assumptions about resource availability and population control made on previous runs, also assumes that land yields are doubled by developments, such as high yielding varieties. Increased food and industrial outputs, to however, increase to levels so high that

Figure 3.51 *The Fifth Run*

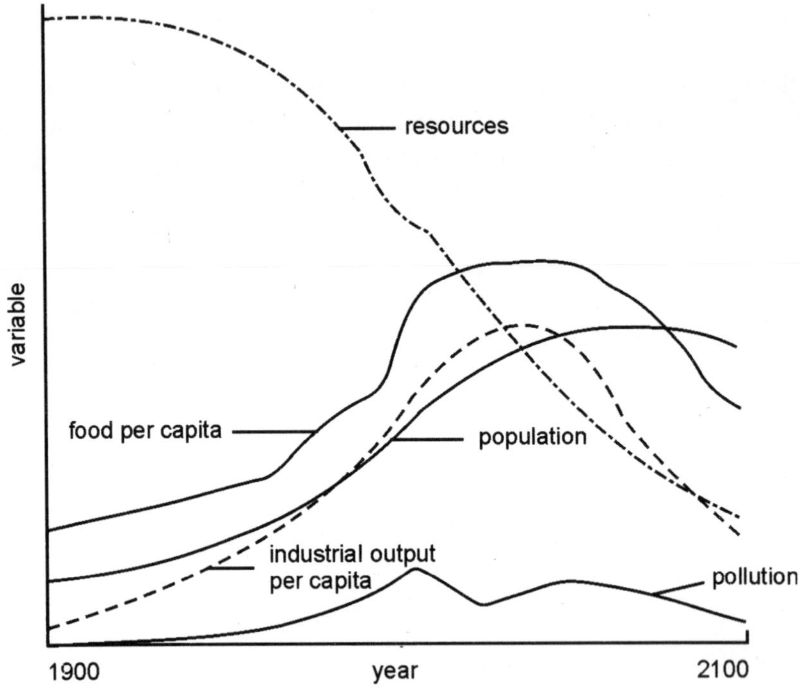

population levels rise despite the controls. This increase in population is again the source of collapse for the same reasons as second and third runs (Fig. 3.51).

The Sixth Run

The sixth run assumes voluntary birth control, which prevents the birth of all unwanted children. This, however, only slows and does not halt population growth only to postponement of the food crisis.

The Seventh Run

The seventh run assumes a combination of all the solutions so far outlined: resource availability, pollution control, food production and population control. The net result is still overshoot and collapse as a result of land revenue and food shortages, resource depletion and excessive pollution leading to a food production crisis and rising death rate. As with previous runs, solutions on one factor often only push the problem on to another (runs 1-4) and technological solutions slow but cannot halt growth (runs 5 and 6) which still on limit to growth calculations, leads to collapse before 2100 (Fig. 3.52).

Figure 3.52 *The Seventh Run*

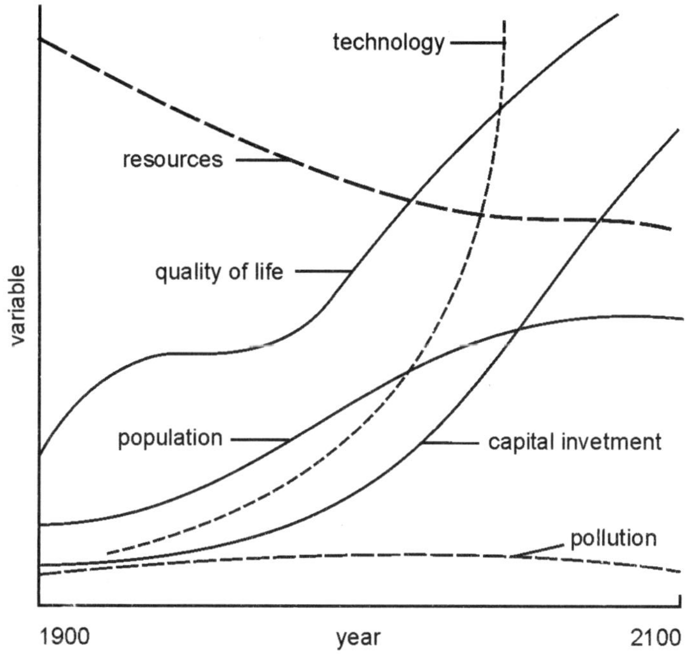

Criticism

'Limits to Growth' model was subjected to intensive scrutiny by scientists in every discipline dealing with population and resources. As a result, every facet of the work has been attacked by one critic or another. And, every time, the same telling points have been made. Many of the criticisms were anticipated and clearly acknowledged in the 'Limits to Growth'.

The authors do believe that their model is very crude, uses educated guesses for many important data. Oversimplified by grouping together many variables into aggregates, and is a preliminary effort to project current patterns—not a prediction of the inevitable.

Observers, however, have pointed many errors in the Forrester-Meadows model:

1. It treats the whole world as a single unit. In reality, the world is divided into rich, poor and not-so-poor countries and regions, each area having its own economic, food, population, resource and social problems.
2. It is extremely crude in that large multivariate subsystems have been lumped together into a single system. For example, many non-renewable resources, which have different and undetermined depletion times, are included under a single variable that assumes a supply large enough to last 250 years at 1970 usage rates. In a similar manner, the pollution variable represents only a long-lived globally

distributed family of pollutants, such as lead, mercury, asbestos, stable pesticides and radioisotopes. Resources supplies, pollution levels, food productions, population growth and industrial growth were estimated from existing knowledge or from educated guesses by experts.

3. It does not take into account the dynamics of resources. Resources are not, they become (Zimmerman). Resources were created as technology improved and as new needs and aspirations developed. For example, coal has existed for a long time but it became a resource once human beings discovered its use.
4. It predicted crisis by the year 2100 and probably within 100 years of the date of publication of report in the early 1970s. The Club and most of the environmentalists now accept that while they think crisis would occur in one of the scenarios outlined by the report if avertive action is not taken promptly. They do not think it will come anything like as soon as this.
5. The most frequently and forcefully put criticism is against the pessimism in the report. The extrapolative method used in the model focuses on physical limits and assumes existing rates of growth and technological innovations. It excludes humans' technological and political capacity to adapt. The conclusions of the model are that there would be crisis of overshoot and collapse whatever way different variables are altered to create changes in the system. The model assumes major technological advances in finding and recycling resources, increasing food productivity and controlling pollution, but does not include technological development or political action. Technology does not increase exponentially on a J-curve, as a result, the model does not assume that technology can solve every problem that might arise. A similar lack of technological optimism was evident in the predictions of Malthus and other 18th and 19th century economists.
6. Systems dynamics and computer modelling methods used in the 'Limits to Growth' model have some inherent disadvantages. Models are simplified versions of reality which pick out particular features and ignore others in order to create a manageable picture to analyze and deal with. The problem is that whatever has been selected in order to create a simplified picture outweighs the imperfections and distortions it creates. Since the 'Limits to Growth' model is a system model, emphasizing interdependency in system, a conglomeration of many tiny exclusions from a model could accumulate and lead to simulation which go far away from what would happen in reality, when the effect to complex details that have been left out in amplified many times.

7. There are problems inherent in any attempt to use modelling techniques for analyzing systems. It is pointed out that the systems analysis and modelling technique used by Meadows are flawed because they are based on inadequate data and empirical information. Moreover, systematic rounding and approximations in numerical values might accumulate and lead to significant miscalculations in the final results. These problems become particularly amplified when extrapolations are made on the basis of the data along exponential rather than just linear lines so that small imperfections in initial numerical values become much bigger when multiplied exponentially.
8. The environmental concern and hostility to growth that is built in the 'Limits to Growth' model reflects the biased interests of the materially well off. An environmental and non-material, which break with growth, development and material well-being, reflects the interests of those who already benefit from the latter. It is not sensitive to the needs of those who would benefit from growth more than environmental protection—the poorer sections in more developed countries and people in less developed countries. Moreover, proposals that emphasize the need for controls on population rather than consumption place the onus of change on the Third World countries with high birth rates than on countries of the North with high consumption, and bolster the idea that concerns for the environment is a luxury of the rich, developed world.
9. The 'Limit to Growth' model analyzes global averages in an aggregative way and is insensitive to regional differences and the specific situation of less developed countries. The trends that Meadows and Forrester fed into their computer runs and from which they extrapolated about the future were based on aggregated world average rates of growth. In other words, they calculated growth across the world but did not distinguish between differences in growth rates in different parts of the world. This is empirically problematic because it does not allow for sensitivity to local conditions, like famine or drought. For inscance, conditions, which may have specific effect on growth in some areas and not others, and so affect local and general growth rates.

Moreover, on prescriptions for change, no growth proposals are seen by critics to avoid real and more complex problems about the composition and distribution of growth. Some forms of growth, for example, may not be environmentally problematic and in need of curbing like manufacturing industry while growth in some services like health and education may be less problematic environmentally.

These criticisms prompted the Club of Rome to sponsor the development of another computer model.

Growth Pole

The concept of 'growth pole' was developed by Francis Perroux—a leading French regional economist—in 1955. He opined that growth is not ubiquitous, rather it occurs at poles, centres and points with variable intensities. He also opined that growth spreads through numerous channels and with variable final effects for the economy as a whole. Perroux and his followers believe that growth pole is capable of rapid economic growth and helps in the reduction of regional inequalities. In fact, the growth poles have a capacity to diffuse the process of industrialization in the neighbouring and distant areas. They also have a sustained impact on the economy of the surrounding areas. They establish the forward and backward linkages and help in the overall development of the region (Fig. 3.53).

Figure 3.53 *Growth Pole, Growth Centre and Growth Point: Forward and Backward Linkages*

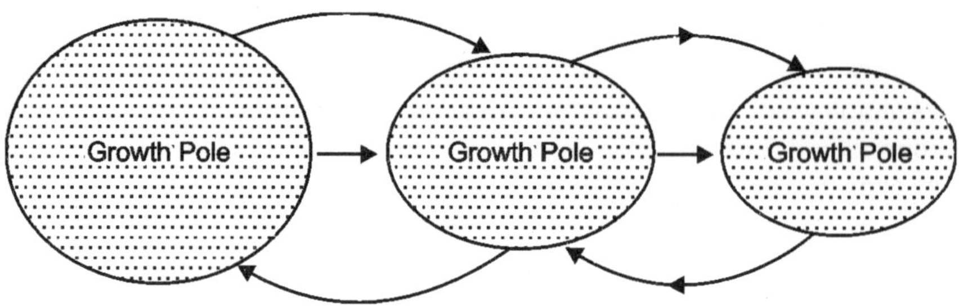

The growth pole concept was subsequently expanded by Bouldeville in 1966 in his book *Problems of Regional Economic Planning.* Later on, Buttler discussed the utility and economic development in his book *Growth Pole Theory and Economic Development.* He was in agreement with Christaller's hexagonal pattern of area of influence.

The basic concept of growth pole is that "development does not appear everywhere and all at once. It appears in points or developments with variable intensities, it spreads along diverse channel and has varying terminal effects for the whole of the economy".

In real world the growth rates vary from one region to another and from one sector to another. This variation is mainly due to: (i) entrepreneurial innovations, and (ii) availability of resources. It has been stressed by these scholars that it is a dynamic concept which explains the inter- and intra-regional variations in the level of economic development.

The population of a growth pole varies from country to country and region to region.

A growth pole, in the opinion of R.P. Misra, K.V. Sundaram and V.L.S. Prakasa Rao, may have a population of about five to twenty-five lakhs, while Perroux and Bouldeville did not give the size of population of a growth pole. At a growth pole, in the initial stage, the basic industries are located. Moreover, there is a dominance of secondary and tertiary economic activities, i.e., financial, educational, medicinal, cultural, technological and industrial sectors.

In India, after the independence, Jamshedpur, Rourkela, Bhilai, Durgapur, Dhanbad, Bokaro and Neyvli were established as the growth poles.

Growth Centres

In the growth pole theory, advocated by Perroux, growth centres are the second-ranking centres identified for economic development. In other words, a growth centres is smaller in size and structure as compared to growth pole. The dominant economic activities in the growth centres are secondary (industries) and tertiary (service) sectors. Generally, the basic industries are missing, they have the dominance of production activities with a focus on consumer goods and services. The population of a growth centre may vary between one to five lakhs within their zones of influence.

These industrial hubs are equipped with the facilities of grain, vegetable and fruit markets, and storage facilities for agricultural commodities. At growth centres, the agricultural inputs like fertilizers, insecticides, pesticides, equipments and other goods of daily requirements are found. In addition to these, there also exist educational, medical, cultural and recreational facilities.

The growth centres are spread in all the developed and the developing countries. In the Northern Plains of India, there are numerous towns and cities which may be placed in the category of growth centres. In the proximity of the national capital city of Delhi, the towns and cities of Faridabad, Gurgaon, Sun-city, Rewari, Bhiwadi, Jhajjar, Bahadurgarh, Rohtak, Sonepat, Panipat, Baraut, Baghpat, Modinagar, Moradnagar, Ghaziabad, Noida, Dadri, Dankaur, Khurja and Hapur may be categorized as the growth centres.

Growth Point

The population of a growth point is less than one lakh in the developing countries. In the developed countries it may be much smaller ranging from one thousand to ten thousand, depending on the size and total population of the country. The growth points render some of the important services to the threshold areas and have the emergence of food processing and other consumer goods industries.

Criticism

The growth pole concept has, however, been criticized on several counts. Some of the important criticisms are as under:

1. The first problem is the identification and selection of growth pole. If an independent or arbitrary decision is made, the resource base may not be sustainable.
2. The nodal points and growth poles develop because of certain favourable geographical locations and not because of selection of arbitrary sites.
3. The problem of threshold-overlapping.
4. Problem about the sectoral composition of growth pole.
5. Problem of establishment of regional network of firms.
6. Problem of social acceptability.
7. Problem of management and monitoring.
8. Problem of appropriate span over which to judge success of failure—say 16 to 25 years may be too long in any social order and political system. In fact, an elected government would like to have positive results within four or five years before the commencement of the next election.
9. Each of the growth poles may thus lead to new socio-economic and ecological problems.
10. The selection of growth pole motivated by political consideration may not be sustainable. It has happened in the case of Baruni and Bongaigaon, which have created new socio-ecological problems.
11. It is not necessary that growth poles and growth centres may remove the regional inequalities, leading to an overall regional development. The impact of Bhilai and Rourkela can hardly be seen in the surrounding tribal areas of Chhattisgarh and Orissa.

The success and failure of a growth pole depends upon the extent of the resource availability and productive demands of the society. The growth pole may thus create new socio-economic problems and may not be helpful in removing the regional inequalities.

References

Ahmed, Qazi, 1965, *Indian Cities: Characteristics and Correlates,* University of Chicago Press, Chicago.

Bhatia, S.S., 1962, 'A Reconsideration of the Concept of the Primate City', *Indian Geographical Journal,* Vol. 37, pp. 23-24.

Blaike, P.M., 1971, 'Spatial Organisation of Agriculture in Some North Indian Villages', *Trans. Inst, Br. Geogr.* 52, pp. 1-40; 53, pp. 15-30.

Buttle, F.H., Hawkins, A.P. and Power, A.G., 1990, 'From Limits to Growth to Global Change: Constraints and Constructions in the Evolution of Environmental Science and Ideology', *Global Environmental Change* 1, pp.57-66.

Chorley, R.J. and P. Haggett, 1971, *Models in Geography,* London: Methuen.

Cole, H.S.D., Freeman, C., Jhoda, M. and Pavitt, K. (eds.), *Thinking about the Future: A Critique of the Limits to Growth*, London: Chatto and Windus.

Dikshit, R.D., 1994, *Political Geography*, second edition, New Delhi: Tata McGraw-Hill Company.

Dunn, E.S., 1954, *The Location of Agricultural Production*, Gainesville, F.L., University of Florida Press.

Farmer, B.H., 1985, 'Christaller Revisited: Description versus Prescription', in Sundram, K.V. (ed.), *Geography and Planning: Essays in Honour of V.L.S. Prakasha Rao*, New Delhi: Concept.

Forrester, J.W., 1971, *World Dynamics*, Cambridge, MA: Wright Allen

Gregory, D., 1981, Alfred Weber and Location Theory. In Stoddart, D.R. (ed.), *Geography, Ideology and Social Concern*, Oxford: Blackwell, pp. 165-85.

Harvey, D. 1974, Population, Resources, and the Ideology of Science, *Econ. Geogr.* 50, pp. 256-71.

Hauner, M. 1990, *What is Asia to Us? Russia's Asian Heartland Yesterday and Today*, London: Unwin Hyman.

Healey, M.J. and Ilbery, B.W. 1990, *Location and Change: Perspectives on Economic Geography*, Oxford: Oxford University Press, ch.14.

Husain, M., 2002, *Systematic Agricultural Geography*, Jaipur: Rawat Publications.

Issard, W., 1956, *Location and Space Economy*, Massachusetts: MIT Press.

Jonasson, O., 1925, 'Agricultural Regions of Europe', *Economic Geography*, Vol. I, pp. 277-314.

Johnston, R.J., D. Gregory and David M. Smith, *The Dictionary of Human Geography*, London: Blackwell.

Knowles, R. and J. Wareing, 2002, *Economic and Social Geography*, Delhi: Rupa and Co.

Kristoff, L.K.D., 1958, 'Political Laws in International Relations', *Western Political Quarterly*, Vol. 21, pp. 589-606.

Losch, A., 1954, *The Economics of Location*, New Haven, CT, Yale University Press/Oxford: Oxford University Press (original German edition 1940).

Mackinder, H.J., 1904, 'The Geographical Pivot of History', *Geogrl. J.* 23, pp. 421-37.

—, 1919, *Reconstruction. London, Constable.: Democratic Ideals and Reality: A Study in the Politics of Reconstruction*, London: Constable.

Marx, K. and Engles, F., 1970, *The German Ideology* (transl. C.J. Arthur), New York: International.

Mayfield, Robert, C., 1967, 'A Central Place Hierarchy in Northern India' in Garrison and Marble (eds.), *Quantitative Geography*, Part 1, Economic and Cultural Topics, Department of Geography, Northwestern University, Illinois, pp. 120-66.

Meadows, D.H., Meadows, D.L., Randers, J. and Behrens, W.W.III, 1972, *The Limits to Growth*, New York: Universe Books.

Minshull, Roger, 1975, *An Introduction to Models in Geography*, London: Longmans.

Morgan, W.B. and R.J. Munton, *Agricultural Geography*, 1972.

Peet, J.R., 1969, 'The Spatial Expansion of Commercial Agriculture in the Nineteenth Century: a von Thunen Explanation', *Econ. Geogr.* 45, pp. 283-301.

Prakasa Rao, V.L.S., 1981, 'Central Place Theory in Retrospect and Prospect', *Annals of the National Association of Geographers of India*, Vol. 1, No. 2, pp. 39-50.

Rafiuullah, S.M., 1965, 'A New Approach to Functional Classification of Towns', *The Geographer*, Vol.12, pp. 40-53.

Rostow, W.W., 1971, *The Stages of Economic Growth: A Non-Communist Manifesto*, second edition, Cambridge: Cambridge University Press.

Shafi, M., 2006, *Agricultural Geography*, Delhi: Dorling Kindersley (India) Pvt. Ltd.

Singh, K.N., 1966, 'Spatial Pattern of Central Places in the Middle Ganga Valley, India', *National Geographical Journal of India,* Vol. 12, pp. 218-26.

Smith, D.M., 1981, *Industrial Location: An Economic Geographical Analysis*, second edition, New York: John Wiley.

Sprout, H. and Sprout, M. 1965, *The Ecological Perspective on Human Affairs*, Princeton, N.J.: Princeton University Press.

von Thunen, J.H. 1914, *Isolated State,* an English Translation of *'Der Isolierte, Staat'* by G.M. Warrenberg (ed.), P. Hall, Oxford: Pergamon Press (originally published in 1826).

Zipf, G,K., 1949, *Human Behaviour and the Principle of Least Effort*, Massachusetts: Addison Weseley Press.

4

Origin of Urban Places

The origin of urban places has been a point of great controversy and the experts of urban geography are not unanimous in their opinions on this issue. There are, however, four broad explanations which have been advocated about the origin of urban settlements. These are described below:

1. *The Ecological Models:* The ecological models typically associate urbanism with the production and concentration of a 'surplus' of some kind through, in particular, the commencing of large-scale irrigation schemes. In the irrigated areas, there was surplus production. In fact, an agrarian society based on large-scale hydraulic constructions, typically irrigation systems, was having surplus production. The effective management of these irrigation works involves an organizational web which covers either the whole, or at least the dynamic core, of the country's population. In consequence, those who control this network are uniquely prepared to wield supreme political power.

The result, Wittfogel contended, was that the state "occupied an unrivaled position of operational leadership and organizational control". It mobilized labour (forced labour raised on a temporary but recurrent, because seasonal, basis) and enforced such an inordinately powerful control over its subject population that it was, in effect, a state stronger than society. The state apparatus, thus, took the form of a *hydraulic bureaucracy*. Wittfogel's thesis has been used to account for urban origins precisely because it is supposed to provide a rationale for centralization and concentration.

2. *Economic Models:* The economic models, typically focus on changing forms of economic integration and, in particular, on the

transition from reciprocity to redistribution, are especially concerned with ways in which such systems of exchange are 'embedded' in non-economic institutions. Most of these models are indebted to Polanyi's substantivist anthropology.

3. Cultural Models: These models typically examine the formative influence of religion on urban genesis. "The religious component is almost alone", Wheatley (1971) argues, "in having left in several of the realms of nuclear urbanism, a more or less continuous success of surviving material traces through ... to fully evolved urban life". In this view, the earliest foci of power and authority took the form of ceremonial centres, with religious symbolism imprinted deeply on their physiognomy and their operation in the evolution of town or iconography of ancient city, which projected images of the cosmic order on to the plane of human experience, where they could provide a framework for social action.

4. Politico-military Models: These models typically conceive of the first city as both fortress and refuge. Many of these models are, of course, compatible with the arguments of previous paragraphs-—the supposed conjunction between 'hydraulic society' and 'Oriental Despotism', for example, or the centring of "political and military power ... first in theocratic and later in monarchical control" (Giddens, 1981, 1985), but they usually go beyond those claims to emphasize the decisive importance of military power through a grid of cities for the creation of empires.

The Size and Spacing of Cities

The urban geographers and sociologists have shown a great deal of interest in the size and spacing of cities. Implicit in such studies is the assumption that there is some short of order or logic underlying the size relationships and spatial distribution of towns. It is assumed that settlements do not grow up in a haphazard or random manner and that a measurable degree of order is to be found in their size and spacing.

In any large region there are many small towns, a smaller number of medium-sized towns and relatively few large cities. This is true for all regions of the world, irrespective of the stage reached in economic development or the degree of urbanization achieved. This regularity is clearly demonstrated when relationships between city rank and city population size are examined.

Rank-Size Rule

The cities in any region may be ranked from largest to smallest according to their population size. Thus, the largest city is ranked No. 1, the second largest No. 2, continuing in this way down to the lower ranking towns.

The rank-size rule, proposed by G.K. Zipf in 1949, attempts to express the relationship between towns in a precise mathematical terms and states that "if all the urban settlements in an area are ranked in descending order of population, the population of the *n*th town will be $1/n^{th}$ that of the largest town". In other words, the population of urban settlements in a region can be arranged in the series of 1, 1/2, 1/3, 1/4, ...1/*n*. This regularity can also be expressed by the formula:

$$Pn = P1/n$$

where P*n* is the population of the town of rank *n* in the descending order and P1 is the population of the largest city. Thus, if the largest city has a population of 2 millions, the tenth ranking town should, according to the rule, have a population of 2,00,000 inhabitants.

Figure 4.1 *The Rank-Size Rule on: (a) normal curve (b) log-normal distribution*

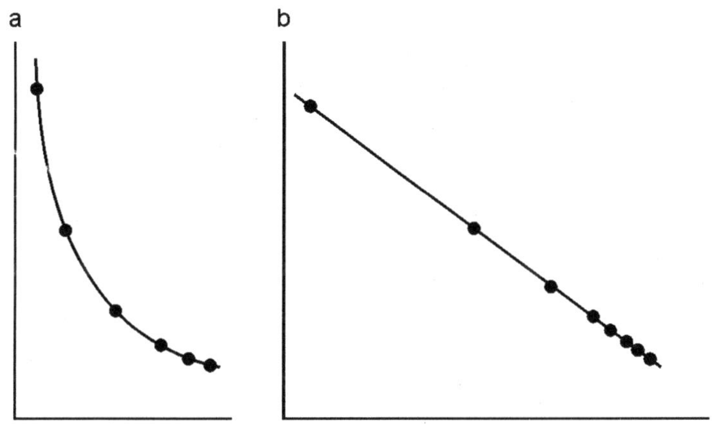

Figure 4.1 above shows in a graph from the theoretical result for the rank-size rule together with the actual relationships between rank and population size for the cities of England and Wales. The use of logarithmic scales on both axes converts a concave curve into a linear or almost linear result. It can be seen that the population of London is greater than the rank-size rule suggests it should be, but that the trend of lower-order towns accord roughly with the rule. Many developing countries show a sharp fall from the largest city (primate city) to the other cities and this is known as primate rule (Fig. 4.2).

Zipf postulates that the size and number of settlements in any nation are governed by two sets of forces, i.e., (i) the forces of diversification, and (ii) the forces of unification. The balance between the two forces results in the regularity of settlement size and number.

Figure 4.2 *Deviations in Rank-Size Distribution*

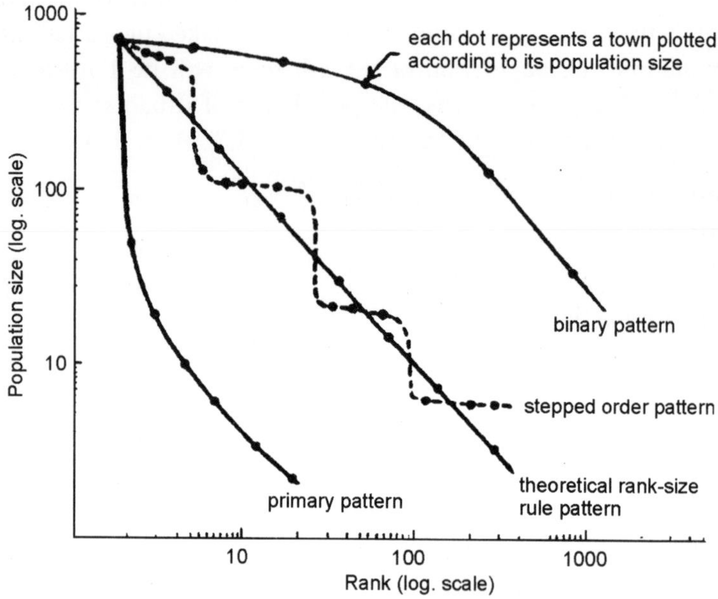

Forces of Diversification

The location of small settlements is generally determined by nearness to the source of raw materials. In such a situation, where primary economic activities predominate, land becomes the basic raw material or resource. Land is tilled by farmers to produce food and other basic necessities of life. A peasant society rooted to the land merges with a large number of village settlements within walking distance of each other. Similarly, apart from agriculture, other primary activities such as mining, fishing and forestry also generate dispersed settlements of small size at regular intervals of distance. As society advances, secondary production makes it possible to locate settlements of greater distances from the source of raw materials. Thus, the settlements specializing in secondary production can be located farther apart, and also be larger in terms of population. Nevertheless, a wide range of secondary economic activities must be located near the source of raw materials so that the costs of transportation can be minimized. Secondary economic activities generate settlements of large size and greater distances apart as compared to primary activities.

Forces of Unification

In contrast to the forces of diversification, the forces of unification result in the emergence of few large settlements. Here, the focus is on tertiary economic activities. Nearness to the market, rather than the source of raw materials, is the determining factor in the location of settlements. The size of market is measured by the population of the settlement itself. Thus, a large settlement in itself constitutes a large market. Tertiary activities, such as education, health and administration, are all consumer-oriented and tend to be concentrated in large cities. In recent times, a wide range of secondary activities have acquired a market orientation (for example, electronic and engineering goods and information technology industries). These secondary economic activities also tend to concentrate in large metropolitan cities. These forces lead to the emergence of a few very large cities.

Rank-Size Relationship in India

The rank-size relationship is absent in India at the national level as the population size of Mumbai, Kolkata and Delhi is very close to each other. Moreover, a great majority of states in India also do not conform to the rank-size rule. In fact, primacy exists in at least 15 out of the 28 states of India and in another eight states (Bihar, Kerala, M.P., Punjab, Orissa, Goa, Arunachal Pradesh and Nagaland) the leading city is only just larger than the second city. In Kerala, the three cities of Cochin, Calicut, and Thiruvananthapuram (Trivandrum) have nearly the same population size; this is also the case with cities of Indore, Jabalpur and Bhopal in Madhya Pradesh, and Ludhiana, Amritsar and Jalandhar in Punjab. Rank-size relationships appear to hold good in the state of Rajasthan. In brief, rank-size rule in India is an exception rather than a rule.

Criticism

The rank-size rule of Zipf is rather rigid and rarely found in an exact empirical fit. Moreover, it attaches great importance to the population size of the largest city in a region or country. The population size of every other settlement depends on the size of the largest city. The application of rank-size rule is difficult because there is no universal definition of city sizes. There are many cities where the built-up area extends outside the administrative boundaries of the city and where many city workers live beyond the edge of the built-up areas.

Looking at the problem of application of rank-size rule, it can be better used for comparative purposes. The rank-size rule is more descriptive rather than explanatory or predictive.

Primate City

The law of primate city was developed by Mark Jefferson in 1939. It is based on the observation that the country's leading city is usually disproportionately larger than others in the system. For example, London is seven times the size of Liverpool, Copenhagen nine times larger than Aarhus, and Mexico City five times the size of Guadalahara. In other words, a primate city has a stratum of small towns dominated by a very

Figure 4.3 *Kathmandu and Thimpu: The Primate Cities of Nepal and Bhutan*

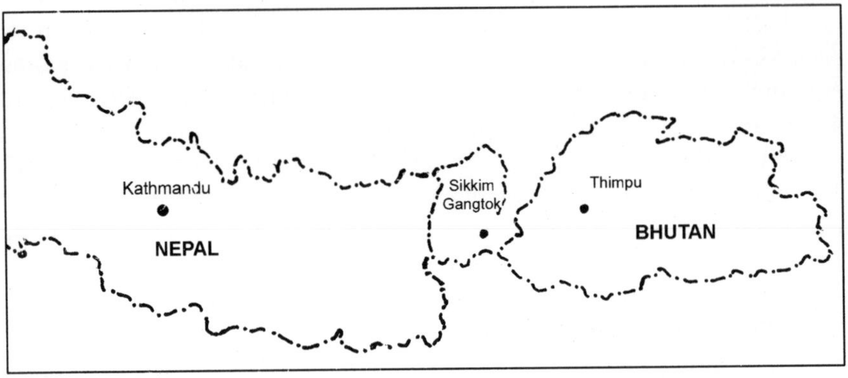

Figure 4.4 *Primate Cities of South America*

large city with an absence of intermediate centres. Jefferson noted that the ratio of the population of the three largest cities is approximately 100:30:20 (i.e., the third largest is one-fifth the size of the largest). According to Jefferson, the largest city has the pre-eminence in infrastructure, economic, social and political affairs.

Primate cities are found almost in all countries—developed or developing. The factors responsible for the primacy and degree of primacy in each one of them vary. The main causes for the development of primate cities are: (i) small territorial extent of the country, (ii) relatively high density of population, (iii) low per capita income, (iv) a high degree of dependence upon agricultural exports, (v) high rates of population growth, and (vi) in many cases, a formal colonial status. Many of these attributes are, of course, typical of developing countries.

The primate cities are mainly found in small countries. For example, Reykjvik in Iceland, Copenhagen in Denmark, Kathmandu in Nepal, Lagos in Nigeria, Lima in Peru, La Paz in Bolivia, Quito in Ecuador, Caracas in Venezuela and Bogota in Colombia (Figs. 4.3 and 4.4). Following factors are responsible for creation of primate cities:

1. Countries with a colonial history also have primate cities. For example, Australia, New Zealand, Malaysia, Sri Lanka, Maldive, Kenya, Ethiopia, etc.
2. Export-oriented economy. A country which is dependent on export of raw materials shows the tendency of primacy.
3. Simple economic and political economy lead to primacy. Countries with small areas and population, low per capita incomes and agriculture as the basic economic activity also have a tendency towards primacy. The cities of Khartoum (Sudan), Addis Ababa (Ethiopia), Mogadishu (Somalia), Kinshasa (Zaire), Nairobi (Kenya), Montevideo (Uruguay), Mexico City (Mexico) are some of such examples.
4. Countries having dual economy (partly labour-intensive and partly capital-intensive) also have primate cities. Dual economies in Indonesia, Malaysia, Sri Lanka and Thailand have been responsible for making Jakarta, Kuala Lumpur, Colombo and Bangkok primate cities.

The agglomeration effect on the growth of cities is a worldwide phenomenon. It has operated in the past and continues to operate even today. All over the world, the leading cities tend to attract the best of human talent and skills. Entrepreneurs are attracted to these cities and with them comes the capital needed for investment in industry. The primate city is perceived as providing the greatest opportunity for employment. The effect of concentration of skills and capital in the leading city is seen in its rapid growth. This also leads to the draining away of skills and capital from a number of smaller cities and towns towards the primate city. These smaller cities and towns stagnate and even decline; they find it difficult to compete with the primate city, and as a result the gap between the primate city and the second largest city tends to widen through time.

The primate city, in course of time, becomes the dominant economic, social, cultural and political nerve centre of a country. The presence of people from all parts of the country makes the primate city a mini-country in itself. Furthermore, since it has the best talent in the country, it symbolizes national achievement. The primate city is, therefore, much more than a premier city—it is, in fact, a symbol of the nation. The existence of primate city signifies strong forces that unify a nation. These forces may be cultural, political or economic, or all of these together.

Primacy in India

There is no primate city in India at the national level at present. The absence of a primate city is partly explained by the large size of the country, its colonial heritage, and weaknesses in the nationalist forces in the country. No city in India exercises dominance over the entire nation.

Figure 4.5 *Regional Primate Cities of North-East India*

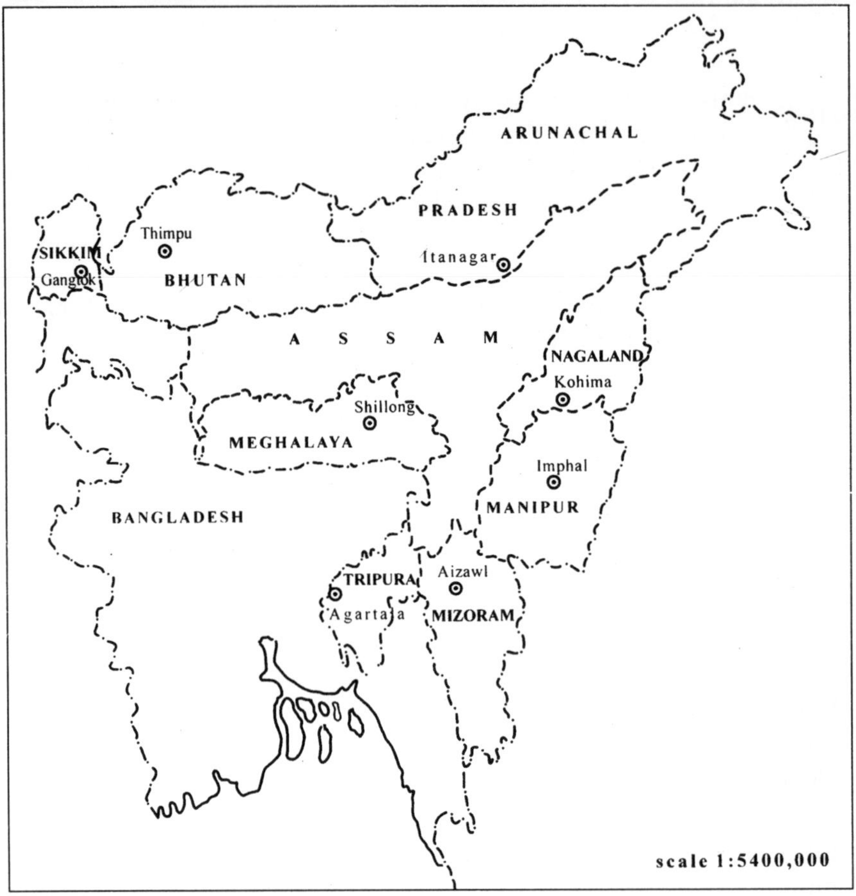

At the state level, however, there are primate cities in fifteen out of the twenty-eight states. Several large states, such as West Bengal, Karnataka, Tamil Nadu, Maharashtra, Andhra Pradesh and Gujarat have primate cities, which have population at least three times the population of the second-ranking city. Among the smaller states, Manipur, Meghalaya, Mizoram, Sikkim and Tripura show a distinct tendency towards city primacy (Fig. 4.5). In the Valley of Kashmir, the city of Srinagar has the status of a primate city (Fig. 4.6). Primate city characteristics are absent in the northern states, from Orissa, and Bihar in the east to Rajasthan and Punjab in the west.

After independence, in India, there has been a distinct tendency, for primacy to increase. This has been because of growing imbalances and increasing concentration of development activities in state capitals. The state capitals, as a matter of fact, provide more employment opportunities.

The law of primate city is still relevant, though the sequence that Jefferson identified is now largely ignored.

Figure 4.6 *Srinagar: The Primate City of the Valley of Kashmir (J&K)*

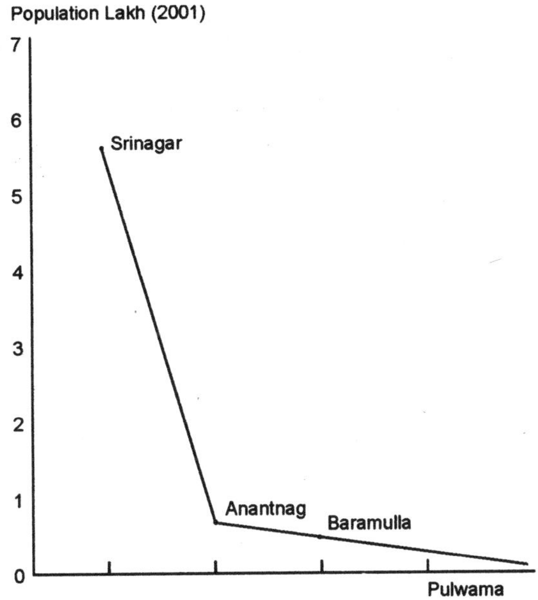

Urban Morphology or the Internal Structure of Towns and Cities

The internal structure of cities and towns is one of the fascinating areas in which the urban geographers are increasingly concentrating. The term 'urban morphology' refers to the physical arrangement or structure of town, its pattern of streets, building blocks, individual buildings, their different functions, densities and lay out. In brief, it is the study of urban land use.

A number of theories and models have been developed by the urban geographers to explain the morphology of urban places. Models of urban land use or structure are of two kinds: (i) partial models, which are concerned with the location of one set of activities, such as residential or industrial land use, and (ii) comprehensive models, which deal with all activities in the urban area and their interrelationship.

Central Business District (CBD)

CBD is the nucleus of an urban area, containing the main concentration of commercial land use (shops, offices, financial institutions, banks, restaurants, places of recreation, and warehouses). It is the most accessible point in the urban area and its peak land value. The CBD contains the densest concentration of land uses and the tallest non-residential buildings within the urban area, and is spatially structured internally, with different uses and categories of use (e.g., jewellery and clothing shops). The main characteristics of a CBD are as follows:

(i) it lies central, at least in terms of its accessibility;
(ii) it has a greater concentration of tall buildings than any other region of the city;
(iii) it normally includes most of the city's offices and largest retail stores;
(iv) it is an area where vehicular and pedestrian traffic are likely to be most concentrated;
(v) it averages higher assessed land values and taxes paid than any other part of the city, and
(vi) it draws its business from the whole urban area and from all ethnic, cultural groups and classes of people (Murphy, 1972).

It has also been said that for out of town people the CBD is synonymous with the city itself. These characteristics of the CBD can be precisely measured. Unfortunately, none of these measures is entirely satisfactory and in practice there is a high concentration of these elements at the heart of the CBD and a gradual diminution of these essential characteristics outwards an indefinite boundary. Indeed, the edge of CBD is best thought of as a zone rather than a boundary. Typically, the CBD merges into a 'blighted' zone of obsolete and deteriorated property which lies between the CBD itself and the surrounding residential districts of the inner city.

According to Murphy, the land valuation data could provide the most objective and widely applicable method of CBD delineation. However, the best means of delimited is probably based on a combination of land valuation data, if available, and land use mapping, whereby the transition from CBD functional areas of the city can be identified. The CBD is, however, a dynamic concept which changes in space and time.

Internal Structure of the CBD

Land use in the CBD shows a tendency towards an ordered arrangement which is delimited by the pattern of land values. The land values decline from the core towards the edge of the CBD. Because of the high land values and high rents of the property, only companies with a large turnover and high profits can afford to conduct their business at the heart of the CBD. Thus, at the core of the CBD, there are departmental stores, major chain stores, supermarkets and headquarters-offices, while the small traders, with only modest profits, are forced out towards the edge of the central area.

In general, the CBD consists of a series of concentric zones, each with a distinct land use. The CBD also has banks, real estate agents, legal offices, and insurance company offices. Major rail or bus routes terminate at the edge of CBD. This can create an area of distinctive land use in the outermost zone with hotels, boarding houses, restaurants and shops, catering for a large transient population. In some cases, small retail stores survive near the heart of the CBD adjacent to large departmental stores by

Figure 4.7 *Connaught Place: A CBD of New Delhi*

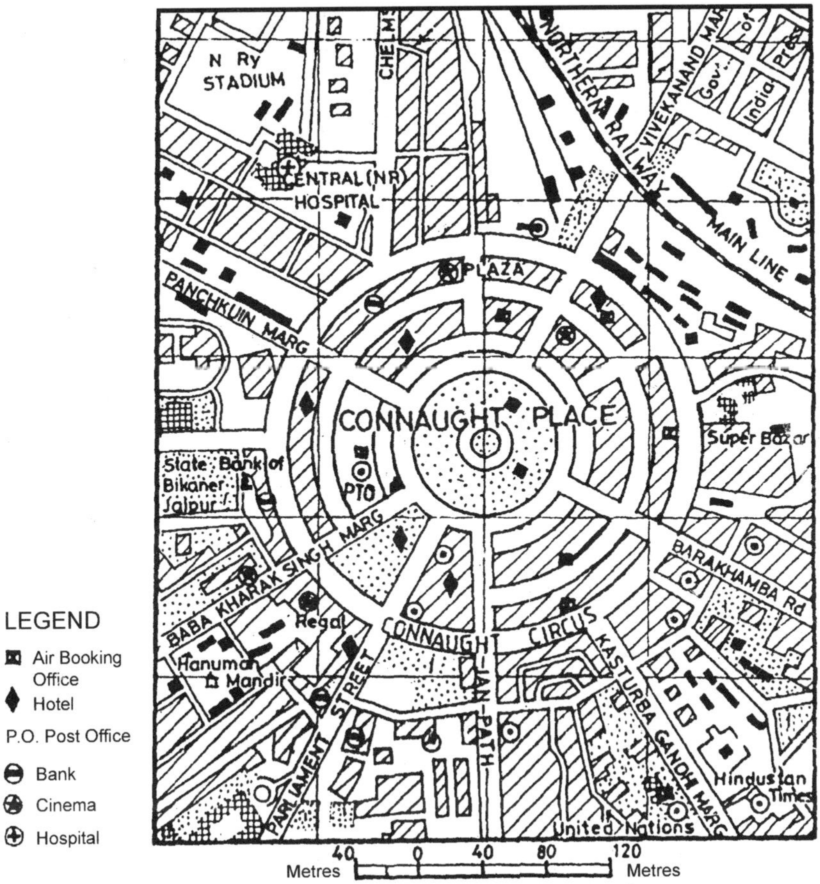

utilizing the drawing power of their larger neighbour or by selling high-grade specialist goods. The Connaught Place of New Delhi is an example of CBD (Fig. 4.7).

The CBD is increasingly under threat due to a combination of traffic congestion, which has led to parking restrictions, and to the growth of out-of-town developments, including superstores. Most CBDs are in relative if not absolute decline, as their characteristic uses are increasingly decentralized to suburban and exurban locations, as with the growth of planned shopping centres and office parks close to major highway intersection.

Theories of Urban Structure

Although each city is unique in respect of the detailed pattern of its internal land use, there is nevertheless a considerable degree of repetition in the broad geographical arrangement of the various categories of urban land use from one city to another. As a result, a number of theories have been formulated which attempt to describe and explain the patterns involved. About the internal structure of cities, three popular models, based on empirical studies in different cultural settings, are important. These theoretical structures are referred as:

(i) The Concentric Zone Model by E.W. Burgess
(ii) The Wedge or Sectoral Model by Homer Hoyt and M.R. Davis
(iii) The Multiple Nuclei Model by C.D. Harris and E.L. Ullman

The Concentric Zone Model

The concentric zone model of urban land use was first proposed by E.W. Burgess (1924, 1927). This is also known as the 'ecological theory of city structure'.

Burgess' main research interest was the determinants of urban social problems, such as vice and crime, and his mapping of their occurrence within Chicago indicated a concentration in certain type areas only, but was later seen to have more universal application. Burgess' model is based on the idea that the growth of a city takes place outwards from its central area to form a series of concentric zone.

Burgess identified a system of five concentric zones, each characterized by a particular type of land use (Fig. 4.8):

Zone I (Central Business District-CBD): CBD occupies the central place and it is the focus of city's commercial, social and civic life. This part is most readily accessible part of the city as all routes radiate here. The heart of this zone is known as the *down town*. It contains departmental stores, smart shops, office buildings, financial institutions, banks, insurance offices, real estate offices, hotels, restaurants, wholesale trade, theatres, recreation places and headquarters of economic, social, civic and political organizations. The land value and rent of space are very high in the CBD.

Zone II (Zone in Transition): This zone surrounds the CBD and it is the zone of residential deterioration. There is invasion of business in this area which led to a decline in residential desirability. Some slums may also develop in this area. There may be some light industries in the inner part of this zone. The owners of property in this zone are not interested in improving the quality of buildings. This area is occupied by lower income groups. They live close to the place of work as they cannot afford high cost of transport.

Zone III (Zone of Independent Workmen's Home): This zone is made up largely of the homes of factory workers and labourers and these are the people who often moved out of the zone II (zone of transition). In India, generally, this area has planned flats. The density and occupancy of this zone is always high.

Zone IV (Zone of Better Residence): In this zone the majority of the middle and upper-middle class people reside. Small businessmen,

Figure 4.8 *The Concentric Theory of Urban Structure: (a) the model; (b) urban areas in Chicago*

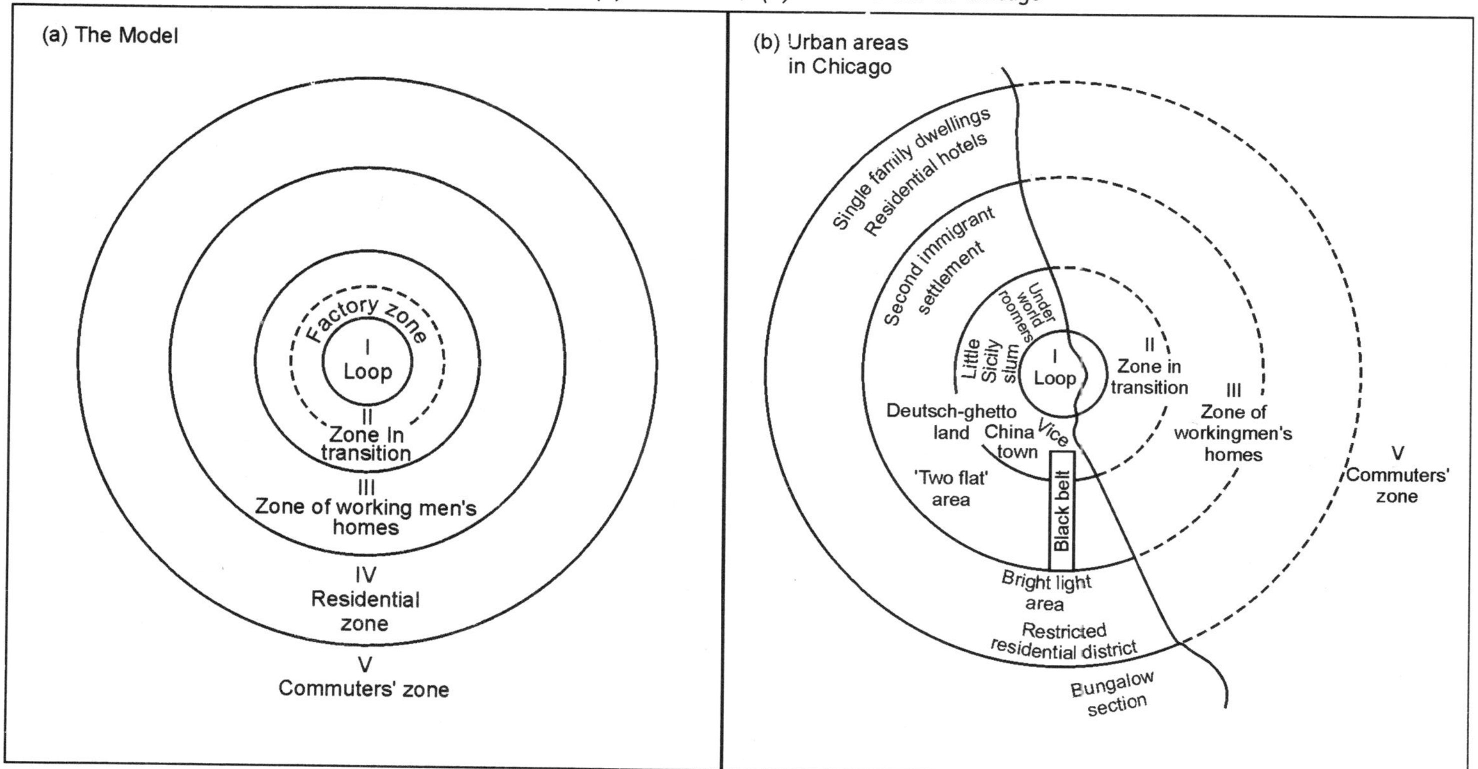

(After E.W. Burgess, 1924)

professional people, better paid government employees and salesmen reside in this zone. This area is dominated by single families.

Zone V (Commuter's Zone): Beyond the city proper are located a small number of small towns which are mostly dormitory towns in that most of the people living here commute to the city every day for work. Most of the people living here work in the CBD. Here are large spacious houses usually having their own lawns etc. and are built at some distance from each other. The cost of land is comparatively low.

As stated above, each of the city zone, according to Burgess, has a tendency to expand outward.

Criticism

1. The Burgess' model suggests a circular shape of the CBD. The CBDs are generally either square, rectangular or linear and not necessarily circular in shape.
2. There is overemphasis on commerce and trade activity and a complete neglect of the industrial land use.
3. The outward expansion of the Zone II into Zone III has not been explained. In fact, how the lower income group people can encroach the outer space of Zone III occupied by rich people.
4. The effect of major transport has been ignored. People prefer to stay close to the roads. Consequently, the space along the roads has high valuation and high rent value.
5. The topographical factors have also been ignored. Concentric zones are not possible in the areas of undulating topography and the cities located near lakes, rivers and other water bodies.

The Wedge or Sectoral Model

The wedge or sectoral model of urban structure was propounded by Homer Hoyt and M.R. Davis in 1939. It was based on the study of 64 American cities.

According to this theory, patterns of urban land use are conditioned by the arrangement of routes radiating out from the city centre which create a sectoral pattern of land and rental values which in turn influences the urban land use pattern (Fig. 4.9).

According to the proponents of this theory, a high-rent residential district in one sector of the city will migrate outwards by the addition of new belts of housing along its outer arc. Similarly, low-rent housing might expand outwards in a different direction. In other words, once contrasts in land use have developed near the city centre, these differences will be perpetuated as the city expands. This idea of a sector-like expansion is an improvement on the earlier concentric model in that it takes into account both (i) the distance and direction of expansion, and (ii) acknowledges the importance of transport routes on the growth of the city.

Figure 4.9 *The Sectoral Theory of Urban Land Use*

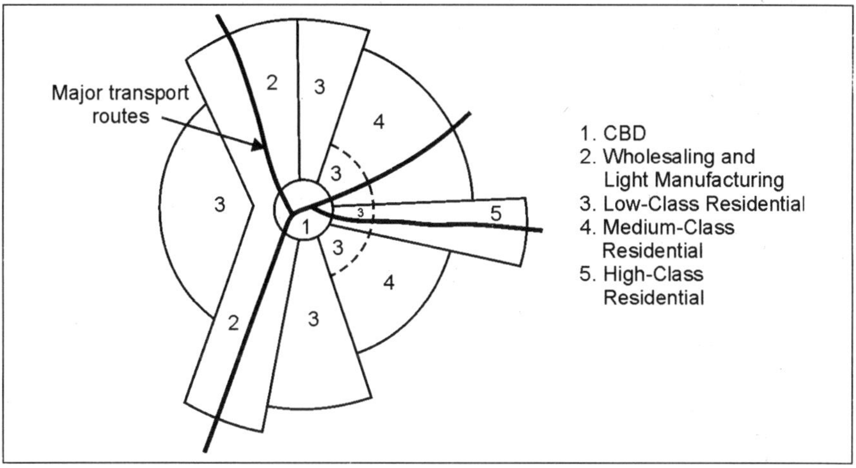

(After H. Hoyt and M.R. Davis, 1939)

The Multiple Nuclei Model

According to C.D. Harris and E.L. Ullman, the concentric and sectoral models have the advantage of an essential simplicity, but actual patterns of land use are generally far more complex and varied, than either model would suggest. Consequently, in 1945, a less rigid model capable of application to a variety of urban patterns was proposed by Harris and Ullman. This was termed the multiple nuclei theory of urban structure.

It was suggested that land use patterns in most large cities develop around a number of discrete centres of nuclei rather than a single centre as described in concentric and sectoral models. The clustering of related land uses around these nuclei in the city creates a cellular structure, the pattern of which will be largely determined by the unique factors of site and history of any particular city. The multiple nuclei theory does not produce a simple model of urban structure appropriate to each and every city, but suggests a number of general principles which are relevant to the land use pattern of most cities (Fig. 4.10).

In this model the term 'nucleus' has been used to refer to any attracting element around which urban growth may take place.

The proponents of this theory emphasized that such multiple nuclei have been present right from the time of origin of cities, presenting as urban growth has filled around them. Metropolitan London is cited as one of the examples with 'The City' and 'West Minister' originating at separate points in the open country—one as a centre of finance and commerce, and the other a political centre.

In fact, with the growth of city, the CBD loses its uniqueness—that it may become just one of the many centres in the city. The high grade activities will abandon it for centres better located to serve the high income

Figure 4.10 *The Multiple Nuclei Theory of Urban Structure*

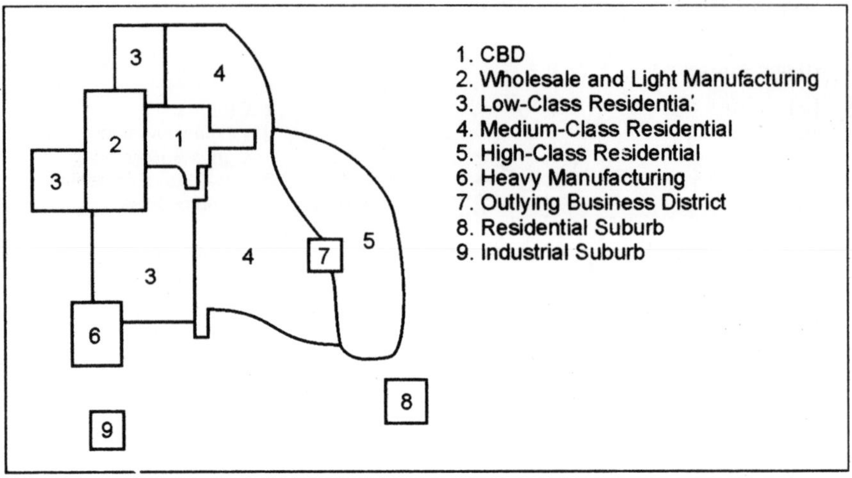

(After C.D. Harris and E.L. Ullman)

areas. The dominant nuclei are proliferating in this manner. Other centres will develop on a regional or specialized basis, thus strengthening the multiple nuclei generalization.

This model is considered to be more scientific which explains the present growth of urban centres both in the developing and the developed countries. This theory goes much farther than the previous two in explaining the modern expanding cities where a number of nuclei, each specializing in particular functions, are evolving.

In brief, none of the three models provides a true explanation of land use in urban areas. Each of the models, however, has some merits.

It must be underscored that urban land use is a dynamic concept. The morphology of a city changes with the change in time and space. The development of transport is constantly changing the internal structure of the urban places.

Internal Structure (Morphology) of Indian Cities

The internal structure of Indian cities is the result of geo-ecological, socio-economic and cultural factors. Consequently, the characteristics and internal structure or urban land use pattern of Indian cities varies from city to city and from one urban place to another. The Indian cities have several focal points, for example, the main market, the railway station, the bus stand, and the professional, educational, medical and recreational institutions. All these focal points are linked with residential areas of rich as well as poor. Religious, linguistic and ethnic minorities have traditionally sought refuge in the city. The Indian cities are, consequently, heterogeneous in terms of population composition.

The internal structure of Indian cities can be examined in terms of (i) layout, (ii) demographic characteristics, and (iii) functions. From the historical point of view, the Indian cities may be divided into the following three categories:

(a) cities of the ancient period,
(b) cities of the medieval period,
(c) cities developed during the British period, and
(d) newly developed planned cities after independence.

Ancient Cities

The cities of the ancient period were the capital cities of the Hindu kings. These cities were surrounded by walls. The centre of the city was occupied by a royal palace. Near the centre and the temples were the residence of high-ranking Brahmins, and ministers of royal cabinet. The Sudras were

Figure 4.11 *Ancient Aryan and Tamil Cities, 600–300 B.C.*

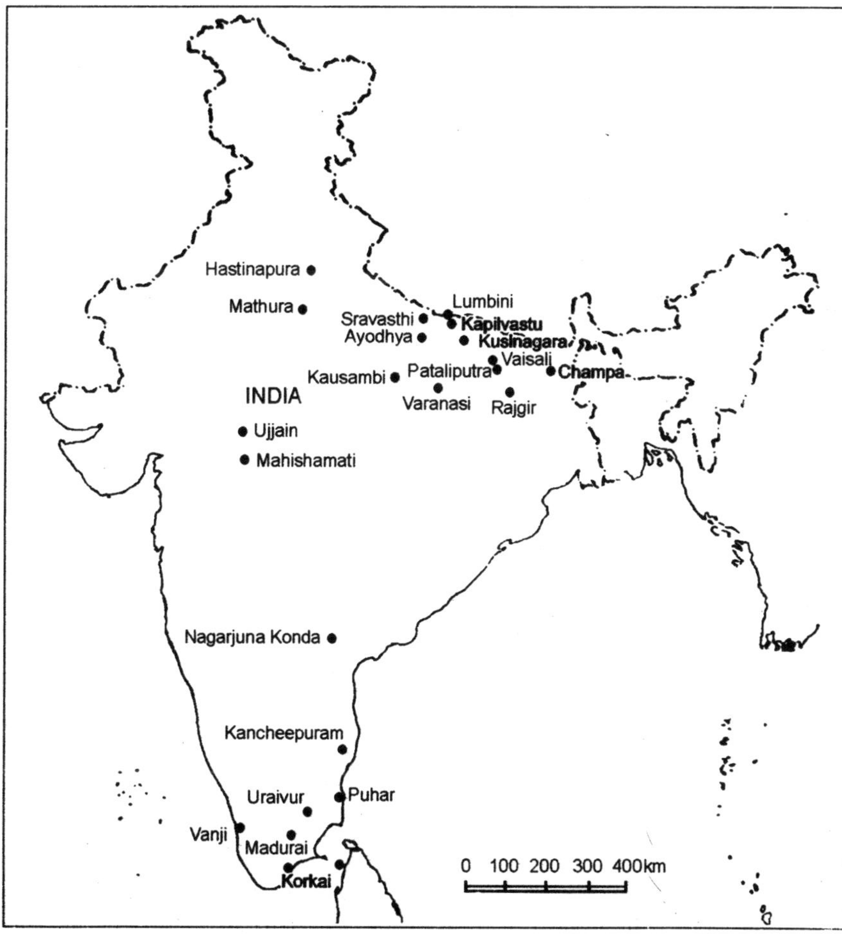

(After R. Ramachandran)

given the southern side. Patliputra (Patna), during the period of Maurya Empire, was one of the most suitable examples of this type of ancient cities (Fig. 4.11).

The impact of Dravidian culture is most marked in cities of south India. In these cities, the temples and tanks acted as nuclei around which grew commercial markets and houses of Brahmin priests and scholars. The rulers built their palaces near the temple. These cities were also surrounded by protective walls. The roads radiated from the centre to the other parts of the cities.

Medieval Cities

During the medieval period of Indian history, the Muslim imprint on the city structure is significantly conspicuous. The Muslims introduced many new elements in the city structure that included fortification, Masjid,

Figure 4.12 *Selected Medieval Indian Cities*

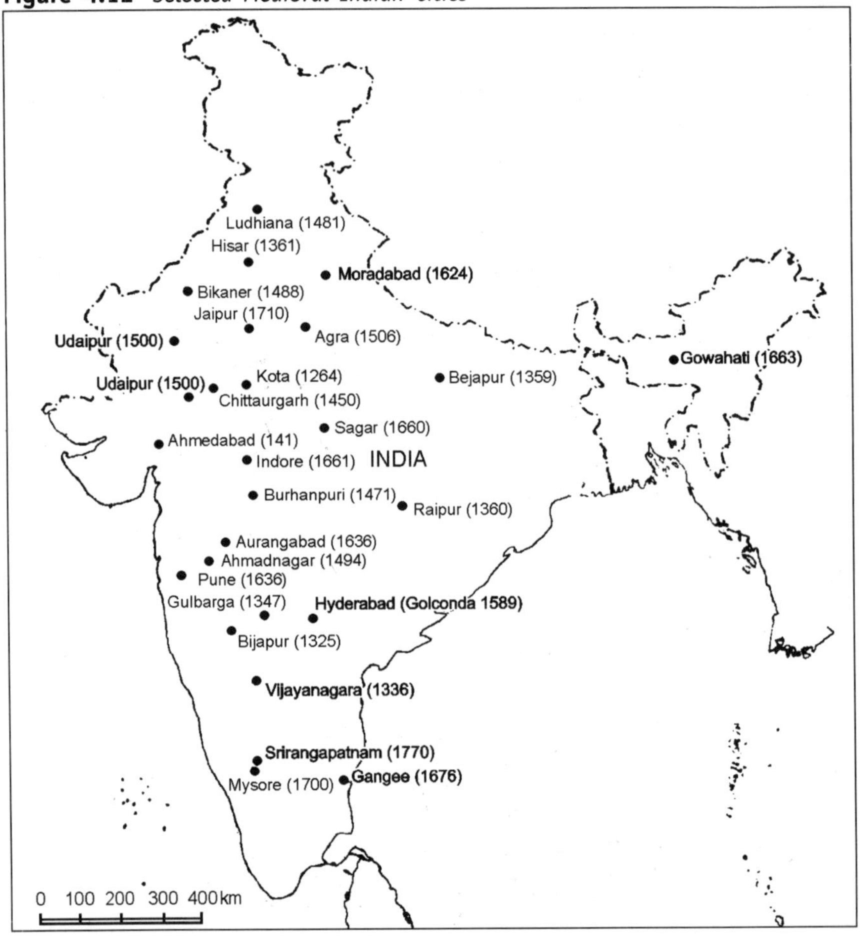

(After R. Ramachandran)

bazaar (market) and a *chowk*, and residential segregation. The most representative city of the Muslim impact may be said to be Shahjahanabad (old Delhi), constructed by Shah Jahan. The city was built on the bank of Yamuna. Its architecture was a fusion of Islamic and Hindu influences. The cities of the medieval period were surrounded by brick walls without a moat. All cities had gates for regulated entries. Every city had a market centre—the main *chowk* (cross-roads) of the city.

Cities of the British Period

The British rulers modified the traditional Indian cities of the ancient and medieval periods. The development of railway lines and increasing commercialization resulted into new patterns in urban land use. Just to have segregated residential areas for the whites, they developed the civil lines and cantonments. The demolition of fortification walls meant

Figure 4.13 *Cities and Towns of the British Period*

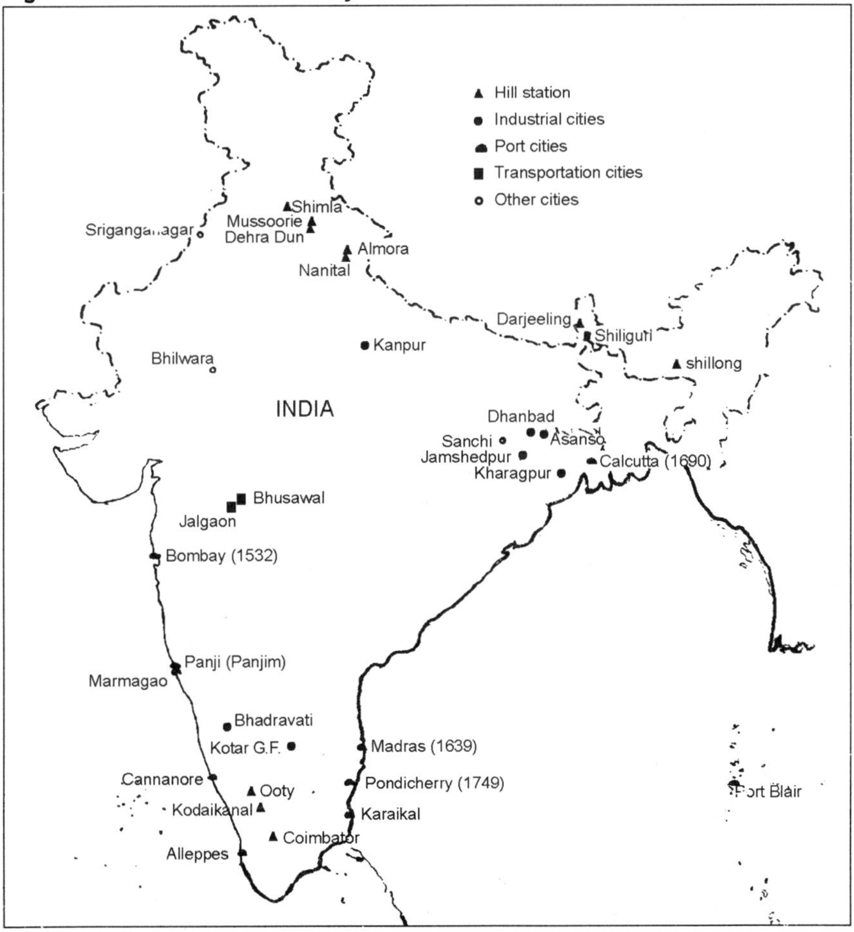

(After R. Ramachandran)

removal of barriers and new extension began to grow along main road and railways. Further commercialization added new nodes in the city, such as Connaught Place in New Delhi. These towns have spacious bungalows, elegant houses, planned streets, Mal Roads with green belts on both sides, clubs for recreation, churches and well maintained gardens, parks and grave-yards (Fig. 4.13).

As a result of the historical processes, the morphology of Indian cities has been transformed substantially.

Cities Developed after Independence

During the post-independence period a number of planned cities have been developed. Chandigarh, Bhubaneshwar, Gandhinagar, Dispur and Bokaro are some of the typical examples of a planned cities (Fig. 4.14).

The city of Chandigarh planned by Le Corbusier was based of the seven principles of roads for fast movement of people and goods. In this city, the concept of sector has been incorporated. Each sector has marketing, educational, medical, recreational and parks facilities. The sector principle has also been applied to Bokaro, Noida, Gurgaon and many new extensions in Bhopal, New Delhi, Navi Mumbai, Jaipur and Indore.

Spheres of Urban Influence (City Region, Zone of Urban Influence)

City and countryside are mutually interdependent. In fact, cities are dependent on their surrounding regions for the supply of vegetables, milk and grain. The industrial cities also depend on the countryside for raw materials of mineral or agricultural origin. In opposite to this, the city is the centre for a wide range of tertiary services, whereas countryside depends on the city for the services. Rural people come to the city daily for medical and educational facilities, for entertainment and for shopping. The city is the focal point for the distribution of a wide range of consumer goods—from clothing and footwear to electronic gadgets. Improved transport and communications have allowed towns to extend their services further into the surrounding rural area and have also permitted urban workers to live far beyond the city limits and travel to work each day. The countryside is trapped in the process of modernization and is, thereby, very dependent on the city for goods and services. A city may either retard or promote growth in its surrounding region. In brief, the modern city is a centre of secondary and tertiary economic activities. Thus, cities and countryside are physically, socially and economically interdependent, affecting each other. The degree of interaction, between the city and the countryside, however, tends to decrease with distance. The exact nature of the economic dependence of any town upon its surrounding regions has been considered briefly in the following pages:

Figure 4.14 *Chandigarh: A Planned Post-Independence City*

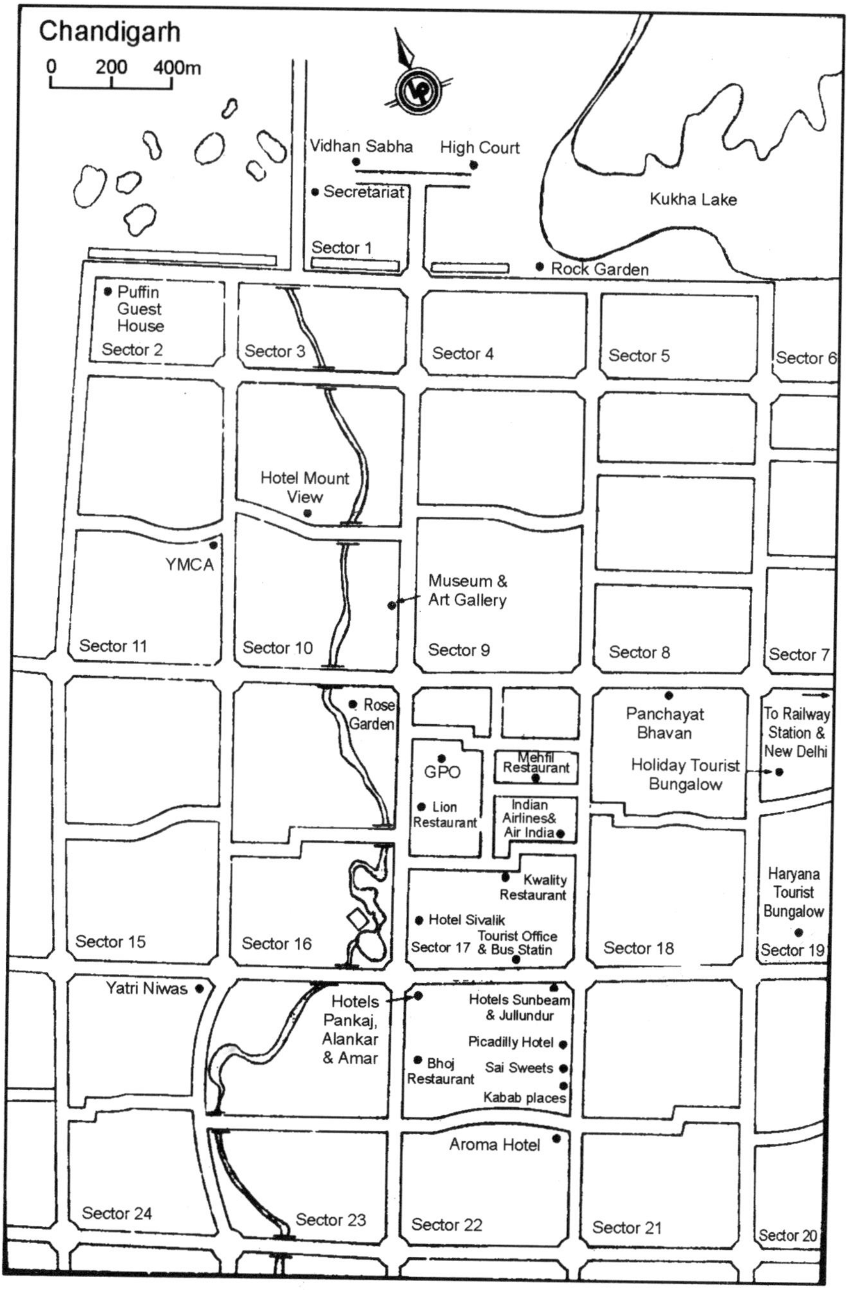

Distance Decay Effect

The intensity of the interaction between a city and its neighbouring towns and villages declines rapidly with distance from the city. Thus, people from nearby villages, within a distance of four to five kilometres, may commute daily to the city for work, shopping, education, medical treatment or entertainment. Farther away, there is a decline in the number of people commuting to the city. Beyond a certain distance, daily movements are replaced by weekly or even irregular movements, until ultimately such movements to the city become rare or absent.

The distance decay effect may be seen in a variety of situations. Apart from the effect on daily or periodic commuting to the city, the effect is also seen in the distribution of goods and services from the city to the surrounding villages. Similarly, the quantum of milk, vegetables, fruits, flowers etc., coming to the city from the villages, is also subject to distance decay. Similar effect may also be noticed in the case of hospital patients and students attending schools or colleges.

The form of distance decay can, in fact, be mathematically specified as a negative exponential curve (Fig. 4.11). This may be expressed as follows:

$$F = aD^{-b}$$

$$\text{or } F = e^{-(a+bD)}$$

where F stands for frequency of interaction and D is distance from the city, while a and b are parameters. The rate of decline with distance from the city is determined by the parameter b. This rate varies within narrow limits, depending on the different purposes of interaction within a region, as well as between city regions, and reflecting differences in the modes of transportation used.

Structure of City Region

The city region is an area around the city over which the city exercises a dominant influence in relation to other neighbouring cities of equal importance. The areas of city influence are contiguous areas around a city from where people commute to the city to obtain certain goods or services. A cinema hall in a city may attract patrons from several villages around the city. The continuous area encompassing all these villages in the area of influence of the city with respect to entertainment through cinema. Likewise, various institutions in the city such as hospitals, colleges, schools, etc., have their corresponding areas of influence.

Methods of Delineation of Areas of Influence

A number of methods have been adopted by the urban geographers to delineate the areas of influence of cities and towns. Under Indian conditions the following four methods have generally been used for the delimitation of city regions. These are: (i) the use of published and unpublished secondary data and information; (ii) the use of primary information through interview of consumers or managers of establishments in the city or town; and (iii) the use of primary information obtained from field interviews of persons from a number of villages in the area surrounding the town.

(i) *Use of secondary data*: In this method, data is obtained from educational institutions (schools, colleges, etc.) Schools and educational institutions generally maintain records of students which include residential addresses of their parents. From this information it is possible to locate the villages or towns from where the students come to schools. The actual location of the villages may be obtained from the Survey of India topographical maps of the area. Similarly, hospitals and banks maintain similar records of their patients and customers. Newspaper data has been effectively used by R.L. Singh and Manzoor Alam to delimit areas of influence of Varanasi and Hyderabad respectively. By using the secondary data, a series of isopleths may be drawn around the city for each service to show the intensity of the service.

(ii) *Use of data from field enquiry in the city*: Areas of influence of a city may be delineated on the basis of data obtained from interviews of persons in the city itself. The interviews may be confined to few persons, such as wholesale dealers, who are requested to give names and locations of their retail clients. The customers themselves may be interviewed briefly in front of shops, cinemas, dispensaries, hospitals, vegetable markets and other establishments. In all field enquiries familiarity with village names and locations is necessary. Adequate preparatory work and reference maps must be prepared for this purpose.

(iii) *Use of data from field survey in the city region*: Field surveys in the surrounding region of a city may be used to determine the nature of the dependence of villages upon the city and to identify the competing cities in the region. The main problem in the collection and use of this data is that there are a number of villages around a city and collection of data may involve more time and resources. Moreover, the villagers may be dependent on more than one town to avail certain services. Moreover, the information obtained from these individuals is subject to personal prejudice and bias and it is always necessary to corroborate information from at least independent interviews in the same village.

Characteristics of Area of Influence

One of the striking characteristics of areas of influence of Indian cities is the near uniformity in terms of actual size of area served by each of the

Figure 4.15 *Area of Influence of Varanasi and Kanpur*

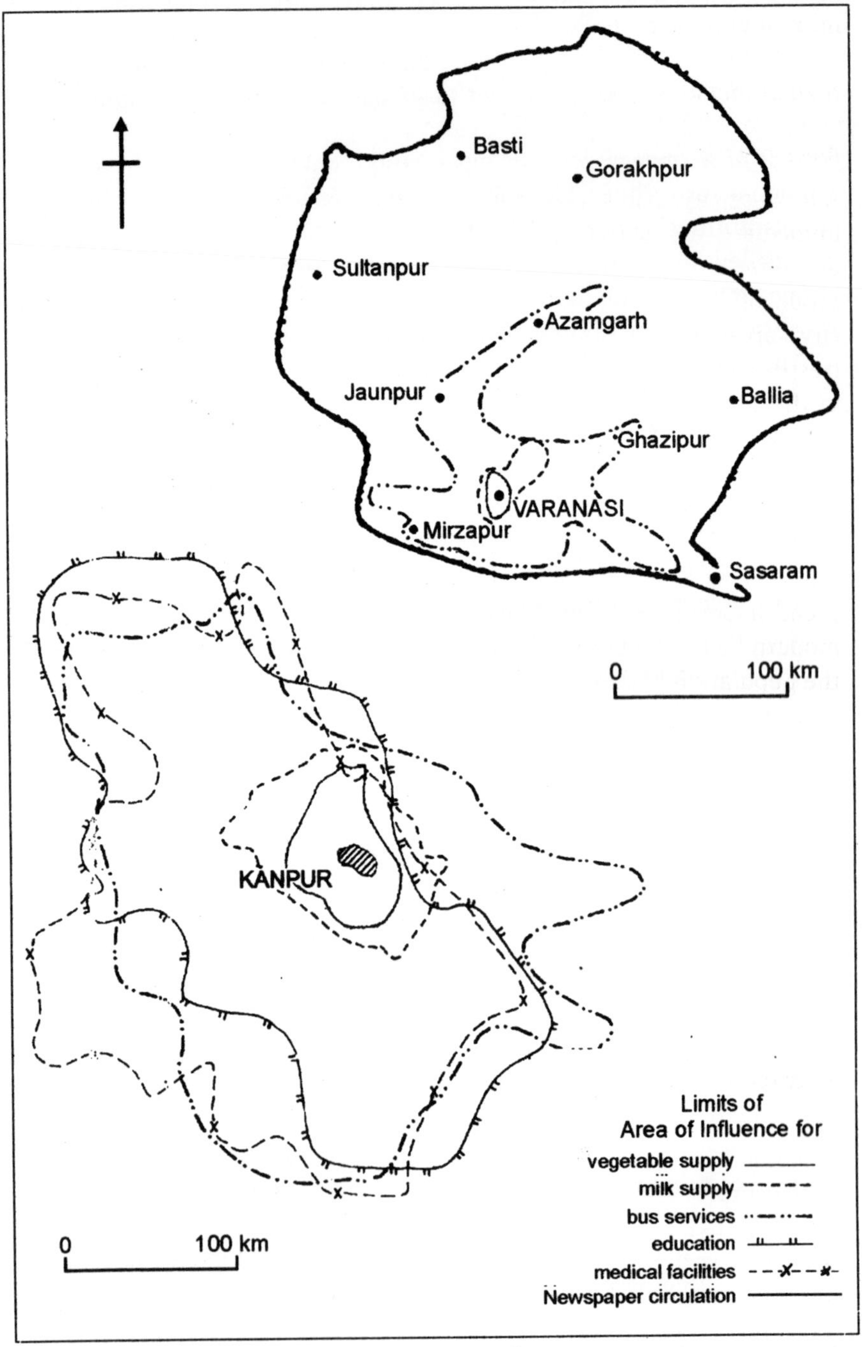

criteria used for the delineation of city region. There is, however, some variation from one service to another. This is the case of the supply of milk, vegetables and circulation of newspapers.

The shape of the area of influence for all services tends to be rather regular and almost circular in most cities (Fig. 4.15).

Areas of City Dominance

A town or city tends to exert a dominant influence over a small area in its immediate neighbourhood. The area of city dominance is almost by definition smaller in extent than the area of influence of any one service or group of services of the same hierarchical level. This is so because the city's areas of influence tend to flow over its own area of dominance and trespass upon the domain of competing cities. An overlap inevitably occurs in the zone of competition. The delimitation of areas of dominance of a city is, therefore, far more difficult than the delimitation of areas of influence.

Implications for Planning

The delineation of areas of influence of individual cities and towns has great importance in the process of planning and development. The modern health and educational facilities may be suitably located to serve the population of the city region more effectively. The extension of health, educational, banking recreation and welfare facilities can be extended to the area of influence of a city region.

City Regions in India

Numerous studies by the Indian urban geographers revealed the existence of four primary metropolitan cities and regions. Since independence, the dominance of primary metropolitan cities have been successfully challenged by newer and more rapidly growing cities, such as Ahmedabad, Bangalore, Hyderabad, Jaipur and Ludhiana. Moreover, all the million and one-lakh cities also show definite signs of dynamic growth.

Metropolitan Regions in India

The colonial cities of Calcutta (Kolkata), Bombay (Mumbai) and Madras (Chennai) have dominated India's economic organization since the early 19th century. To this list one should add Delhi, which became the national capital in 1911. Delhi's importance in economic and regional development has been spectacular increased since independence, and at present it overshadows the colonial cities in the national spatial economy. The four metropolitan cities between them account for over 80 per cent of the richest people in the country. They also attract the greatest volume of traffic in terms of goods and people. Their area of dominance includes

Figure 4.16 *Metropolitan Regions in India*

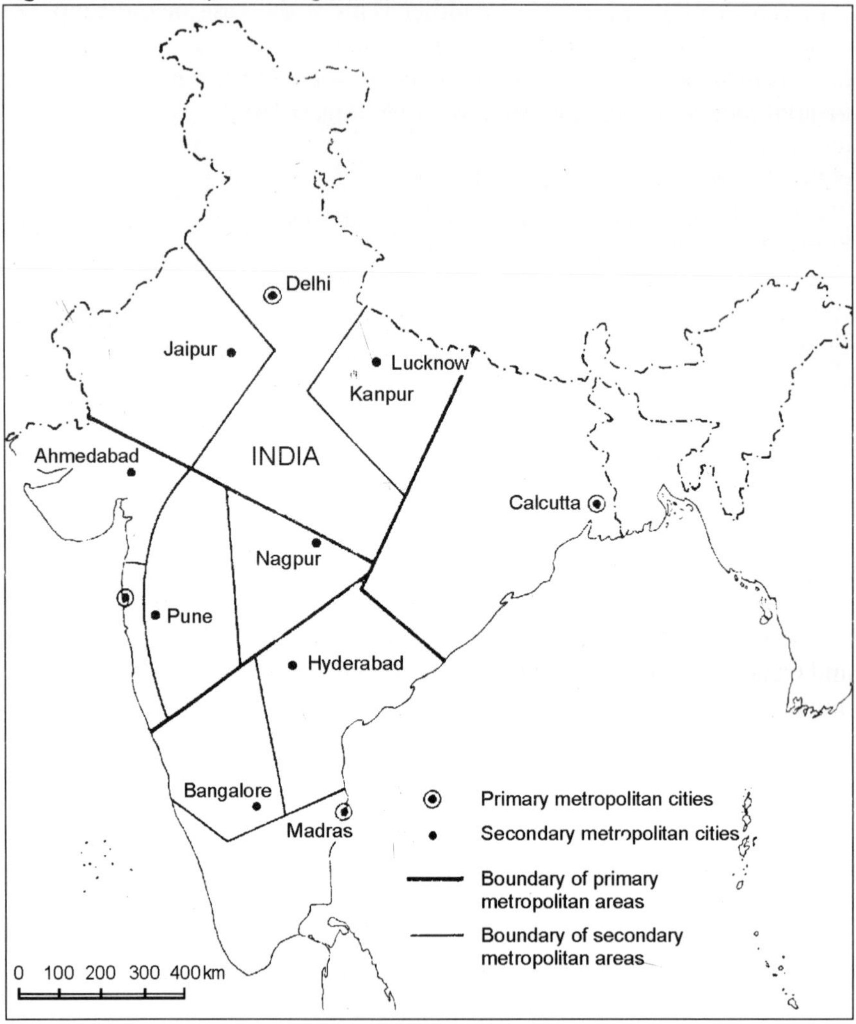

(After R. Ramachandran)

several states. The area of influence of the four metropolitan regions has been demarcated on the basis of: (i) air travel, (ii) railway passengers and commodity flow, and (iii) long distance telephone calls (Fig. 4.16).

The Delhi metropolitan region covers the northern states of Jammu and Kashmir, Himachal Pradesh, Punjab, Haryana and Rajasthan. The western part of Uttar Pradesh, Uttrakhand and most parts of Madhya Pradesh also form the part of the Delhi region. The Calcutta region includes all the states of North-East India (Arunachal Pradesh, Assam, Manipur, Meghalaya, Mizoram, Nagaland, Sikkim, Tripura). Bihar, Jharkhand and Orissa also form part of this region.

The eastern part of Uttar Pradesh and Chhattisgarh form part of the Calcutta region. The Bombay metropolitan region sprawls the states of Maharashtra, Gujarat and Goa and the union territory of Daman and Diu. In addition, it includes the south-western portion of Madhya Pradesh, including the city of Indore, and Malwa region. The Madras metropolitan region includes the four southern states of Tamil Nadu, Kerala, Andhra Pradesh and Karnataka and the southernmost part of Orissa and Madhya Pradesh—the area linked by the Vizag-Jagdalpur railway line (Fig. 4.16).

The city region concept provides an ideal framework for regionalization of the country into a set of hierarchical territorial divisions. Delineation of city regions in India is a tedious job, mainly because of:

(i) clustering of one-lakh cities, and
(ii) absence of one-lakh cities over vast areas in certain parts of the country.

The clustering of one-lakh cities may be found around some metropolitan areas, such as Delhi, Coimbatore, Patna, Agra, Kanpur, Lucknow, Jhansi, Meerut, etc. In the case of clustering cities become complementarity.

The city regions in India, by and large, are much larger than an average district. In the more urbanized states of Tamil Nadu, Maharashtra and Gujarat, the majority of the districts have a city within their territorial limits, while in northern and eastern India, the city regions encompass several districts. The size of the city, therefore, bears an inverse relationship with the level of urbanization. The hilly and mountainous areas of the country have few urban places.

At the meso-level, Srinagar in the Valley of Kashmir, Shillong in Meghalaya, Agartala in Tripura, Aizawl in Mizoram, Imphal in Manipur and Gangtok in Sikkim are all primate cities in their own way.

City Region and Planning

The city region has often been recommended as a territorial unit for regional development planning. The city region has a number of advantages as unit for development planning, which are as under:

(i) The city, being the only focal point, has an easy access to all places within the region. Even, at the initial stage, its influence in terms of tertiary services extends to all parts of the region.
(ii) Economic and social planning can take advantage of existing links between the city and the countryside.
(iii) Development generally takes place in an uneven manner in which the city may become an initial nuclei of development.
(iv) Cities play an important role in bringing about changes in agriculture, rural education and health.
(v) At present, cities have become the new centres of modern industry. The forces of urban and industrial agglomeration help in the development of the city and also lead to the industrialization of

countryside. Thus, the city region provides a convenient and sound spatial framework for such an approach to regional planning.

City Planning

From the Third Five-Year Plan period (1961-66), master plans have been prepared for all the one-lakh cities of India. The city master plans generally use the terms like City Planning Area, City Planning Region or Metropolitan Planning Region.

Rural-Urban Fringe

The concept of rural-urban fringe was propounded by R.J. Pryor in 1968. It is a zone of transition between the continuously built-up urban and suburban areas of the central city and the rural hinterland. The rural-urban fringe has also been defined as "the area of transition between well-recognized urban land use and the area devoted to agriculture". This definition is universally applicable.

In the opinion of Blizzard and Anderson, the rural-urban fringe is that "area of mixed urban and rural land use between the point where full city services cease to be available and the point where agricultural land use predominate". In the context of India, the definition of the rural-urban fringe may be as under:

"The rural-urban fringe is an area of mixed rural and urban populations and land uses, which begins at the point where agricultural land use appear near the city and extends to the point where villages have distinct urban land uses or where some persons, at least, from the village community commute to the city daily for work or other purposes" (Fig. 4.17).

Figure 4.17 *The Rural-Urban Fringe*

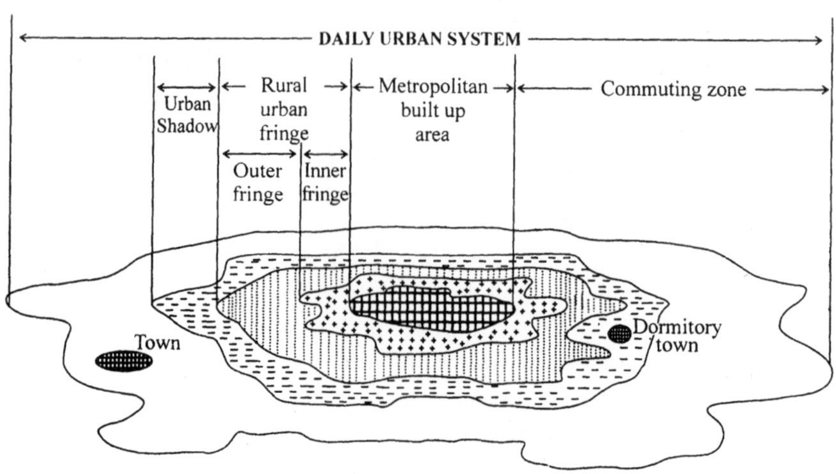

The area displays a changing mix of land use, social and demographic characteristics and is an area in which the large-scale urban amenities are located. In other words, it is a transition zone between the city and its suburb, and the countryside. Certain types of land use are characteristics of this zone. For example, garden centres, farm houses, horse-riding stables, golf courses, sewage works, airports, vegetable and flower fields. The rural-urban fringe is neither truly urban nor truly rural.

Characteristics

The main characteristics of the rural-urban fringe are as under:

1. The rural-urban fringe is known for the airports, golf course, horse-race course, sewage treatment plants, large-sized godowns, and farm houses.
2. Agricultural land use in general is intensive and the arable land is devoted to perishable commodities, e.g., vegetables, fruits, flowers and dairy products.
3. Encroachment of residential and industrial estates. This is the area into which the city is physically expanding.
4. The size of holdings and farms is generally small.
5. Social amenities and public utility facilities are inadequate.
6. There is constant change in the land use. The land shifts from agriculture to the secondary and tertiary economic activities.
7. The builders develop residential colonies in the rural-urban fringe.
8. This is often been called as the 'Green Belt' or Green-Field Site'.
9. The planners generally allocate land for schools, colleges, hospitals, cremation grounds and cemetery.
10. The rural-urban fringe attracts mobile middle class residents.
11. People living in the fringe areas commute daily to their place of work in the city.
12. The rural-urban fringe is the a problem area from the point of view of administration. It suffers from the administrative problems and there is nobody responsible for the management of complex problems. Often, the criminals escape the administrative boundaries after committing a crime in the neighbouring administrative areas.

Evolution of Rural-Urban Fringe of Indian Cities

During the medieval period, the walled cities in India stood apart from the surrounding rural areas. The limits of the physical city were then clearly defined by the walls, the moats and other protective structures around the city. The gates, few in number, provided the only regulated points of entry to and exit from the city. Inside the walled city lived an urban class of people engaged in non-agricultural occupations, and in the villages outside the city, there lived rural people who were primarily engaged in agriculture, animal husbandry, forestry and fishery. The city and countryside were clearly divided by a conspicuous boundary. Even, where walls were absent, the boundary between the traditional Indian city and the rural villages was abrupt and clearly defined.

The origin of the rural-urban fringe is a rather recent phenomenon around Indian cities, though its occurrence around the western cities was observed long ago. It has been asserted that the rural-urban fringe was non-existent even around the largest metropolitan cities in India before 1950. During the British period, a number of villages around existing towns and cities were totally relocated or dislocated in order to obtain space for the construction of cantonments. City and town expansion throughout the British period was invariably confined to the development of new cantonments and civil lines. They developed the civil lines and cantonments in the vicinity of the old towns.

The post-independence period has witnessed a radical transformation of the urban scene. In particular, during this period, the one-lakh cities and million-cities began to grow rapidly. This rapid increase in the size of cities led to urban encroachment in the rural areas. The villages in the periphery of the city, which had hardly any administrative or political clout, were an easy target for the manipulative tactics of the new urbanites—both rich and poor. The rural people were helpless against the money power of the new industrial and commercial elite. This resulted into the presence of urban land use within the rural areas surrounding the rapidly growing cities.

The growth of industries, commerce, administration and institutions of learning, arts and health generates jobs for the rural population. For those who want to continue with farming, the rapidly growing city provides an expanding market for vegetables, fruits, milk etc. Consequently, the rural people change their lifestyle. Thus, we have the emergence of the semi-urban society—transitional phase between the rural and urban societies.

The rural-urban fringe is not only an Indian phenomenon; it has also its counterparts in the developed countries of the world. The delineation of the rural-urban fringe has been attempted by many of the urban geographers. The bases for the demarcation of rural-urban fringe adopted, by the Indian geographers, are as under:

(i) Density of population.
(ii) Population growth in the preceding decade.
(iii) Females per thousand of males—800 or less.
(iv) Proportion of workers in agricultural activities—50 per cent or more.
(v) The outer limit of the city bus services or local train services.

The rural-urban fringe may be in the form of concentric circle, rectangular or linear. It may extend for about 30 to 40 km along the roads and railways. Green belt surround the city and forms a part of the rural-urban fringe. There may also be satellite towns in the rural-urban fringe, and in due course of time there may emerge new towns and urban countermagnets.

The rural-urban fringe offers the greatest challenge to the urban planners. It is an area of rapid change in land utilization and in population characteristics. Some of the major problems of rural-urban fringe are as under:

1. The rural-urban fringe is often described as the garbage and sewage dump of the city. All obnoxious land uses relevant to normal life of urban areas are pushed towards the city's outer limits, and often into fringe villages. The classic examples of such land uses are garbage dumps, water works, sewage disposal tanks and farms, burial and cremation grounds, airports, timber yards, brick kilns, etc. Industries emitting obnoxious gases or generating chemical effluents are allowed to locate in the city's peripheral fringe villages.
2. The people of the fringe villages dispose of their land for non-agricultural purposes. Urban land developers, industrialists, businessmen and others who wish to invest black money are all attracted to the fringe village.
3. The fringe areas suffer from concentration of landownership, speculation on land, and rapidly rising land values. All these require urgent regulation. Land is the most vital asset in urban areas. The fringe land can generate a lot of capital which can be utilized for the development of infrastructure in the urban area as well as in the urban fringe.
4. The fringe zone, in recent times, has been used for the relocation of slums (*jhuggi-jhopri*), which are uprooted from the city. This attitude is basically unfair. While the fringe may have to be developed for the physical expansion of the city, this expansion ought to be well-planed, and the fringe area must be made as an attractive area.

Green Belt

The green belt is a part of the rural-urban fringe. The green belt aims to control the physical expansion of large cities. A green belt of 2 to 5 km in width should encircle the entire city. It acts as barrier, preventing further physical expansion of the city and preserves the quality of the environment around the city. It has picnic spots, parks and other recreational areas, where the city people spend their leisure time. In the green belt the land is mainly devoted to vegetables, orchards and social forestry. The green belt concept is very attractive planning proposition and, in fact, all city master plans prepared in India since 1960 have included the green belt in their physical land use plans. Unfortunately, in most of the plans of

Indian cities and towns, the green belts have been encroached upon by urban residential structures and industries.

At present, urban geographers are applying statistical techniques to delineate for accurate zones of rural-urban fringes. The rural-urban fringe may be delimited by joining the outer limits of all the five boundaries mentioned above.

Satellite Town

A satellite town is a town designed to house the overspill population of a major city, but located well beyond the limits of that city, and operating as a discrete, self-contained entity, theoretically independent of the central city. For example, most of the early new towns were satellite towns of London. In India, the satellite towns, in fact, are the suburbs of the major metropolitan cities like Mumbai, Kolkata, Delhi and Chennai. The people living in the residential suburbs, however, identify themselves with the metropolitan city and even claim to be living within the city. The postal and telephone departments consider the suburbs as part of the metropolitan city and treat them in the same way as central areas of the city. These suburbs (satellite towns), however, do not receive the benefits of any of the metropolitan services, nor do they pay taxes to the city.

Figure 4.18 *Satellite Towns of Delhi*

The term 'satellite town' is at times used to designate a place that is located at a distance from the city, but linked to it through the daily commuting people. Suburbs are closer to the city, while satellite towns are located farther away. Satellite towns are generally located outside the rural urban fringe, while suburbs are an integral part of the rural-urban fringe zone.

In India, most satellite towns are purely residential in character.

People living in these suburbs (satellite towns) commute to the city for work and are in fact dependent upon the city for most services such as higher education, health and shopping. There are a few suburbs where large and medium scale industries provide employment locally. In some cases, the industrial establishments themselves provide housing for their employees. In this case, we have a self-contained township. Such townships with residential, commercial and institutional areas are often designated as 'satellite townships'.

Satellite towns differ from suburbs mainly on account of the presence of industrial and other establishments which provide an employment base in the settlement itself. A good example of satellite town is Faridabad, located about 25 km south of Delhi. In many cases, satellite towns generate reverse commuting from the city to the satellite town.

Census Definition of Urban Areas

From 1901 onwards, the Census Organization in India has set up its own criteria for identifying urban places. The criteria have, however, changed from time to time. The Census Organization of India gave its definition of urban places in the censuses of 1971 and 1981. The census is by far the most important source of information on urbanization in India and, therefore, an understanding of the census definition of urban places is important to the students of urban geography.

The 1981 Census defined an urban place as (a) any place with a municipality, corporation, or cantonment or notified town area; or (b) any other place which satisfied all the following criteria:

(i) a minimum population of 5,000,
(ii) at least 75 per cent of the male working population in non-agricultural activities, and
(iii) a population density of at least 400 persons per square kilometre (i.e., 1,000 persons per square mile).

The census criteria are, however, conservative as well as vague. Some of the vagueness in the definition are as under:

1. The census prescribes a lower population limit of 5,000 persons for an urban place. According to the census, there are 2,000 places with over 10,000 or more persons, which were identified as rural. In the Ganga-Yamuna Doab, there are numerous villages, which had a population of nearly 50,000 persons.
2. Another census criterion relates to the density of population, with all places having a gross density of population of 400 persons or more per square kilometre qualify as urban place. This is indeed a low density value in Indian context as the average density of population of India is 324 persons per square kilometre. A much higher value, of

around 1,000 persons per square kilometre, would be more appropriate in the Indian situation.

3. A third criterion suggested by the Census Organization would classify as urban places having 75 per cent or more of their male working population engaged in non-agricultural activities. This criterion is based on assumption that agricultural activity is unsuited to an urban environment. However, according to the census data, at least 25 per cent of the census towns have agriculture as the dominant (most common) activity. Further, female workers are excluded from the criterion and it is not very clear why this has been done.

A rather simple way out of this dilemma is to recognize urban places on the basis of: (a) a minimum population of 5,000 or more, and (b) the presence of a compact settlement within the limits of the revenue village. A compact settlement with a population of 5,000 or more persons is bound to have a commercial core to serve its own population and the population in the smaller settlements around it. If we accept the simple definition of an urban place, the total number of urban places in India would increase substantially.

Conurbation

The term 'conurbation' was coined by Patrick Geddes to describe a built-up area created by the coalescence of once separate urban settlements, initially through ribbon-development along the main inter-urban routes. With greater urban sprawl the term has now largely been replaced by concepts such as daily urban system, megalopolis, metropolitan area and metropolitan labour area. In a conurbation, there is substantial number of daily commuters.

Conurbation and metropolitan areas, having their origins in USA, are used to describe very large urban settlements. The first and the largest conurbation

Figure 4.19 *USA: Conurbation*

of the world may be observed in the eastern part of USA. In this region, there is a coalescence of the million cities of Boston, New Haven, New York, Philadelphia, Baltimore, Washington D.C., New Port, Columbia and Columbus (Fig. 4.19).

In India, a gigantic conurbation is developing between Agra, Mathura, Hodal, Faridabad, Delhi, Panipat, Ambala, Chandigarh and Kalka. It is expected that within the coming two decades, the cities of Agra, Mathura, Hodal, Ballabgarh, Faridabad, Delhi, Sonipat, Panipat, Karnal, Kurukshetra, Ambala, Chandigarh and Kalka will form a continuous urban settlement having the merger of several million cities along the National Highway 1, which may be the largest conurbation of the world (Fig. 4.20).

Figure 4.20 *Emerging Conurbation in North India*

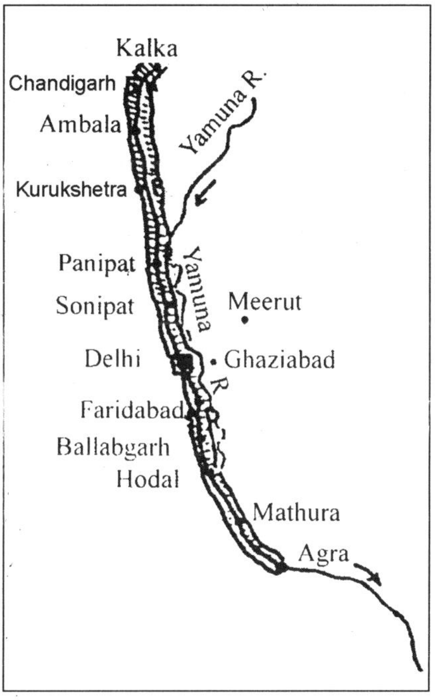

Problems and Prospects

The conurbations and metropolitan areas have their own merits and demerits as given below:

(i) Most of the metropolitan cities of India are growing at an alarming pace and there is less development of infrastructure. Consequently, life in conurbations is not enjoyable.

(ii) The unplanned growth of population in conurbations is resulting into more urban slums and squatters. The population of slum dwellers in India has already crossed the mark of 150 million.

(iii) There is decay in the quality of environment, and the conurbations are characterized by poverty, unemployment, insecurity and crime.

(iv) Owing to the overlapping of area of influence, there is a serious problem of administration and control of crimes.

(v) Huge size of urban settlements leads to traffic congestion, scarcity of water, electricity and environmental pollution.

Despite these shortcomings, the metropolitan areas provide more employment to the skilled, semi-skilled and unskilled workers. Moreover, the conurbations integrate the rural-urban cultures and thereby, help in solving the problems of housing as a large section of the workers commute to their suburbs and rural areas.

Rural-Urban Continuum

The concept of 'rural-urban continuum' was developed by R.E. Pahl, in 1966. It suggests a continuous gradation of ways of life between the two poles of truly rural community and truly urban society. The concept also suggests that the way of life is best understood in terms of the type of settlement that people live in, and it has been used both to distinguish between urban and rural extremes and as a theory of social change which emphasized the transformations from one pole to another.

In the developed countries, rural-urban contrasts have all but vanished; around 90 per cent of their people now live in urban areas and the remaining about 10 per cent live in isolated farm houses and hamlets. The majority of those who live in these 'rural' habitations are not engaged in agriculture, for the labour force engaged in farming is only around 5 per cent of the total labour force. Moreover, even the isolated farm houses have all the modern amenities such as piped water supply, telephones, electricity, etc. Moreover, the automobile provides the people easy access to the cities. In contrast, the absence of these facilities is common to both urban and rural areas in India.

In India, the villages as well as cities are socially stratified on the basis of ethnic, religious and linguistic origins, *jati* or *varna*. Social customs, particularly those relating to marriage, are governed by nearly the same rules, irrespective of the size of the settlement. The number of children per household or even the size of the household does not differ markedly between city and village. Piped water supply, sewage and garbage disposal are not available to great majority of Indian city dwellers and in this respect, they are, if anything, worse off than rural folk. Huts and fragile dwellings are as much a part of the Indian city as of the Indian village. Large parts of cities are occupied by hutted slums. The proportion of people below the poverty-line is nearly the same in the smaller settlements as in the larger urban settlements. Further, there is a gradual and continuous increase in settlement size from hamlets to million cities and there are no abrupt breaks in the frequency distribution of settlements according to their size. Thus, according to this view, dividing settlements into two categories, rural and urban, is arbitrary, as there is no sharp dividing line between the two. On the contrary, it may be asserted that settlements fall along a continuum without any distinct dividing line.

In India, the processes of Aryanization, Persianization, and Westernization have brought about sharp differences in social behaviours of the rural and urban population. Westernization has influenced cities much more than villages, though villages have not been unaffected by it.

The contrasts in lifestyle become greater and more self-evident when large cities are compared with the smallest settlements. From this the rough contours of the rural-urban dichotomy emerge: the city and the village stand out as two distinct entities, deserving recognition as such.

Urban Sprawl

A term, often used pejoratively, to describe the unplanned extension of relatively low density urban land use into rural areas, usually alongside main roads. In almost all countries of the world, especially in the developing countries like India, cities are growing at the expense of surrounding agricultural land and the fringe areas are continuously being shifted outward. In both the developed and the developing countries, the wealthier classes of town dwellers are constantly moving from the crowded centres of the cities to the more pleasant suburbs where they can build larger houses.

Historically, suburbs have grown first along major roads leading to the towns as 'ribbon settlements'. Such sites are the first to be developed because of their accessibility, but soon the demand for suburban homes causes the land between ribbon settlements to be brought, built on and made accessible by the construction of new roads. In this way, towns are continuously growing and in some areas, the suburbs of a number of neighbouring towns merge to form an almost continuous urban growth, called conurbation.

Urban sprawl, along the main roads, emerged from a built-up area, may too create conurbation. Locations on such routes usually offer relatively cheap but accessible sites for business, trade from passers-by. Within built-up areas, the term is often used to describe strings of commercial land use along the main roads leading to major centres, with the same advantages.

Sprawl implies little control of land sub-division, so that the conversion of plots for urban uses may create enclaves of agricultural land. Farmers in such areas may suffer negative externalities through the impact of neighbours on their land uses. Planning legislation, especially in heavily populated countries, often includes policies to reduce sprawl, as with green belt.

In the developing countries, like India, the metropolitan and other million and class one cities are sprawling along the national highways and major roads. The Delhi-Chandigarh, Delhi-Hardwar via Meerut, Delhi-Dehradun, Delhi-Agra, Delhi-Jaipur, Delhi-Rohtak and Delhi-Moradabad sprawls are quite conspicuous, which have emerged during the last three decades. Ultimately, these sprawls may achieve the form of urban conurbations. Similar sprawls may also be seen along Mumbai-Pune, Ahmedabad-Vadodra, Kolkata-Asansol and Chennai-Coimbatore roads.

Slums

An area of poor housing, often characterized by multi-occupancy and overcrowding. Schools are poor, sanitation poor and inadequate, and surroundings generally polluted. Slum area often exhibits high concentration of drug abusers, alcoholics, criminals and vandals. The Indian slums are known by different names in different cities. For example, *Bastees* in Kolkata, *Cheri* in Chennai, *Jhopadpattis* in Mumbai and *Jhuggi-Jhopries* in Delhi.

The slums are characterized by poverty, disease, malnutrition, dislocation, pollution, crime and corruption. The unhealthy environment of slums and sub-standard dwellings offer an important habitat for a wide range of insects and rodents (fleas, cockroaches, bugs, mosquitoes, flies, rats and rodents), which transmit a number of diseases. The poor quality of drinking water becomes the cause of typhoid, dysentery, diarrhoea, malaria, jaundice, cholera, yellow fever, dengue, and conjunctivitis.

The unplanned growth of urban places has resulted in an acute shortage of housing in the country. The gap between the demand and availability of housing has widened in the last three decades.

In India, slums are created when squatters illegally occupy land, either on the edge of a built-up area or in the interstices of existing development, as alongside railway lines, ponds, drains, and public lands. Such an occupation may be entirely unplanned and piecemeal, but most slum settlements are the result of planned invasion of land which neither private owners nor the state are likely to resist. In India, the slum areas have grown rapidly especially in the past four decades.

Slum Population in India

The Indian cities have been subject to an ever burgeoning population, and as a result, the problems of finding space and housing for all have intensified. Slums have become an inevitable part of the major Indian metropolises. The proportion of the population in towns and cities living in slums has been increasing over the years, and has risen from 17 per cent of the urban population in 1981 to about 24 per cent in 2001. According to the census of 2001, 607 towns reported slums (the state of Maharashtra has the largest number of slum dwellers, about 11 million). The second highest slum population is in the state of Uttar Pradesh, which has over 7 million, followed by West Bengal. Delhi is estimated to have 5 million people living in slums, accounting for about 40 per cent of the city's population. The slum population of India is likely to be 100 million by 2015.

Sex ratio in the slums was 1,032 in Pondicherry and 707 in Chandigarh (2001), while the literacy rate of slum dwellers was 88 per cent in Shillong and 55 per cent in Chandigarh. The slum dwellers are mainly dependent on secondary and tertiary economic activities.

The slums lack a basic infrastructure—public utilities such as electricity, running water and sewage and garbage removal—and much of the housing is of a poor quality. People in slum areas live under the most deplorable socio-economic and health conditions, with little or no access to effective social and health care services, potable water, sanitation facilities and are, therefore, more vulnerable to epidemics and developmental challenges such as HIV/AIDS, cholera and dengue fever. Their low socio-economic status, low level of education, and high fertility and mortality all indicate that they need special attention in terms of public health, family planning, and reproductive health programme. Special plans and strategies are therefore required to improve the quality of life of the urban slum dwellers.

Squatter Settlements (Shanty Colonies)

An urban development, usually predominantly residential, on land neither owned nor rented by its occupants, is known as a squatter settlement. Such settlements are mainly found in Third World countries, where the conventional housing markets are unable to cope with the demands produced by rapid growth of urban places. They are created when squatters illegally occupy land, either on the edge of a built-up area or in the interstices of existing development (along the railway lines). Such an occupation may be entirely unplanned and piecemeal, but most squatter settlements, like slums, are the result of planned invasion of land which neither private owners nor the state are likely to resist.

Squatter settlements are erected without official permission (unauthorized on land which squatters do not own). They are usually located on the periphery of the cities and are constructed of any easily available and cheap materials such as cardboards, metal sheets, planks, straw, mats and sacks. There are no minimum sanitation standards, the construction is primarily uncontrolled and the areas tend to lack the essential services of an urban environment such as roads, water, electricity and sewage. Educational and medical facilities are severely limited. This creates many health and other problems.

Policies to deal with squatter settlements vary. For sometime, the city authorities tried to bulldoze the settlements. In some cities, the authorities have recognized the value of their informal migrants, and have attempted to up-grade squatter settlements by self-help schemes, by granting legal title, or by provision of sites with limited services around which migrants can construct their houses.

For many governments squatter settlements were for long seen as major irritants, creating not only substantial blots on the landscape and potential health hazards for the wider population but also possibly containing radical 'social movements'. From the 1960s on, however, some housing specialists started arguing that squatter settlements provide a sensible resolution of the housing in rapidly growing, relatively poor countries. The conventional housing market cannot cope with the explosion of demand, and scarce capital is better invested elsewhere in the economy. Thus, squatter settlements provide a solution which works, allowing people to invest in housing and its improvement as their circumstances allow. In some countries, the squatter settlement movement has even been encouraged both by assisting groups in the search for land on which to establish their communities and by providing a basic infrastructure—a piped water system, a basic drainage network, and an electricity grid, for example. In some cases, even basic dwellings (core rooms—a kitchen and one bedroom) are provided, which the occupants can extend when money is available. However, such policies are condemned by others as ways of ideologically sustaining class differentials within divided societies.

Ghetto

Ghetto, a part of a city, not necessarily a slum area, occupied by a minority group. The term was first used for the enforced concentration of Jews into specific residential areas in European cities from the Middle Ages, but has now spread to include other ethnic groups in unofficial ghettos, especially black minorities in USA. Lifestyles within the ghetto differ distinctly from those of 'host' population and the prejudices of the host confine the sub-group to particular locations. Although ghettos are characterized by social disadvantage, most ghettos display a spread of socio-economic groups and the better-off may move to the affluence of the 'gilded ghetto'.

Urbanization in India

There has been a steady increase in the size of urban population in the country since independence. A study of the urbanization process since the beginning of this century reveals a significant increase in the size of its urban population, the number of urban centres, the level of urbanization since 1911, and a rapid rise since 1951. From 1951 to 1991, India's urban population more than tripled from 62 million to 217.2 million. The number of urban centres (towns and urban agglomerations) swelled from 2,843 in 1951 to 3,768 in 1991—an increase of 32.5 per cent.

The proportion of the population living in urban areas (level of urbanization) increased from 17.3 per cent in 1951 to about 28 per cent in 2001. The annual rate of growth of the urban population during 1991-2001 was about 5 per cent.

Trends in growth of population in the six categories of urban areas reveal that the lion's share of the increase in population since 1981 has been in the class I cities. The proportion of the Indian urban population living in class I towns is steadily growing towards two-thirds of the total. Class V and VI towns (below 10,000 population) have actually experienced a net loss of population during the decade of 1991-2001. India had 23 million cities in 1991 which increased to 35 in 2001. Thus, the urbanization process in India has essentially been the growth of large towns and metropolitan cities, at the cost of smaller towns and urban centres. While the pace of urbanization in the country as a whole has been slowing down, the pace for metropolitan cities has been accelerating.

Classification of Indian Cities and Towns

Cities and towns are numerous and too varied in their characteristics to be understood without meaningful categorization. Towns have been classified from the origin, site, situation, morphological and functional points of view. The classification of urban places may be based on: (i) physical characteristics, i.e., the geographical setting of towns, e.g., hill stations, river view, lake view, sea front; (ii) classification based on climate, e.g., hot and humid summer sites, coastal cities, pleasant weather cities, plateau towns and cities, inland cities; (iii) historical classification, e.g., the ancient cities, medieval cities, British Indian cities, and post-independence cities; (iv) classification based on religion and ethnicity; and (v) economic or functional classification.

In India, towns are classified into six population sizes as follows:

Class I: Population of 100,000 or more
Class II: Population between 50,000 and 99,999
Class III: Population between 20,000 and 49,999
Class IV: Population between 10,000 and 19,999
Class V: Population between 5,000 and 9,999
Class VI: Population below 5,000

Functional Classification

Classification of towns is an interesting area in which the urban geographers have been increasingly probing. In India, there are numerous towns and cities and they vary in their characteristics. For a proper understanding of cities and towns, it is imperative to classify the towns on the basis of their functions. The term 'function' refers to an economic activity in a town, whether at the primary, secondary or tertiary level.

A categorization of cities and towns according to the functions (economic activities) they discharge is known as functional classification of towns. In reality, there is no mono-functional town as any one city may perform a number of functions.

Some of the important scholars who developed new techniques and contributed in the classification of towns and cities are Griffith Taylor, Lewis Mumford, Aurousseau, C. Harris, H. Nelson, Victor Jones, Ullman,

Duncan Reiss, Moser, Scott, Richard Forstall, Qazi Ahmad, S.M. Rafiuulah, S.S. Bhatia, Asok Mita and Ashish Bose.

Urban places may be classified on the basis of qualitative assessment of their characteristics and roles, or alternately, on a quantitative analysis of the people and their occupations. Cities and towns may be classified on the basis of the types of economic activities (functions) carried on within the city. Certain economic activities have acquired greater significance than others. The presence of an iron and steel industry or textile industry lends a certain name and character to the city. Some cities are known for their administrative status as state capitals or district headquarters, others for their educational institutions, and yet others for trade or transport activities. Cities in general show a remarkable degree of specialization in one or more specific activities, such as trade, transport, industry, administration and housing. It is, therefore more meaningful to classify cities on the basis of detailed categories of primary, secondary and tertiary activities.

Data on economic activities in urban places are rather meagre. The main source of information on this aspect is the Indian census. In 1961 and 1971 censuses, data for all towns and cities are provided for nine industrial categories of workers. This classification of workers closely corresponds to the Standard Industrial Classification of workers at the first digit level. The categories are:

I Cultivation
II Agricultural labour
III Forestry, fishing, plantation, mining and quarrying, etc.
IV Household industry
V Manufacturing other than household industry
VI Construction
VII Trade and commerce
VIII Transport, storage and communication
IX Services

The industrial categories of workers include primary, secondary and tertiary economic activities. The first three categories (viz., cultivation, agricultural labour) represent primary activities, categories of household industry and manufacturing represent the secondary economic activities, while categories pertaining to construction, trade and commerce, transport, storage, communication and services represent tertiary activities.

Asok Mitra, a former Registrar of the Census of India, attempted a classification of all towns and cities common to 1961 and 1971 censuses. As a first step in his classification, he grouped the seven industrial categories of workers (excluding the first two categories representing the agricultural sector) into three broad groups to derive three major functional types as under:

1. Manufacturing town
2. Trade and transport town
3. Service town

According to Mitra, on the basis of functions, Indian cities and towns may be classified in the following categories.

1. *Manufacturing Towns*: Manufacturing towns are generally located in the areas where raw materials are available, especially on lines of transportation lines. In manufacturing towns there is a segregation of houses and business establishments; the officers quarters may be grouped at one place and that of the labourers at the other. Ahmedabad, Jamshedpur, Rourkela, Bhilai, Bokaro, Durgapur, Khetri, Gajrola, Modinagar, Modipuram, Coimbatore, Chitranjan, Papernagar, Ulhasnagar, Vishakhapatnam are some of the examples of manufacturing cities.
2. *Administrative Towns*: The main function of administrative towns is to administer the country/state/district or a specific territory. It includes not only the capital cities of countries, but all the centres of provinces, states, districts and other administrative divisions of the country. For example, New Delhi, Mumbai, Kolkata, Chennai, Hyderabad, Bangalore, Patna, Lucknow, Jaipur, Dispur, Bhopal, Kohima, Shillong, Chandigarh, Shimla, Dehradun, etc.
3. *Industrial Towns (Production Centres)*: The United Nations defines manufacturing as "the mechanical or chemical transformation of inorganic or organic substances into new products, whether the work is performed by power-driven machinery or by hand, whether it is done in a factory or in the worker's home, and whether the products are sold wholesale or retail". This is a wider definition, however, and 'manufacturing' is more usually restricted to modern manufacturing industry with its sophisticated products and elaborate structures. Modern manufacturing employs a large proportion of the working population, and its vast output has enabled the industrialized countries to develop a distinct and very favoured type of economy. In industrial and manufacturing towns, there is segregation of houses and establishments, the officers quarters may be grouped at one place and that of the labourers at the other. The leading manufacturing cities, which show clear specialization, are Ahmedabad, Surat, Jamshedpur, Sindri, Ulhasnagar, Ludhiana, Salem, Sholapur, Rourkela, Bhilai, Dhanbad, Bokaro, Coimbatore, Kanpur etc.
4. *Collection Centres*: The mining towns, fishing ports and lumbering centres fall under the category of collection centres.
   (i) *Mining centres*: There are numerous metallic, non-metallic, precious stones and energy resources, which are obtained from mines. The places which have these mines may be small settlements, serving a particular mine like Zawar mines near Udaipur

(Rajasthan), Bailadela (Chhattisgarh), Digboi (Assam), Koyali (Gujarat), Raniganj and Jharia, etc. In the markets of such settlements one may find mining equipments and special clothes and accessories needed for miners.

(ii) *Fishing ports*: They are the base for small boats which go to sea everyday or may be used by large vessels, which stay at sea for days or weeks, but they have many features in common. They are usually grouped fairly closely round the harbour and have facilities of landing, storing, cleaning, drying, packing and despatching the fish, as well as canneries, fertilizer plants, fish feed or freezing plants in some cases. In many fishing ports, other related activities such as boat building, net making or repairing and fishing equipments are also found.

(iii) *Lumbering towns*: The main functions of lumbering towns are to collect and partly process the wood and they often have saw mills and joinery. Some urban places may also have plants for making pulp and paper. The timber towns of Kathgodam, Kotdwar, Haldwani (Uttranchal), Dimapur (Nagaland), Kalka (Himachal Pradesh) and Siliguri (West Bengal) are some of the examples of lumbering towns. Apart from the saw and paper mills, the lumbering towns have tree/plants nurseries and research centres.

5. *Trade Centres*: Throughout the history trading routes have been the important factors, which helped in the development of urban places and cities. The cities which have a high degree of specialization in trade are Vijaywada, Vishakhapatnam (Andhra Pradesh), Siliguri (West Bengal), Katihar (Bihar), Chandigarh, Dehradun (Uttrakhand), Bhubaneshwar (Orissa), Shillong (Meghalaya) and Agartala (Tripura). The major seaports like Mumbai, Okha, Kandla, Chennai and Calicut can also be included in this category. Desert crossing, mountain crossing (passes) and river crossing are also the sites on which the urban places develop.

6. *Cultural Centres*: Towns which perform cultural functions like educational institutions, religious places and art centres are known as the cultural cities. For example, Shanti Niketan, Nalanda, Magadh, Allahabad, Varanasi, Hardwar, Triupati, Madurai, Badrinath, Kedarnath, Ajmer, Amritsar, Roorkee, Bodh Gaya etc. In such cultural centres religion occupies an important place and have many religious buildings (places of worship), *Dharamshalas* (residential accommodation for pilgrims), shops selling religious books, flowers, joss sticks, etc. There is an elaborate arrangement to accommodate thousands of pilgrims. Such places often have subsidiary functions as tourist centres.

7. *Resorts*: The urban places, which cater the recreation needs of people, are known as resorts or recreation towns. These resorts may be based on health-giving water (hot springs), seaside recreation, mountain climbing, cultural attractions, historical monuments, sports facilities, national parks and attractive landscape.

   Resorts and recreational towns have many hotels to accommodate visitors. They also provide sport and recreational facilities, such as golf courses, swimming, pools, trekking, rock climbing, skating and skiing. Moreover, they have numerous entertainment facilities, such as theatres, cinemas, night clubs, massage centres and children parks. They also have good range of shopping facilities as most of the tourists like to buy souvenir, postcards and clothing and accessories of sports. The old age people also prefer to stay there to pass their retired life. In some British seaside resorts almost half of the population is over 65 years of age. Dehradun is also known as the 'city of gray head (retired persons) and green hedges'. Srinagar, Gulmarg, Pahalgam, Sonmarg, Kokarnag, Yushmarg (Kashmir), Shimla, Kullu, Manali, Dalhousie (Himachal Pradesh), Mussourie, Nainital, Almora, Ranikhet, Bhawali, Bageshwar (Uttrakhand), Mt. Abu (Rajasthan), Darjeeling (West Bengal), Gangtok (Sikkim), Panchmarhi, Amarkantak (Madhya Pradesh), Mahabaleshwar, Matheran (Maharashtra), Kodikkanal, Udagamandalam (Tamil Nadu) and Shillong (Meghalaya) are some of the important hill resorts of India. Seashore resorts are attracting large number of national and international tourists. Consequently, these cities are growing much faster. Panaji, Kovalam etc. are the examples of such seaside resort cities.
8. *Services Cities*: Kanchipuram, Nagarcoil (Tamil Nadu), Sagar (M.P.), Mughalsarai, Varanasi (U.P.), Bhagalpur (Bihar) and Imphal (Manipur) are some of the service cities. Transportation is a major specialization of the cities in the coal mining, namely, Dhanbad and Asansol. The railway cities of Kharagpur and Bhusawal are good examples. Anamalai Hills of Tamil Nadu specialize in plantation (tea) industry, while Kolar gold fields have specialization in mining.
9. *Defensive Towns*: During the medieval period, most of the towns and cities used to be developed on the strategic sites of military importance. Forts and garrisons used to be constructed at strategic places. The defensive towns have barracks, cantonments, training facilities for the armed forces, airfields, and harbours for warships. Vishakhapatnam, Mhow, Khadakwasla, Udhampur, Halwar, Adampur and Jamnagar have been developed as defensive towns.
10. *Residential Towns*: In the post-independence India, some of the cities have been developed to accommodate the growing population of metropolitan cities. In residential towns, most of the land is devoted

to residential flats, often in the form of multi-storeyed buildings. These towns are very well connected with the metropolitan cities by roads and railways. In the National Capital Region of India, Shahdara, Shahibabad, Loni, Shakarpur, Ghaziabad, Ghazipur, Rohini, Dadri, Greater Noida, Badarpur, Surajkund, Gurgaon, Sanpla and Bahadurgarh are some of the examples of such towns.

11. *Towns of Diversified Functions*: As stated at the outset, none of the towns is mono-functional. Over space and time, the functions of an urban place increase giving it the form of a multi-functional city. The cities of Mumbai, Kolkata, Delhi, Chennai, Hyderabad, Jaipur and Patna may be put in this category.

Urban place is a dynamic concept. Numerous towns and cities have changed their sites in the past. For example, Delhi has shifted at several sites, e.g., Indraprastha, Surajkund, Mehrauli, Siri, Chirag-Dilli, Tughlakabad, Shahjahanabad (Old Delhi) and New Delhi.

Problems of Urbanization

Although India is not highly urbanized as only about 28 per cent of the total population is living in the cities and urban places, but that 28 per cent represents nearly as many as the entire population of the United States. The rate of urban growth is on the upswing. People by the hundreds of thousands are arriving in the already teeming cities, swelling urban India by about 5 per cent annually, almost three times as fast as the overall population. Urban growth is the cause of numerous problems, some of which are due to the increase in population and many due to the physical expansion of the cities and towns. The rapid increase in urban population is the cause of concern for planners and administrators. The urban problems may be classified under two broader categories: (i) internal problems, which affect the city area and its inhabitants, e.g., shortage of space, shortage of residential accommodation, transport bottleneck, paucity of clean drinking water, pollution, disposal of city garbage and sewage, supply of electricity, maintenance of law and order and control of crimes; and (ii) external problems, which have their impact on the area and people of its fringe area and inland. Some of the important problems of urbanization in India have been presented briefly in the following paragraphs:

(i) Problem of Space and Residential Accommodation

Urban growth in India after independence has been tremendous at least in respect of absolute number. The area and size of some of the metropolitan cities, like Mumbai, Kolkata, New Delhi, Chennai, Bangalore and Patna have increased at an unparallel scale. These cities are overcrowded with large population and high density of population. Cities constantly require more space for their growth. This demand for space is met by colonizing fringe areas. But, sometimes, due to physical constraints, the expansion of the city is obstructed. The island nature of Mumbai and occurrence of salt water in the eastern periphery of Kolkata are causing such problems. The inadequacy of space leads to enormous increase in land values and high rents which make the living of poor people difficult and miserable. All the million cities of the country are facing this problem of scarcity of space and overcrowding. The land prices in all the cities have gone up, making residential accommodation beyond the reach of ordinary workers. The density of population in the slums and shanty colonies is enormous. The slums dwellers inevitably outnumber the new homes provided by city housing authorities.

According to one estimate, there is an annual shortage of about two million houses in Indian cities. This has forced low income group people to live in slums or occupy footpaths and road pavements. The number of such slums and pavement dwellers is rising by leaps and bounds in metropolitan cities, like Mumbai, Kolkata and Delhi.

(ii) Inadequacy of Social Amenities

The unemployed migrants to the city are not in a position to afford the rent of approved locations. Their poverty aggravates the problems of providing social services such as drinking water, sanitation and sewage disposal. With the greater concentration of people in the cities, the city governments are hard pressed to provide all human beings the services that would fulfil the basic needs of the residents. The social amenities like housing, running water, sewage systems, electricity supply, garbage disposal systems, educational, medical, entertainment and recreational facilities are inadequate.

(iii) Unemployment

Unemployment is the state of being involuntary out of work. The unemployment rate is the number of unemployed as a percentage of the total population of working age. An unemployment level of under 3 per cent is thought of as a natural rate as people change their jobs. In India, the rate of unemployment in the urban places, both for the skilled and unskilled workers, is progressively increasing. According to one estimate, nearly 25 per cent of the workers are unemployed in the urban areas. The high rate of unemployment and underemployment many a time leads to high-tech crimes. In cities like Mumbai, Delhi, Patna, Kanpur, etc., unemployed youths indulge in kidnapping, abduction, rape, extortion, snatching, pick-pocketing, theft, robbery and murder.

According to the sociologists, unemployment and underemployment have been the main causes of crime in urban areas. The slum and squatter settlements are especially infested with unemployed criminals who in due course of time become habitual offenders.

(iv) Problem of Transport

Transport bottleneck and traffic congestion are the major problems of most of the Indian cities and urban places. The larger a town grows, the more important its functions become. Consequently, more number of people are likely to work or commute for shopping there. As the town becomes larger, the people living within the built-up area need to travel from one area of the town to another. Many of the outsiders bring their cars and vehicles or travel by public transport. Wherever trade is important, commercial vehicles such as vans and lorries also add to the traffic.

In each city, because of narrow roads and increasing number of vehicles, especially in the CBDs, the smooth flow of traffic is obstructed resulting into traffic congestion. The traffic congestion is especially observed during the peak office hours, i.e., in the morning and evening. Traffic congestion is also felt in the small shopping centres which grow in the suburbs. Such congestion becomes greater when the centre is built up in tall skyscraper blocks whose offices sometimes employ thousands of workers, because at the end of office hours everyone wants to leave the building within a short space of time to make their way home. This puts tremendous pressure on public transport and causes journey to take much longer time than it normally takes.

Traffic congestion is also seriously felt in the old towns. The narrowness of the streets, which were built long before the days of motorized transport, and the lack of parking facilities, help to create congestion. A multiplicity of narrow streets, sharp corners and the waiting time required to join lines of traffic may slow down movement and create greater congestion.

If the number of vehicles are allowed to increase at the present rate without widening and upgrading of roads, the whole transport system of major cities is bound to collapse.

(v) The Energy Crisis

The energy crisis is the most serious problem in most of the metropolitan, class I and small urban centres. On energy depends the industrial growth, transportation efficiency and human comfort. In almost all the urban places there is load-shedding. The peak power demand in the metropolis is increasing day by day and the power situation is not geared to meet it.

(vi) Inadequacy of Water Supply

Water is the first and foremost requirement in our daily life. In fact, water is life and man cannot subsist without water. The average per head per day consumption of water in Kolkata is 250 litres, in Mumbai, 175 litres, and in Delhi 80 litres in comparison to 1,200 litres in Los Angles and 1,050 litres in Chicago. The scarcity of water in the urban places may be appreciated from the fact that Chennai, Hyderabad, Shimla, Jodhpur, Udaipur, Bikaner and Rajkot get only about an hour supply of water every day. Many small towns have no running water supply at all and are dependent on such sources as personal tubewells, pumping sets, handpumps, open wells or even rivers which have some storage of water in pools in the summer. The scarcity of water is acute in the national capital also which receives part of its water supply from Ramganga river (U.P.), about 180 km away from the city. The Yamuna shrinks near Delhi during the summer months and the colour of water is brackish, polluted by the chemicals discharged from the factories and the sewerage from the residential areas. Though the central and state governments provide financial assistance in the form of loans and grants to Municipal Corporation to improve the water supply conditions, the supply has not improved much.

(vii) Environmental Pollution

Environmental pollution is the other serious problem of all the million, industrial and class I cities of India. It is not only air pollution caused by smoke emitting from factories, cars, vehicles and houses, water and noise pollutions are equally serious. The scarcity of dumping grounds makes rural-urban fringe less healthy and less pleasant to live in. Noise pollution is quite serious in the vicinity of airport. Pollution originating in towns can also have adverse effect on the surroundings. The problem of garbage disposal (hazardous plastic, metal and packaging) is thus quite serious in most of the Indian cities. Unfortunately, most of the garbage is dumped into the rivers. The three metropolitan cities—Kolkata, Mumbai and Chennai—continue to discharge a major part of their sewage into the sea. In Delhi, more than half of the total inflow of sewage goes into river Yamuna, leading to its pollution. In Delhi, there are numerous drains and nallahs carrying brackish water in the course of which, there are slums and *jhuggi-jhopris*. The water from drains and nallahs merges into the Yamuna and get mixed with other industrial effluents and chemicals which are regularly discharged into them. Occasionally, the heavily polluted water of Yamuna flows into the taps of residents of Delhi. Local governments lack the resources and technical and managerial expertise to deal with the trash and garbage crisis of an increasingly consumerist behaviour of the urban people.

(viii) Social Problems

In the urban places, especially in the major cities and towns, people inmigrate in search of jobs and economic benefits. Usually, they come alone to cities leaving their families back in villages. In the urban areas, they are exposed to the glitz and insensitive people who have gross disregard for the feelings and sentiments of the fellow citizens. In such a situation, the immigrants face a cut-throat competition for jobs. The lack of job opportunities and unemployment have resulted in the breaking down of social values. It precipitates problems such as juvenile delinquency and crimes, emotional depression, nervous strain, mental illness, suicide, insecurity, etc. The increasing graph of crimes creates more problems for administration.

(ix) Problem of Urban Crimes

Increasing urban crimes are disturbing the peace and tranquillity of the modern cities. Material culture, growing consumerism, selfishness, stiff competition, lavishness, increasing socio-economic disparities, rising unemployment and loneliness are some of the main reasons of this menace. It is not only the poor, deprived and slum dwellers who are prone to such crimes but many youngsters from well-off families also indulge into crimes, e.g., prostitution, alcoholism, drug addiction, crimes against females and property, social unrest and political unstability. The problem becomes complicated when such criminals get protection from politicians, bureaucrats and elite class of the urban society. Unfortunately, some of the criminals have even reached to respectable political positions using their money and muscle power. It has been observed that in general the northern Indian cities and the metropolitan cities are more prone to heinous crimes.

Town/Urban Planning

Cities and towns are the important creations of man on the earth surface. These are human agglomerations showing a lot of economic, social and cultural diversities. Unfortunately, each of the Indian cities is characterized by numerous problems. These problems can be solved only through proper urban planning, which make the cities livable, clean, safe and healthy not only for the present generation but also for the future generations. Tjeerd Deelstra (1989) has identified the following ingredients in a good city:

1. Good cities are livable places where growing old can be a pleasurable experience, where children can live and learn.
2. Good cities incorporate nature, in simple and complex form. This contributes to balanced urban climates, wildlife and habitat. Nature in cities can contribute to food production, the recycling of organic waste, education and recreation.
3. In good cities, a lot of activities is going on. There is continuous and multiple use of space.

4. Good cities are safe. Pedestrians and cyclists are given priority. Women, children, senior citizens and handicapped can get around easily.
5. Good cities are clean and healthy. They have good air and water and healthy buildings in which it is good to stay.
6. Good cities combine centralized infrastructure with decentralized systems. Management of flow of water, materials and energy is optimal and attained to natural life-cycles.
7. Good cities have great experimental value. Activities, movements and flows are visible. This enhances emotional, behavioural and cognitive growth of cities, and facilitates their engagement with urban life and the environment.
8. Good cities have past and future, identifiable in their built form as well as in their social and cultural life.

Principles of Urban Planning

Some of the broad principles of urban planning are as follows:

1. *Urban Decentralization*: Under this principle decentralization of industries and residential areas is encouraged through the development of new suburban towns to bring down overcrowding in large metropolitan cities.
2. *Garden Suburbs and Garden Cities*: The development of gardens, and green belts are imperative to provide the cities a clean and healthy environment.
3. *New Towns*: This principle suggests the formation of new planned towns outside the municipal limits of the city. Such towns should be equipped with all urban facilities so as to attract industrial and business houses. In India, New Delhi has been planned on the basis of this principle.
4. *Expanded Towns*: These towns attract workers, artisans, businessmen and industrialists from neighbouring cities and, hence, play a significant role in easing out the burden of population.
5. *Urban Renewal*: This is an important concept of urban planning. It helps in planning and improvement of land use pattern in the old existing cities through the development of master plan for 10 or 20 years.
6. *Slum Clearance*: Under this principle two-pronged strategies are being adopted to improve the conditions of urban slums. These slums are removed and slum dwellers are provided alternative residential accommodation in new planned colonies free of cost or at highly concessional rates. Wherever such slum clearance is not possible, basic civic amenities like drinking water, electricity, garbage disposal, sewerage, roads, etc. are provided to improve the environmental conditions of the slums.

7. *Rehabilitation versus Redevelopment*: This principle is applied for improving the housing conditions in the old cities. It may involve improvement in the old buildings or construction of the new ones at their place after demolition.
8. *Traffic Segregation*: This principle is applied to remove transport bottlenecks and to improve urban traffic. It offers various solutions, like widening of roads, construction of flyovers and subways, bridges, metros, underground railways and parking facilities.
9. *City Centre Redevelopment*: Under this principle attention is focused on the improvement of CBD.
10. *Planning for Future Cities*: This principle focuses attention on new towns and cities keeping in mind the future needs of the city region. This may be done through development of satellite towns, linear cities, dispersed cities or radial cities.

Town Planning in India

The geographers are not only concerned with past and present distributions and interactions, but also with the future. Because of the resource problems arising from population growth, and complexities of modern urbanized and industrialized societies, it is becoming increasingly necessary to adopt suitable strategies for urban planning. Planning has been described as "an ordered sequence of operations, designed to lead to the achievement of either a single goal, or to a balance between several goals".

In India, the practice of urban planning is quite old. The remains of Mohenjodaro, Harappa, and Lothal belonging to the Indus Valley civilization bear testimony to urban planning. During the Vedic period cities like Ayodhya, Hastinapur, Mathura, Vaishali, Patliputa, Sanchi, Kaushambi, Kanchipuram, Madurai, etc., were built.

Mughal rulers also took great interest in city planning. Cities like Agra, Fatehpur Sikri, Srinagar, Shahjahanabad (Delhi), Jaipur, Moradabad, Hyderabad, Bareilly, Rampur, Shahjahanpur, Aligarh, Saharanpur, Muzaffarnagar, Panipat, Tughlakabad, Aurangabad, Adilabad, Ahmadnagar, etc., were developed during the Mughal times.

The British regime gave opportunity for adopting Western model in city planning in which greater attention was focused on administrative and military towns, especially cantonments, railway colonies and civil lines. Mumbai, Kolkata, Lucknow, Ambala, Jalandhar, Kanpur and Allahabad were given special attention. In December 1911, New Delhi was made the capital of India by the Britons.

Since independence many planned cities like Chandigarh, Bhubaneshwar, Gandhinagar, Itanagar, Dispur, Bokaro, Kandla, Paradwip, Haldia, etc., have been built using latest techniques in city planning.

National Urbanization Policy

The main objective of the urbanization policy is to develop an urban-rural continuum replacing the existing urban-rural dichotomy. The national urbanization policy gives first priority to urban housing in which both public and private sectors have been invited to contribute. Here public sector's responsibilities have been limited with the improvement of slums, and providing cheap housing facilities to weaker sections of the urban society.

Recently, a major change in national urbanization policy has been observed towards the slums. The policy puts greater emphasis on the improvement of slums rather than on their demolition or removal. For this purpose slum beautification programmes have been launched in several cities to provide the basic civic amenities.

Under the Urban Land Ceiling Act, 1976, the state governments have been authorized to fix maximum ceiling for urban land. State governments and local bodies have also been granted right to acquire any private property for public use after paying adequate compensation.

The urbanization policy also puts greater emphasis on the development of infrastructural facilities in small and medium-sized towns and cities.

Under the policy incentives have been given to industrial houses and entrepreneurs to establish new industries in small, medium and intermediate towns. Such facilities include tax concessions, factory site at cheaper rate, infrastructural facilities free of cost or at concessional rates.

The urbanization policy makes obligatory even to small, medium and intermediate towns to prepare their master plans. The management of all the towns and cities has been given to the respective state governments. There is a plan to divide country into urban regions so as to provide greater opportunity for the solution of local and regional urban problems. Provisions have also been made for research and policy formulation on urbanization and urban development.

References

Adams, R. Mc.C., 1966, *The Evolution of Urban Society*, Chicago: Chicago University Press.

Ahmad, Qazi, 1965, *Indian Cities: Characteristics and Correlates*, The Department of Geography Monograph Series, University of Chicago.

Alam, S.M., 1965, *Hyderabad-Secunderabad (Twin Cities): A Study in Urban Geography*, Bombay: Allied Publishers.

—, 1980, India: Metropolitan Systems and their Structure, *Technical Bulletin No.1*, Centre for Urban Research, Osmania University, Hyderabad.

—, 1980, Distortion in Settlement Systems of India, *Geographical Review of India,* Vol. 42, No.4, pp.305-22.

Ali, S.M., 1966, *The Geography of Puranas,* Bombay: Peoples Publishing House.

Ashish Bose, 1973, *Studies in India's Urbanisation 1901-71,* New Delhi: Tata-McGraw Hill Pub.Co.

Bhatia, S.S., 1956, Historical Geography of Delhi, *Indian Geographer*, Vol. I, pp.17-43.

Bhatia, S.S., 1962, A Reconsideration of the Concept of Primate City, *Indian Geographical Journal,* Vol.37, pp. 23-24.

Berry, B.J.L., 1967, *Geography of Market Centres and Retail Distribution*, N.J.: Englewood Cliffs.

Bhattacharya, B., 1979, *Urban Development in India*, Delhi: Shree Publishing House.

Bose, Asish, 1980, *India's Urbanisation: 1901-2001*, New Delhi: Tata-McGraw Hill.

Breese, G., 1966, *Urbanization in Newly Developing Countries,* N.J.: Englewood Cliffs.

Brush, J.E., 1968, Spatial Patterns of Population in Indian Cities, *Geographical Review,* No. 58.

Burgess. E.W., 1924, The Growth of the City: An Introduction to Research Project, *Publ. Am. Sociol. Soc.* 18, pp. 85-97.

—, 1927, The Determination of Gradients in the Growth of the City, *Publ. Am. Sociol. Soc.*, pp 21, 178-84.

Carter, H., 1983, *An Introduction to Urban Historical Geography,* London: Edward Arnold, pp. 1-17.

Census of India, Towards Population and Development Goals, UNFPA for United Nations System in India, Oxford University Press.

Chand, M. and V.K. Puri, 1983, *Regional Planning in India,* New Delhi: Allied Publishers Ltd.

Davies, W.K.D., 1968, The Morphology of Central Places: A Case Study, *Ann. Asso. of American Geographers,* Vol. 58.

Deelstra, T., 1989, The Resourceful City, Management Approaches to Efficient Cities Fit to Live in, *Proceedings of the MAB: II Workshop,* Amsterdam, September 1989.

Dickinson, R.E., 1964, *City and Region: A Geographical Interpretation*, London.

Geddes, P., 1915, *Cities in Evolution*, London.

Gibson, M.S. and Langstaff, M.J., 1982, *An Introduction to Urban Renewal,* London: Hutchinson.

Giddens, A., 1981, *A Contemporary Critique of Historical Materialism*, Vol.1, Power, Poverty and the State, London: Macmillan.

Gopi, K.N., 1978, *The Process of Urban Fringe Development: A Model*, Delhi: Concept Pub. Co.

Gottmann, J., 1961, *Megalapolis: the Urbanized Notheastern Seaboard of United States*, New York.

Habib, Irfan, 1982, *An Atlas of the Mughal Empire: Political and Economic Maps with Detailed Notes, Bibliography and Index*, New Delhi: Oxford University Press.

Hardboy, J., 1989, *Squatter Citizen: Life in the Urban Third World*, London: Earthscan.

Harris, C.D., 1943, A Functional Classification of Cities in the United States, *Geographical Review,* Vol. 33.

Harris, C.D. and Ullman, E.L., 1945, The Nature of Cities, *Ann. Am. Acad. Pol. Soc. Sci.* 242, pp. 7-17.

Harris, C.D. and Ullman, E.L., 1959, The Nature of Cities, in Mayer, H.M. and Kohn, C.F. (eds.), *Reading in Urban Geography*, Chicago: Chicago University Press, pp. 277-860.

Harvey, D., 1973, *Social Justice and the City*, London: Edward Arnold/ Baltimore, MD: Johns Hopkins University Press.

Hodder, B.W., 1973, *Economic Development in the Tropics* (2nd edn), London: Methuen.

Hoyt, H., 1939, *The Structure and Growth of Residential Neighbourhoods in American Cities,* Washington DC: Federal Housing Administration.

Husain, M., 2005, *Human Geography*, (3rd ed.), Jaipur: Rawat Publications.

Hussain, Irshad M. and N.A. Siddiqui, 1982, Urban Encroachment of Rural Land Use: A Case Study of Saharanpur City (1950-73), *The National Geographical Journal of India,* Vol. 28, parts 3 & 4, pp. 186-96.

Johnston, R.J., 1971, *Urban Residential Patterns: An Introductory Review,* London: George Bell.

Johnston, R.J., D. Gregory, and David M. Smith, 1994, *The Dictionary of Human Geography*, (Third Edition), Oxford: Blackwell Publishers Ltd.

Jones, E., 1966, *Towns and Cities*, New York.

Knowles, R., Wareing, J., 2002, *Economic and Social Geography Made Simple*, Delhi: Rupa Co.

Lloyd, P.C., 1979, *Slums of Hope: Shanty Towns in the Third World*, London: Penguin.

Mayhew, S., 1997, *Oxford Dictionary of Geography*, Indian (second edition), Oxford University Press.

Mumford, L., 1938, *The Culture of Cities*, New York.

Pahl, R.E., 1966, The Rural-Urban Continuum, *Sociol. Rur.* 6, pp. 299-327.

Pahl, R.E., 1965, Urbs in Rure, London School of Economics, *Geographical Papers, No. 2.*

Park, R.E., Burgess, E.W. and Mackenzie, R.D. (eds.), *The City*, Chicago University Press.

Pathak, C.R., Spatial Variation in Urban Industrial Growth in India, *Indian Journal of Regional Science,* vol. 7, No. I, pp. 1-10.

Prakasa Rao, V.L.S., 1983, *Urbanisation in India: Spatial Dimensions*, New Delhi: Concept Pub. Co.

Rafiullah, S.M., 1965, A New Approach to Functional Classification of Towns, *The Geographer,* Vol. 12, pp. 40-53.

Rao, M.S.A., 1970, *Urbanisation and Social Change: A Study of Rural Community on a Metropolitan Fringe,* New Delhi: Orient Longman.

Robinson, G., 1990, *Conflict and Change in the Countryside,* London: Belhaven.

Roy, B.K. (Direction), *Census Atlas: National Volume,* Census of India, plates 35-45.

Sack, R.D., 1980, *Conception of Space in Social Thought: A Geographic Perspective,* London: Macmillan/Minneapolis: University of Minnesota.

Siddhartha, K., S. Mukherjee, 2005, *Cities, Urbanisation and Urban Systems,* Delhi: Kisalaya Publications.

Singh, R.L., 1955, *Banaras: A Study in Urban Geography,* Banaras: Nand Kishore & Bros.

Stokes, C.J., 1962, A Theory of Slums, *Land Economics, 38,* pp. 127-37.

Sundaram, K.V., 1977, *Urban and Regional Planning in India,* New Delhi: Vikas Pub.

Tiwari, R.C., 2004, *Geography of India,* Allahabad: Prayag Pustak Bhawan.

Wheatley, P., 1971, *The Pivot of the Four Quarters: A Preliminary Inquiry into the Origins and Character of the Ancient Chinese City,* Edinburgh: Edinburgh University Press/Chicago: Aldine.

Wheatley, P., 1972, Proleptic Observations in the Origins of Urbanism. In Steel, R.W. and Lawton, R. (eds.), *Liverpool Essays in Geography,* London: Longman, pp. 315-45.

Zipf, G.K., 1949, *Human Behaviour and the Principle of Least Effort,* Massachusetts: Addison Wesley Press.

5

Rural Settlements

After food, shelter is the most important need of man. Men construct houses and develop settlements to protect themselves against the vagaries of weather and to enjoy social life. In fact, settlement is man's important step towards adaptation in his physical environment. The habitations in which the people are associated with primary economic activities are known as the rural settlements. The oldest rural settlement may be as old as the origin of agriculture—the domestication of plants and animals—most probably around 8000 B.C. or 10,000 B.P. (before present).

Human Settlements

The reliable reasons for the establishment of human settlements are not known. The origin of human settlements is shrouded in history. The site, growth and development of human settlements are closely influenced by the availability of water, fertility of soil, micro-climate, accessibility, elevation, forest and mineral resources. Depending on the quality and nature of resources, the settlements may be temporary or permanent. Settlements, unless planned, are rural in origin which may become urban in due course of time. Thus, settlements may be divided into: (i) unstable settlements, and (ii) stable settlements.

Unstable Settlements

About 1 per cent of the total population of the world is nomadic and nearly 5 per cent still without houses. The hunters, food gathers, pastoralists, and the transhumants have essentially movable dwellings. For example, the Gujjars and Bakarwals of the Jammu and Kashmir and the Gaddis of Himachal Pradesh and Utrrakhand construct *Kothas* (residential accommodation) and *Bandis* (cattle-sheds) as their seasonal shelters. The unstable settlements are, however, confined to the mountainous regions, deserts and semi-deserts, swampy lands and areas inundated by floods.

Stable Settlements

Settlements in which most of the people are engaged in agriculture, hunting, food gathering, cattle-keeping, dairying, ranching, forestry, mining and fishing are known as rural settlements. In the cultivated areas, a rural settlement is often called an agricultural workshop. It cannot be separated from the land whose use it ensures. Its type, shape and pattern are generally in accordance with the kind of work, the agricultural techniques, and the way the soil is utilized.

Most of the rural settlements in India are stable and permanent. The rural areas are dominated by open countryside, extensive land uses. Most of the Indian settlements are rural as about 72 per cent of the total population is rural in character.

Rural Settlements

Functions

As stated at the outset, people living in the rural settlements (villages) are generally engaged in the cultivation of crops and domestication of animals. Apart from agriculture and dairying, the rural settlements perform several other functions also. Its religious places (temples, mosques, gurdwaras, churches, etc.) with one or two shops in the centre have some meeting place too where people pass their leisure time and exchange their views.

In the rural settlements where fishermen live, the main economic activity is fishing. Such villages are located on the banks of rivers, lakes and sea coasts. The fishermen look more to the water rather than to the land to earn their livelihood. Although the main occupation of these villages is fishing, but this may be combined with some agricultural activity.

There are numerous rural settlements in the forest areas in which the dominant economic activity is gathering of forest products and lumbering. Such villages are generally small in size and may be found in the Himalayas and hilly tracts of different parts of the country.

The fishing, mining and lumbering rural settlements are like agricultural settlements in which there are a few shops and a few small-scale industrial units. In all the rural settlements the secondary and tertiary activities are almost absent.

Classification of Rural Settlements

Rural settlements can be classified on the basis of their size, distribution, layout, functions, types, patterns or design. In general, the rural settlements may be grouped under two categories: (i) agglomerated, nucleated, clustered or compact settlements, and (ii) dispersed or isolated in which houses are isolated, away from each other. A number of intermediate types of settlements may be identified between the given two extremes. The major types of rural settlements are as under:

1. Compact or nucleated settlements
2. Semi-compact or semi-nucleated settlements
3. Hamleted settlements
4. Dispersed settlements

Compact or Nucleated Settlements

A compact or nucleated rural settlement comprises of houses, cattlesheds, farm structures, and other buildings such as temples, mosques, churches and gurudwaras, ponds, burial and cremation grounds surrounded by agricultural fields and community land.

The compact settlements are mostly found in the highly productive alluvial plains and valleys. In fact, compact settlements are the characteristic features of the Northern Plains from Punjab in the west to West Bengal in the east. Elsewhere, they are also found in Assam, Tripura, coastal Orissa, Mahanadi basin in Chattisgarh, Karishna and Kaveri deltas, Maidan districts of Karnataka, and Andhra and Rayalseema regions of Andhra Pradesh. These are either rectangular or amorphous in shape in which streets criss-cross at different angles forming *galis* (lanes). For the origin and development of compact settlements, factors such as fertility of soil, security from enemies and wild animals, and need for mutual cooperation are mainly responsible.

The agricultural communities in these productive plains have settled firmly on the soil and dependent on the cultivation of crops and dairying for their sustenance. During the Neolithic period, when the first rural settlement (hut) was developed by the cultivators, prehistoric man was incapable of combating with nature individually and had little control over nature. For the cutting of forests and clearing of land by fire to develop fields, he had to cooperate with others. Such type of cooperation, in due course of time, helped in the development of compact settlements (Fig. 5.1).

Agricultural communities with primitive techniques of cultivation often needed to unite in order to surmount great material difficulties, for instance, to protect themselves from wild beasts who often had them at their mercy, and above all to cooperate in tilling the soil. In the forests and valleys of Siwaliks and other mountain ranges and ridges, many compact settlements developed to defend against wild animals and enemies.

In the Sutlej-Ganga plain, one more characteristic feature of the compact settlements is that they have a high degree of segregation of the upper castes (*Swarn-Jati*) and the lower castes (Sudras). Each caste occupies a specific locality in the rural settlement. In general, the upper

Figure 5.1 *Compact Settlements*

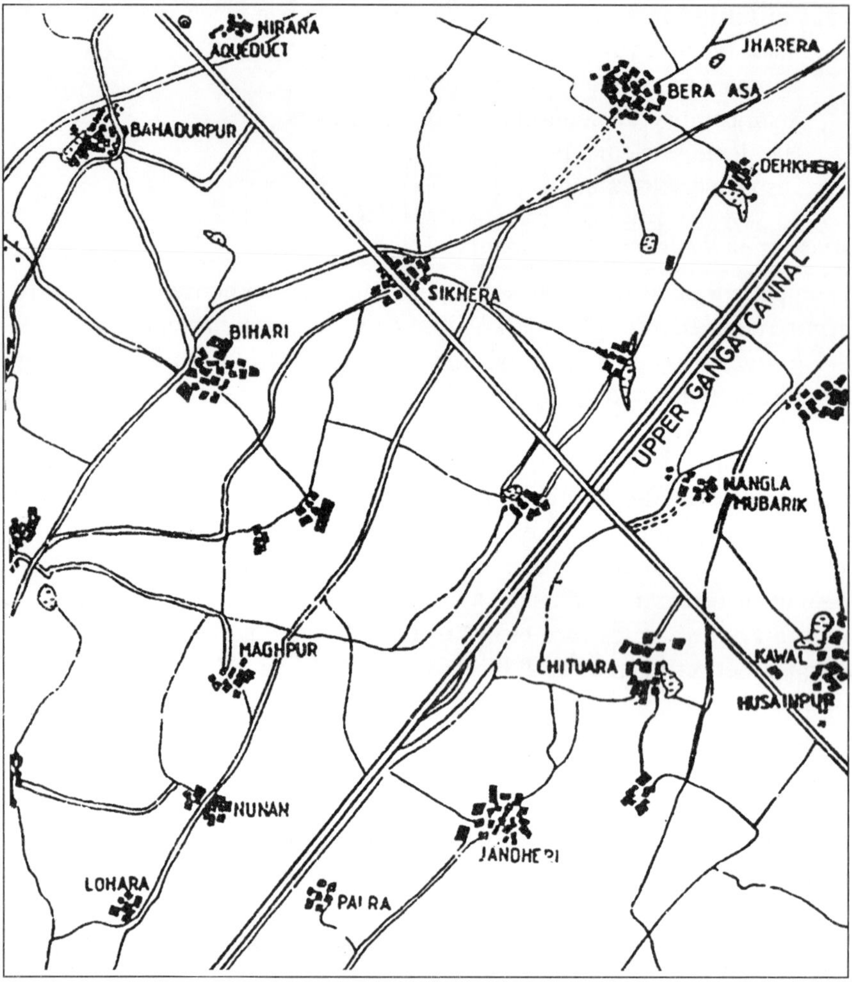

Source: Survey of India, Sheet No. 53 G15.

castes (Brahmins, Kshatriyas) reside in the western and northern sides which are considered as the superior directions being ruled by the deities, Varun and Kuber, respectively (Fig. 5.2). In the eastern side, ruled by the deity Indra, reside the Vaishyas (agriculturists and artisan classes), while the southern side having the deity Yam is occupied by the lower castes (Sudras). The few shops are always near the centre of the village. The religious place is also located in the centre of the settlement. In the irrigated areas of Ferozpur (Punjab) and Suratgarh, Hanumangarh and Ganganagar (Rajasthan), there are planned villages. These new compact settlements are generally rectangular in shape with an open space where stands the temple, mosque, gurdwara and shops (Fig. 5.2).

Figure 5.2 *Caste Segregation in an Indian Village*

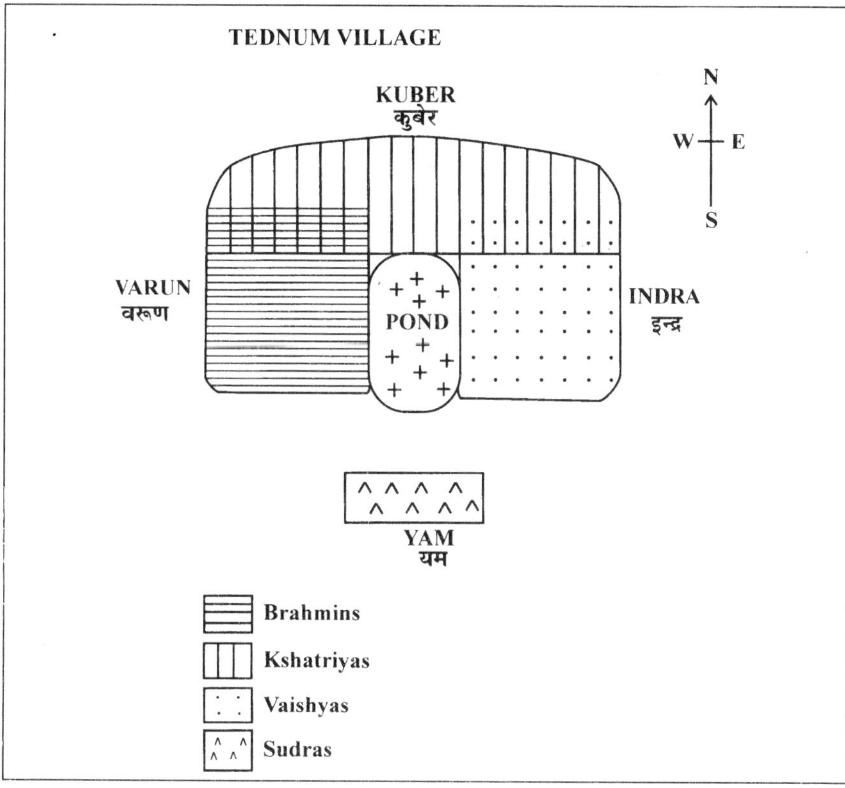

Apart from the productive cultivated areas, compact settlements are also found in many of the hunting and fishing communities. Big villages are still numerous along the Brahmaputra and Hooghly rivers, where fishing requires cooperative management of the weirs and traps on the river. In the areas where natural resources are poor and scanty, the villages are small and scattered.

Intensive cultivation, which clearly represents the most scientific use of soil, has given rise to the greatest number of large compact settlements. The history of settlements very often shows that with the increase of population in the agricultural communities and more demand for food resulted into intensification of agriculture. The rural settlements become compact to accommodate more people over minimum space. When there was not adequate space to construct houses or the fields were at long distance from the settlement, farmers might have started to build their houses in their fields. Such settlements are known as *Purvas* and *Majras* in Uttar Pradesh and Punjab and *Dhani* in Rajasthan and Haryana.

Compact settlements at the tops of hills and ridges of Nagaland, Mizoram and Manipur are the result of the practice of head-hunting in the past. From the security point of view such compact settlements were developed.

Semi-compact or Semi-nucleated Settlements

Semi-compact or semi-nucleated settlement is a transitional phase in the growth of compact settlement. The semi-compact settlements represent either an outgrowth from the compact settlement due to the increasing pressure of population or the inhabitants of late immigrants who came to earn their livelihood by working as agricultural labourers. Semi-compact settlements may also be the result of low-caste people segregated by the social taboos, like untouchability. As the new technology is adopted, and the population increases, the scattered settlements start taking shape of semi-compact settlements. For the growth of semi-compact settlements no

Figure 5.3 *Semi-compact Settlements*

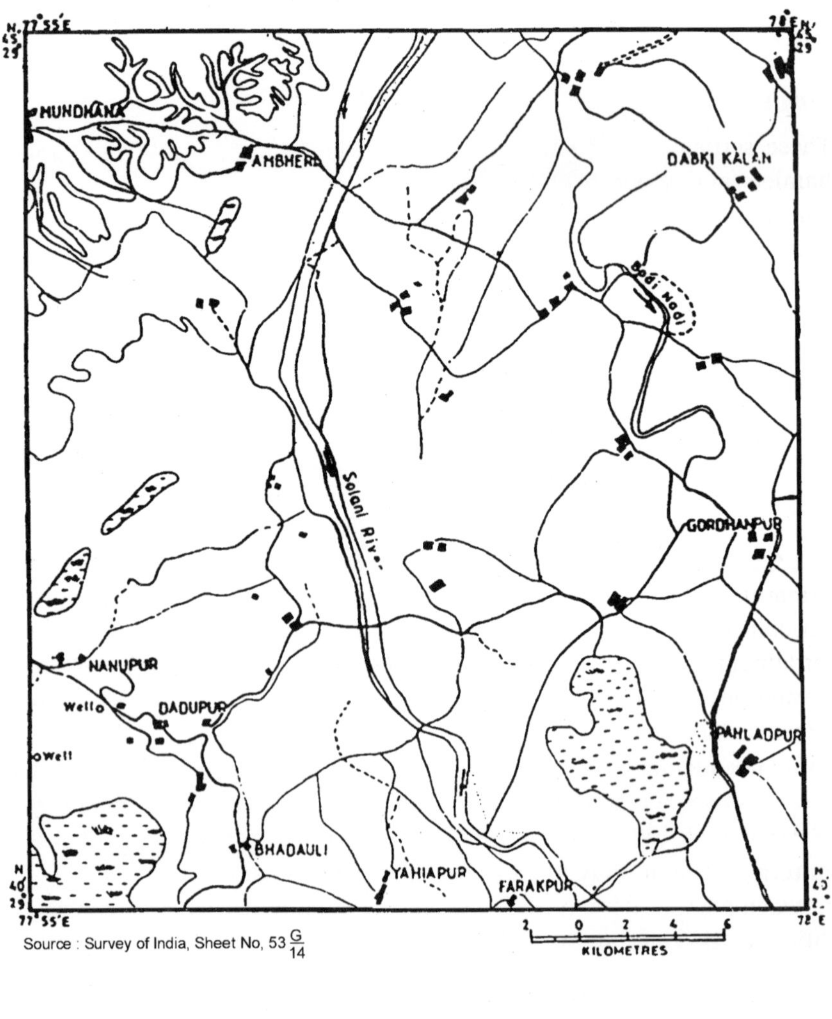

Source : Survey of India, Sheet No, 53 $\frac{G}{14}$

simple environmental explanation is acceptable. The reasons for the emergence of semi-compact settlements in the semi-arid areas differ from that of the humid regions, and in the marginal productive lands to that of the highly fertile intensively cultivated lands. The only generalization about the semi-compact settlements is that in the arable lands and fertile plains, the primary type of settlement was small and loosely nucleated and, in fact, lacked a pattern. Increase in population can cause village to grow and the number of houses increases. These houses start occupying the open spaces and lead to semi-compact settlement, which ultimately acquires the shape of a nucleated compact settlement (Fig. 5.3).

Figure 5.3 shows a rural landscape having several semi-compact settlements. Such settlements are well marked in the north-western parts of the Ganga-Yamuna doab, Khadar plain of the Ganga and its tributaries, and the ravine lands of the Yamuna river. In the east of Aravallis, in the state of Rajasthan, and in the hilly tracts of Madhya Pradesh, Arunachal Pradesh, Meghalaya, Manipur, Mizoram, Tripura, Sikkim and Siwaliks, one may observe numerous semi-compact settlements.

Hamleted Settlements

These settlements are characterized by the presence of several smaller hamlets and separate individual habitations, spreading over the entire *Mauza/Basti* (settlement). In these loosely connected settlements, street pattern is generally lacking and the straggling residences are separated by intervening cultivated fields. Such settlements generally occupy transitional zone between compact and isolated or dispersed settlements. In India, they are well marked in the Ganga-Ghagra doab, Trans-Ghagra plain, Kashmir Valley, Doon Valley and the Vidhyans. On an average the number of hamlets is 5 or more and in exceptional cases may reach 50 or even more (Fig. 5.4).

Dispersed Settlements

Dispersed settlements are generally found in the areas of extreme climates, hilly and mountainous tracts, thick forests, grasslands, poor agricultural lands and in the areas of extensive cultivation. The dispersed settlements are relatively recent in origin. In India, dispersed settlements started developing mainly during the medieval period. Such settlements may also be seen in the *khadar* (new alluvial plains which are frequently inundated) lands (Fig. 5.5).

Patterns of Rural Settlements

Pattern or design of a rural settlement has been defined as the relationship between a house to another house. Or, in other words, the spatial organization of houses in a village is known as its pattern. The pattern of

Figure 5.4 *Hamleted Settlements in a Mountain Valley*

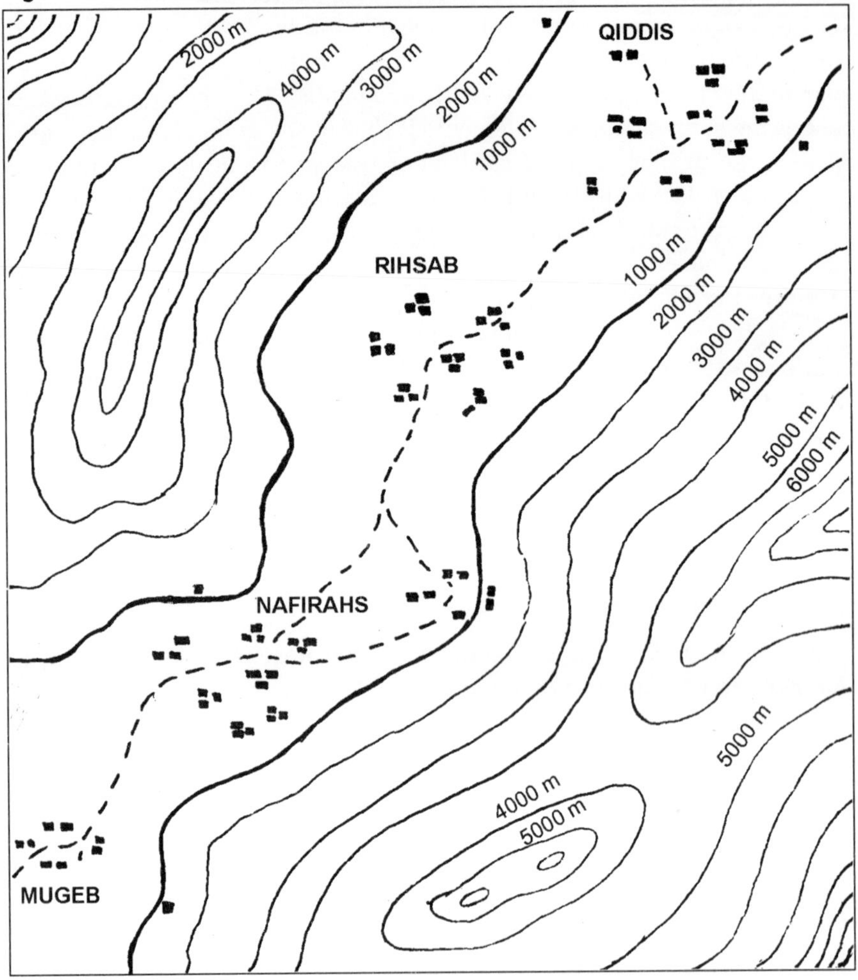

settlement may be easily identified by reading and observing a large scale map, like that of the topographical sheets, published by the Survey of India, Dehradun.

The term, 'pattern of settlement' deals with the compact and semi-compact settlements only as each of the dispersed and scattered settlements has its own shape and design.

The rural settlements have different shapes and sizes. The site of the village, and the surrounding topography and terrain, influence the shape and size of the village. In fact, the pattern of rural settlements is the result of a series of adjustments to the environment which have been going on for centuries. Moreover, the socio-cultural and economic factors like caste structure and functional needs of the people also have a close bearing on their shape and size. In the valleys and mountainous areas, the pattern of

Figure 5.5 *Dispersed Settlements*

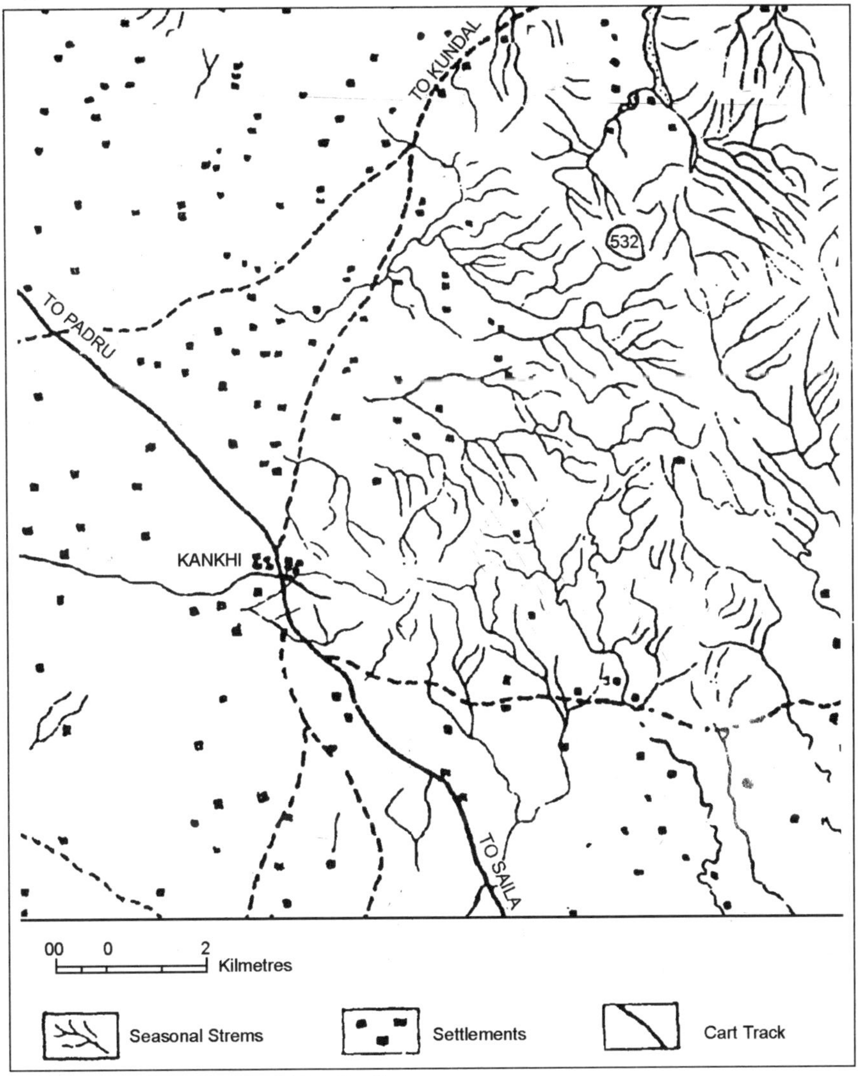

Source: Survey of India, Sheet No. 45 c 7.

rural settlements is generally linear, while in the fertile alluvial tracts their shape is rectangular, and near the ponds and lakes circular or semi-circular. In exceptional cases the settlement pattern may be nebular and amorphous. The rural settlements may be broadly classified under the following categories:

1. Rectangular pattern
2. Linear pattern
3. Circular and semi-circular pattern
4. Star-like pattern

5. Triangular pattern
6. Nebular pattern
7. Terraced pattern

Rectangular Pattern

Most of the rural population in India live in rectangular pattern of settlements. The houses in rectangular settlements are generally attached to each other, and the lanes are straight. Such settlements are found in the Northern Plains of India (Fig. 5.6). The well-planned settlements in the Indira Gandhi Canal catchment area, especially in Suratgarh and Ganganagar districts of Rajasthan are rectangular in shape.

Figure 5.6 *Rectangular Pattern*

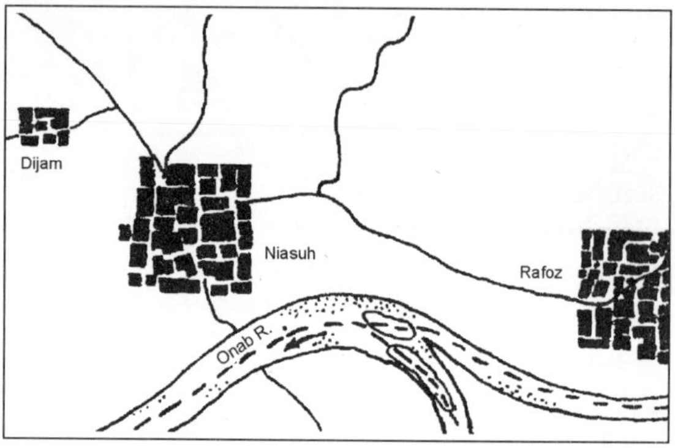

Linear Pattern

Linear pattern is the other most important pattern of rural settlements. In the linear settlements, houses are arranged along either side of the river, road, railway line, river or canal. Such rural settlements also evolve along the edge of a valley, especially in the mountainous areas above the flood level or along the coast.

The development of linear settlements in the hilly and mountainous areas is largely controlled by the terrain and topography. Along the river banks and the sea shore, the flood and high water level influence the origin and spread of linear settlements. Such settlements are well spread in India along the roads and in the valleys of rivers in the hilly tracts (Fig. 5.7).

Figure 5.7 *Linear Pattern*

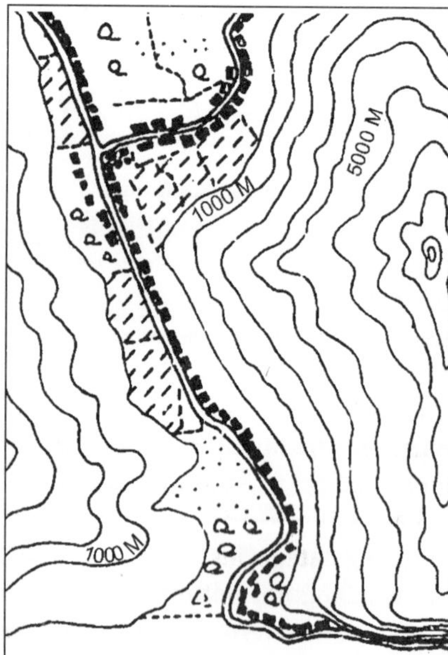

Linear Settlement in a Side-Valley of Kashmir (J & K)

Circular and Semi-circular Patterns

The fishermen and salt producers develop their settlements along the sea coasts and salt lakes. Since the people prefer to stay near the water bodies, they construct their houses along the coasts. Such settlements acquire the circular or semi-circular shapes. In the vicinity of crater lakes and on the levees of ox-bow lakes are found such settlements (Figs. 5.8 and 5.9). These people earn their livelihood from the water either by catching fish, water nuts, grasses, or by providing services to picnic goers and aesthetic lovers.

Figure 5.8 *A Circular Pattern: Uhsba Village*

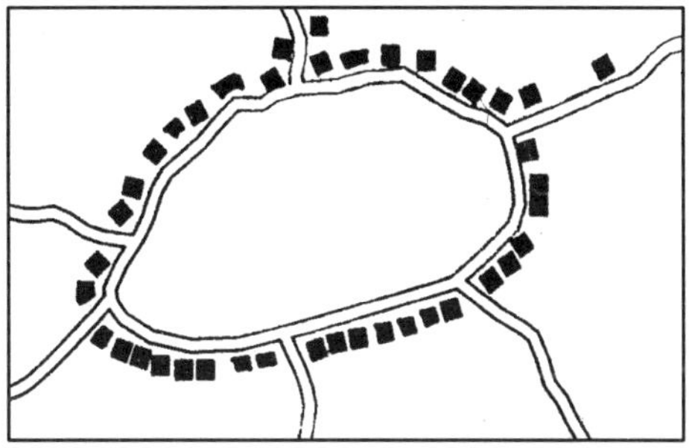

Figure 5.9 *Semi-circular Pattern: Abih Village in Siwaliks*

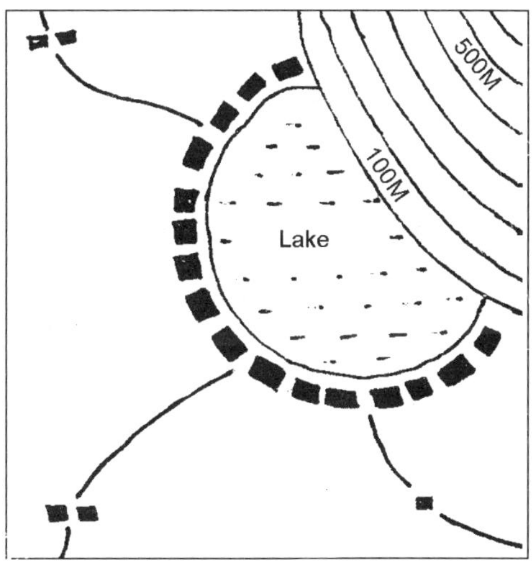

Star-like Pattern

The star-like settlements develop on the sites where several metalled or unmetalled roads converge. In the star-shaped settlements, houses spread out along the sides of roads in all the directions. This pattern is common to both villages and towns and is caused mostly by new development, spreading out along the major roads. This type of settlement is the characteristic of the countryside of the Great Plains of India, comprising Punjab, Haryana, Uttar Pradesh and Bihar (Fig. 5.10).

Figure 5.10 *Star-like Pattern: Abih Village*

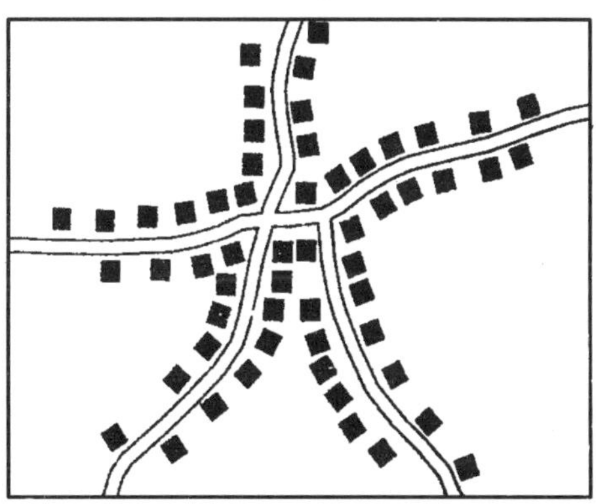

Triangular Pattern

Triangular pattern of rural settlements generally

develops at the confluence of rivers. The lateral expansion of houses at the confluence is constrained by the rivers. Consequently, the settlement acquires a triangular shape (Fig. 5.11).

Figure 5.11 *Triangular Pattern: Naarbij Village*

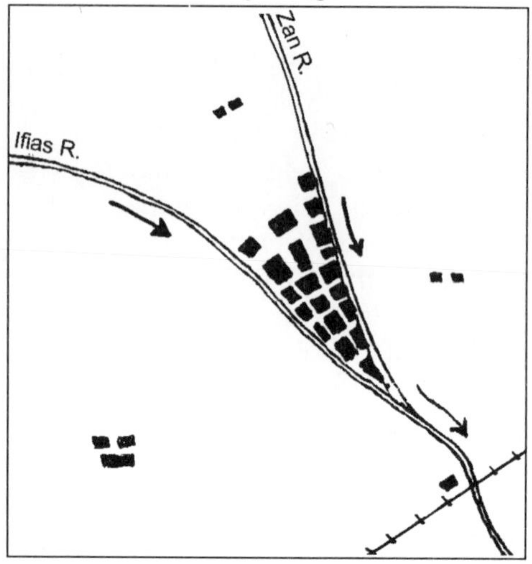

Nebular Pattern

When the shape of a settlement resembles to a nebula, it is known as a nebular settlement. The arrangement of roads in such a settlement is almost circular which ends at the central location or the nucleus of the settlement. Generally, the size of nebular settlements is small and they develop around the house of the main landlord of the village or around the temple, mosque, church or gurudwara. They are also found at the top of hillocks in Uttrakhand, Himachal Pradesh and Vindhyan mountains. There are several villages of this type in the upper Ganga-Yamuna doab, Uttrakhand and Himachal Pradesh (Fig. 5.12).

Figure 5.12 *Nebular Pattern around a Hill Aheedam Village*

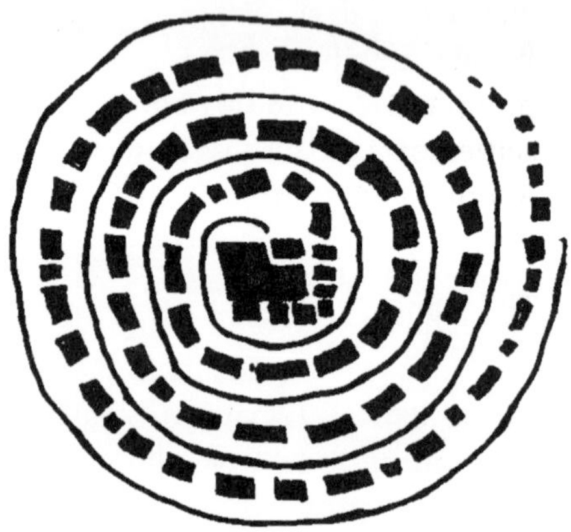

Terraced Pattern

These settlements are found in the hilly and mountainous areas. In the hills and mountains, terraces are cut into the slopes. In the terraces cultivation of crops is done with great skill. The cultivators construct their huts and houses in their terraced fields, resulting into a staircase like pattern of rural settlements (Fig. 5.13).

Planned Rural Settlements

Most of the rural settlements and their shapes are the results of geo-historical and socio-cultural factors. However, in the recent decades,

Figure 5.13 *Terraced Settlement: Sadargam Village, District Tehri*

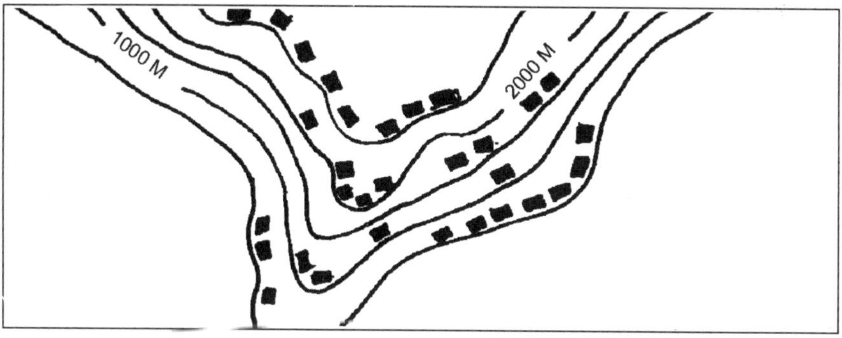

attempts are being made to develop planned rural settlements. In specific regions rural settlements are planned to give each family a reasonably good income, a better standard of living and to make the rural life more enjoyable. Examples of planned settlements are found in the sparsely populated parts of the country and the catchment of rivers on which multi-purpose projects have been constructed and rehabilitation of people has been planned. The main advantages of the planned rural settlements have been given as under:

1. People are ensured a reasonably good income. There is adequate provision of roads, electricity and drinking water. In the unplanned settlements, this is less likely to be the case.
2. Another important advantage of the well-planned rural settlements is that they ensure a judicious utilization of land.

The planned rural settlements are, however, not free from disadvantages. Some of the disadvantages of planned villages are given below:

1. In the early stages of planned settlements, the settlers have little freedom of action and the schemes do not provide an immediate solution of their problems as perhaps they thought they would. Some settlers become dissatisfied and leave the scheme but most realize its long-term advantages.
2. Settlements are planned for a given number of families, based on the capacity of the land to support them at a given income. Naturally, when they have children, not all the young people will be able to live in the same areas. To overcome this problem, people will either have to move away from the area in future generations or will have to make the original settlement schemes the basis for trading and rural industries. If no alternative jobs are created, land schemes will only postpone the 'drift to the towns' for a generation, instead finding a permanent solution.
3. The cropping pattern adopted may not be useful in the long run. Generally, in planned rural settlements, cash crops are grown and

most of such crops, are soil exhaustive. The merits of diversification and rotation of crops which enhance the fertility of soils by fixing nitrogen, will not be available.

The demerits of planned settlements may be overcome by adopting a flexible and pragmatic agricultural land use planning as the number of people increase.

Rural Settlements in India

India is often called as a country of villages. In fact, about 72 per cent of its population (2001) is living in rural settlements, largely engaged in primary economic activities. There are about 5,81,000 rural settlements in India. These settlements occupy alluvial plains, river banks, flood plains, natural levees, hill slopes, valley floors, river terraces, piedmont areas, wet points (water points), flat-hill tops, spurs, ridges and coastal lands.

In India, the average areal size of a village is 5.6 sq. km., which inhabits an average population of 1,085. The average areal size of rural settlements in Pondicherry (1.9 sq. km.) is the smallest, followed by West Bengal (2.3 sq. km.), Uttar Pradesh and Bihar (2.5 sq. km.). Contrary to this, Jammu & Kashmir with 34 sq. km. and Arunachal Pradesh with 23 sq. km. has the largest area per settlement.

The average village density is about 18 villages per 100 sq. km., which varies from a minimum of 3 villages per 100 sq. km. in Mizoram and Jammu & Kashmir to 53 in Pondicherry and 42 in West Bengal.

The rural population density (average 191 per sq. km.) also exhibits great variations being a minimum of 9 persons in Arunachal Pradesh to 710 persons in Lakshdweep. High density areas of rural population are found mainly in the alluvial plains of Sutlej-Ganga-Brahmaputra rivers and the deltas of Mahanadi, Godavari, Krishna and Kaveri.

Rural Settlements in Himalayas

Himalayas, having parallel ranges in the form of Siwaliks, Lesser Himalayas and Greater Himalayas, have different types of rural settlements at different altitudes. The rural settlements of the Himalayas, excluding those in the valleys, may be classified under the following categories: (i) hamleted or semi-sprinkled, (ii) dispersed, and (iii) isolated homestead.

Hamleted type mainly occupies low-lying valley with regular stretch of fairly level land, like that of the valley of Kashmir, the Dun Valley, the Kullu Valley etc. The small settlements of valley floors are nucleated. The dispersed type of rural settlements are found in patches and generally they are bigger in size, while the isolated homestead are found in the higher altitudes and alpine pastures. At the high altitudes there are seasonal houses and huts of the transhumants (Gujjars and Bakarwals). An interesting feature of the rural settlements in the Himalayas is that they are mostly located on the southern slopes, and generally below 3,500 metres above the sea level.

In the Eastern Himalayas, the settlements are generally small and widely dispersed due to steep slope, more undulating topography, heavy rainfall, dense forest cover and multiplicity of tribes with different languages and dialects, customs and traditions. Here, compact and well-organized villages are found in the northern zone of Indo-Tibetan culture: the lower zone exhibiting Assamese impact, while the middle zone is characterized by dispersed settlements.

In Meghalaya, the villages of Khasi tribes are located along the hill slopes near the water bodies, which range from isolated homestead to dispersed and compact settlements. In Nagaland, because of the past tradition of head-hunting, villages have been developed either at the hill tops or at the sides and spurs between 1,200 to 2,000 metres. Each Naga village has 20 to 120 houses. The houses are arranged in parallel rows through which passes the main lane (road).

In Manipur, Kukis, practising shifting cultivation (*jhuming*), build their houses on flat-topped ridges like that of the Nagas. The Lushais (Mizo) prefer either a valley along the stream or the hill top and flat ridge for the establishment of their villages.

Rural Settlements in the Great Plains of India

The Northern Plains of India present mixture of types and patterns of rural settlements. Rural settlements in Rajasthan plain are small, compact and sparsely distributed. The availability of water is the main criterion for the selection of site of a rural settlement.

In the arid and desert areas of Barmer, Jaisalmer and Bikaner, where, there is predominance of sand dunes, settlements are confined to the oases, especially near the water points. But, in the northern Rajasthan and to the east of Aravallis, the rural settlements are large and compact. After the introduction of the Indira Gandhi Canal a number of compact settlements have developed in its command area. In the highly irrigated plains of Punjab and Haryana the villages are almost uniformly distributed. These settlements are compact settlements but in the *khadar* area the settlements are generally semi-compact.

Over the Ganga-Yamuna plain almost 60 per cent of the population lives in medium to large sized villages. In the *Tarai* belt of Uttrakhand and Uttar Pradesh, rural settlements are unevenly distributed due to high percentage of forests, marshy lands and seasonal floods. In the Ghar (Bhabhar) region settlement sites follow the drainage lines and the nature of slope. In the areas of Bhangar (areas of older alluvium) the settlements are generally compact and closely packed surrounded by mango orchards.

In the middle Ganga plain the distribution and patterns of rural settlements are largely influenced by alluvial morphology. Areas of east

Uttar Pradesh and west Bihar are marked by small-sized but closely-spaced villages. Semi-compact settlements are typical of the Ganga-Ghagra doab. The Mithila plain exhibits wide variation in settlement pattern and types: linear in lower Gandak valley dispersed in submontane tract of Champaran, and linearly oriented along the levees of the dead-channels or ox-bow lakes in Purnea district. The south Bihar plain has more compact settlements than its northern part.

In the lower Ganga plain, hydrological characteristics have played a dominant role in determining the types and patterns of rural settlements. Here, scattered villages are common in Rarh plain, Duars, and Sundarban delta.

In the Brahmaputra valley villages are generally agglomerated, aligned along the river levees and transport arteries. Moreover, the villages are smaller in size in which houses are separated by bamboo fences. Machan type of houses on wooden pillars are constructed in low-lying and flood-prone areas where boat is the only means of transport during rainy season.

Rural Settlements in the Peninsular India

Rural settlements in the peninsular part of the country again exhibit mixed types depending upon the nature of relief, soil fertility, water supply and socio-economic development. Throughout the hilly tract of the Aravalli region huts are widely dispersed within the revenue village land. In the dissected hills of Mewar, Marwar and Alwar, isolated farmsteads are dotted in the long narrow valleys, while the districts of Jaipur, Sawai Madhopur and Bundi are characterized by compact and semi-compact settlements.

In the highly dissected and ravine tracts of Bundelkhand, large compact settlements occupy the favourable and protected sites, while badlands are marked by semi-compact and dispersed settlements.

The Chotanagpur plateau region shows great variations in the types and patterns of rural settlements. Here, the Rajmahal tract is characterized by semi-compact settlements. However, compact settlements have been developed in the fertile areas of the Baghelkhand and Chhattisgarh region. The lava plateau region of Maharashtra, owing to its rich soils, has favoured the clustered settlements.

In south India, large compact and widely dispersed villages are the characteristic feature of the northern part of Karnataka and Rayalseema of Andhra Pradesh. Between the Kaveri and the Tungbhadra are the numerous tanks along which the settlements are generally compact.

Rural Settlements in the Coastal Plains

The eastern and the western coastal plains are dotted with several fishing villages of different shapes and designs. In the Mahanadi delta area, high escarpments provide ideal location for rural settlements to safeguard against floods. In the Godavari, Krishna and Kaveri delta regions intensive farming activity has promoted the compact or nucleated settlements, which are generally located along the canals and on high ground levees.

The Malabar coast is dominated by large compact villages but coconut and cashew plantations have encouraged the growth of isolated dwellings. The coastal plains of Gujarat are marked by nucleated settlements of medium to large size. The Saurashtra region is characterized by small villages with long inter-village distances. In the semi-arid region of Kutch one may observe a few semi-compact settlements at long intervals hardly containing a room or two.

Morphology of Rural Settlements of India

The morphology of rural settlement mainly deals with the layout, plan and internal structure of the settlement. It not only views settled area in terms of physical space but also identifies its various components in respect of socio-economic space, which has a direct bearing in controlling the arrangement of buildings, patterns of streets and fields and functional characteristics of settlements in general. In an old settled country like India, it assumes considerable importance as it helps in understanding the socio-cultural structure of the villages, their economic and sanitary conditions as well as their response to new innovations (Tiwari, 1984).

The morphology of a settlement has two main components: (i) ground plan, and (ii) built-up area. The ground plan includes mainly buildings, street patterns, arrangement of buildings and cultural artifacts like temple, mosque, church, *chaupal, chowk*, residence of landlord and shops etc. Here, the process of change is comparatively slow. On the contrary, built-up area undergoes frequent changes. The main part of a village consists of the following:

(a) Homogeneous part consisting of field, grazing lands and community land.
(b) Circulatory part consisting of village roads, streets, *chak*-roads and lanes.
(c) Central part having the built-up area of the village.
(d) Special part marked by temple, school, *panchayat ghar, chaupal, morung*, place of recreation, etc.

An Indian village, though recognized as the smallest revenue-cum-administrative unit, generally consists of aggregate of residence, the inhabitants of which have certain relations, and some kind of union or bond of common government. The built-up area consisting of clusters of houses is surrounded by cultivated fields and linked with *kutcha* (unmetalled) and *puccka* (metalled) roads or village footpaths. At times there are more than one inhabited site within the territorial limits of one revenue village interspersed with cultivated fields, pastures and village groves. The main site, generally occupying the central location and inhabited by early village settlers (mostly belonging to upper castes and zamindars), former nucleus of the village, is known in northern India as *Khas Gaon*. Other features around it are small hamlets named after the dominant castes residing therein (mostly low castes, landless labourers, village artisans or emigrants from the main village). These small hamlets are locally called by different names such as *purva, pura, tola, nangla, majra, patti, dhani,* etc. The main village and its outlying hamlets, though physically detached from each other, function as an integrated unit under the old *jajmani* system.

The general pattern of land use includes multi-cropped highly productive zone, locally called as *gouhan, goira,* or *goind,* around the inhabited site, followed by relatively less fertile area of low productivity. The built-up area, forming the nucleus of the settlement, attracts the maximum concentration of socio-economic activities and transport movements.

References

Ali, S.M., 1966, *Geography of the Puranas*, New Delhi: Peoples Publishing House.

Bose, Ashish, 1980, *India's Urbanisation 1901-2001*, New Delhi: Tata McGraw Hill.

Chisholm, M., *Rural Settlement and Land Use* (3rd edn), London: Hutchinson.

Husain, M., 2002, *Human Geography*, Jaipur: Rawat Publications.

Knowles, R., J. Wareing, 2002, *Economic and Social Ge*ography, New Delhi: Rupa Co.

Prakasa Rao, V.L.S., 1983, *Urbanisation in India: Spatial Dimensions,* New Delhi: Concept.

Ramachandran, R., 2001, *Urbanization and Urban Systems in India,* New Delhi: Oxford University Press.

Sundram, K.V., 1977, *Urban Regional Planning in India,* New Delhi: Vikas.

Tiwari, R.C., 2000, *Geography of India*, Allahabad.

6

Population and Migration

Population

The study of human population, their growth, distribution, density, composition and migration is an important area in which the population geographers are increasingly probing. In the initial phase of human history, the human population grew at a snail speed. The hazardous climatic conditions, the migratory character of hunters and gatherers, poor nourishment and poor technology were all unfavourable for the growth of human population. During the last 10,000 years, the growth of world population accelerated at two distinct times. The first time was about 10,000 B.P. (before present) or 8000 B.C., when man started the domestication of plants and animals (the beginning of agriculture). This abrupt change in human economic activity is known as the *Agricultural Revolution*. The assured supply of food from agriculture and settled life provided better nourishment. The human body became more resistant to

Figure 6.1 *Growth of Population since the Neolithic Period*

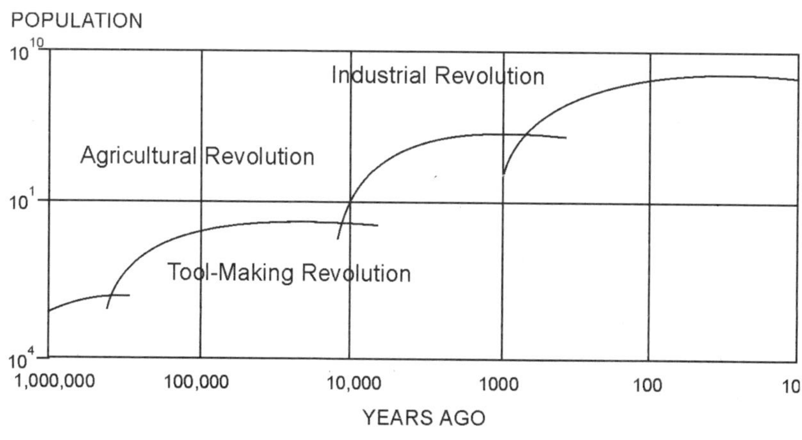

the adverse weather and climate. Consequently, the human population started growing at a steady pace. The second time when population started increasing at a faster pace, was 1779, when man harnessed fossil fuel and developed steam engine. This development is known as *Industrial Revolution* in Britain and Europe (Fig. 6.1).

The agricultural and industrial revolutions mentioned above were important ecologically as there is a close relationship between population, ecology and environment. In fact, if population number rise or fall, everything on the earth surface is liable to change. Before discussing the spatial distribution, density and dynamics of population, it is important to present a brief account of the growth of human population during the pre-historic period.

Population Prior to Agricultural Revolution

The population of our ancestors a few million years ago was confined on both the sides of equator in Africa and the number was only a few lakhs. By the time our ancestors started agriculture, the information started passing from generation to generation. The transmission of knowledge about the domestication of plants and animals resulted into the decline of death rate. The increased food supply led to relatively better nutrition standard, greater resistance to diseases, increased longevity of life and hence growth in human population.

As stated at the outset, the domestication of plants and animals was started around 8000 B.C. There is, however, no direct evidence to know the size, population growth pattern, distribution and density of population during that period. The experts of population have, however, estimated the population of the past on the basis of indirect or circumstantial evidences. For example, agriculture was unknown before 8000 B.C.; prior to that all human groups made their living by hunting and gathering. Not more than 52 million square kilometres of the earth's total area could have been successfully utilized in this way by the early ancestors. From this and the population densities of the hunting and gathering tribes of today, it has been estimated that the total population in 8000 B.C. was about 5 million people (Durand, 1973). They lived in bands of fewer than 50 people; the men being primarily hunters, whereas the women providing most of the food through plant gathering and roots collection. Nowhere on the earth was the population large. Typical densities were around four persons per hundred square kilometres. There were large uninhabited areas and each band usually lived in relative isolation from others. Such low densities are not surprising because the economies of bands were based on a form of extensive land use. In other words, they obtained their food from large areas without expanding much labour per unit of land.

It was the period of high and fluctuating birth and death rates. The high death rates restricted population growth. It is probable that population growth was also restrained by deliberate control through restricted sexual intercourse, abortion, infanticide, and use of plants with contraceptive powers. Slow growth for a band meant that its size did not rapidly exceeded the carrying capacity of the territory being used for hunting and gathering.

The distribution of population, prior to the development of agriculture, covered only parts of the old continents (Africa, Asia and Europe). By 8000 B.C., hunters and gatherers had migrated from Africa throughout Europe and Asia and from South-East Asia to Australia. They crossed the Bering Strait and moved southward in the continents of North and South America. Only Antarctica was the uninhabited continent. The dispersal of modern man prior to the Neolithic period has been shown in Figure 6.2.

Table 6.1 *World Growth of Population*

Year	*Population (in million)*	*No. of years to add one billion people*
8000 B.C.	5	—
1 A.D.	200	—
1000 A.D.	300	—
1650 A.D.	500	—
1850 A.D.	1000 (1 billion)	80
1930 A.D.	2000 (2 billion)	32
1960 A.D.	3000 (3 billion)	13
1975 A.D.	4000 (4 billion)	12
1987 A.D.	5000 (5 billion)	10
1999 A.D.	6000 (6 billion)	13
2070 A.D.*	9000 (9 billion)	—

*Projected figures.

Sources: 1. *United Nations Population Reference Bureau*, 1998.
2. *The World Guide—See the World from a Different Angle*, 10th ed., Oxford: New Internationalists™ Publications Ltd.

It may be seen from the Table 6.1 that the population at various points of time from the onset of the first agricultural revolution (8000 B.C.) until the first scanty census data recorded in the 17th century A.D. has also been estimated. It is thought that the total human population at the time of Jesus Christ's birth (1 A.D.) was about 200 million, which became about 500 million by 1650 A.D.

It may also be observed from the Table 6.1 that in 1850 A.D. the world's estimated population was 1000 million which became 2000 million in 1930. In 1960 the population was 3,000 million, which rose to 4,000 million in 1975, reaching 5,000 million in 1987 and 6,000 million in 1999. Thus, the stupendous growth of human population took place

Figure 6.2 *Dispersal of Early Hominids*

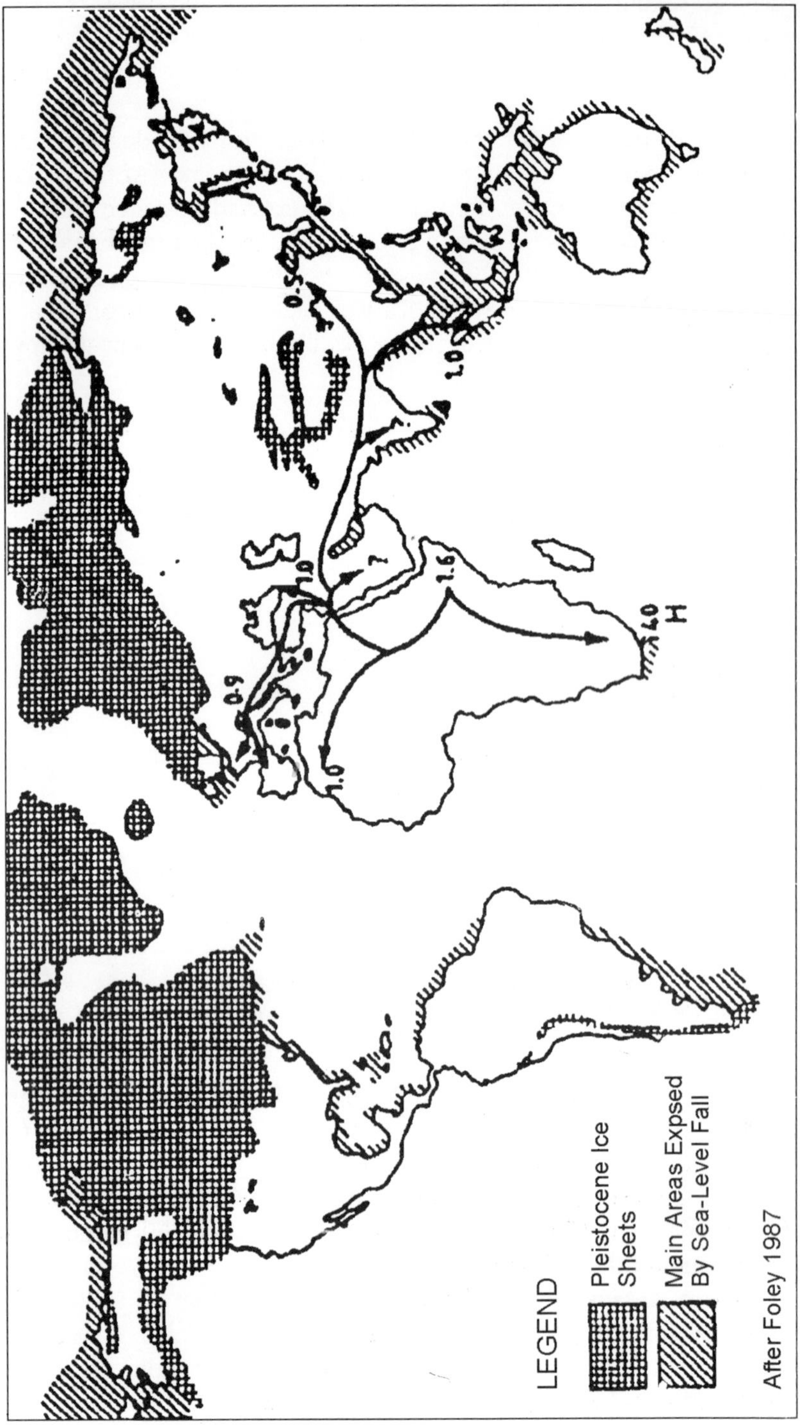

Figure 6.3 *The Rate of Innovation Through Human History*

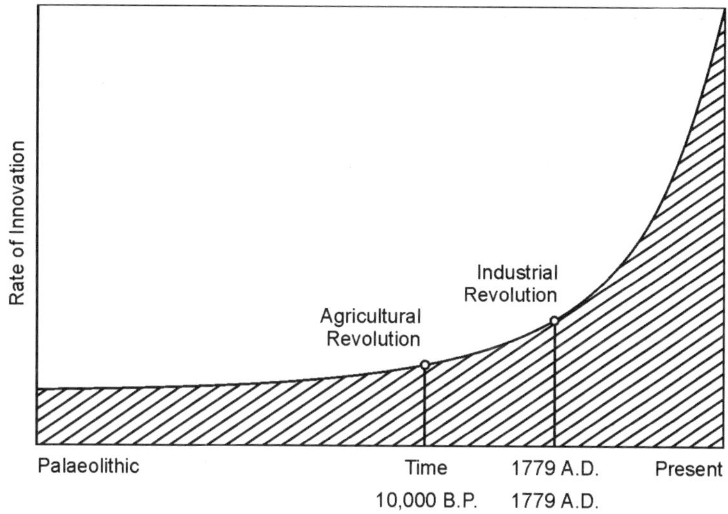

between 1850 A.D. and 1999 A.D. The doubling period of world population has been given in Table 6.2.

Table 6.2 *Doubling Period of World Population*

Period	*Population*	*Time in which population doubled*
8000 B.C.	5 million	—
1650 A.D.	500 million	1500 years
1850 A.D.	1000 million	200 years
1930 A.D.	2000 million	80 years
1975 A.D.	4000 million	45 years
2025 A.D.*	8000 million	26 years

* Projected figures.

Sources: 1. *United Nations Population Reference Bureau*, 2001.
2. *The World Guide*, 10th edition, Oxford: New Internationalists™ Publications Ltd.

It may be seen from Table 6.2 that the population on an average doubled once after every 1,500 years during the Neolithic period. The next doubling, i.e., from 500 million to one billion, took 200 years and from 1 billion to 2 billion reached in 80 years, while 4 billion mark was touched in only 45 years. If the present growth of population may continue, the population may be 8 billion in 2030 A.D.

Thus, the world had 3 billion people in 1960, 5 billion in 1987, and 6 billion in 1999. It, is estimated that by 2070 the world population will reach a peak of 9 billion, after which it will start to decline, reaching approximately 8.4 billion in 2110 (Fig. 6.4).

Figure 6.4 *World Population Growth (8000 B.C. to A.D. 2000)*

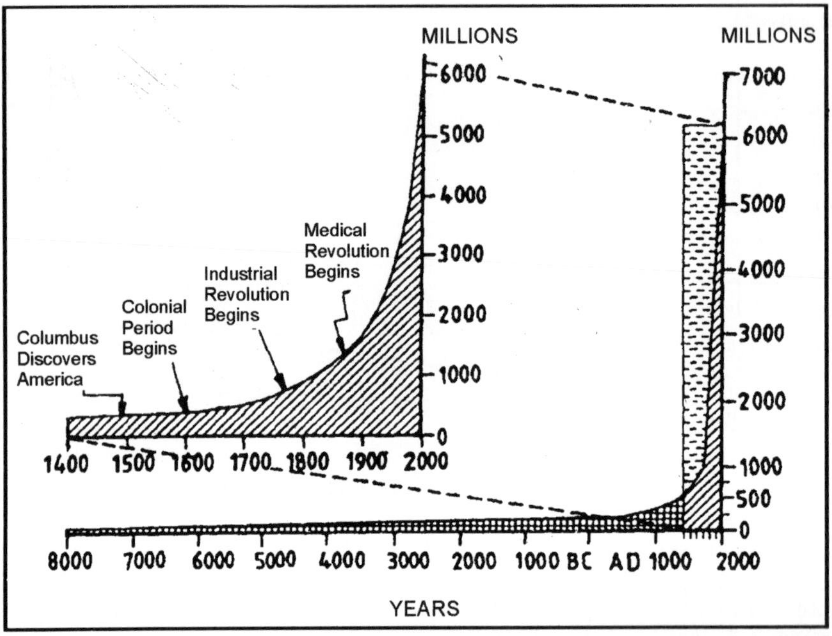

According to the UN annual report entitled *State of World Population, 2003,* global population growth slowed down, in part, due to efforts made in the last 30 years to raise awareness on the dangers of overpopulation. Although the report states that the growth will continue to drop during the next decades, high birth rates in the past determine a demographic inertia, which generates approximately 90 million births per year.

The UN report also states that although in several European countries people over 65 years of age (senile population, senior citizens or retired people) make 20 per cent of the population, in 2050 this will amount to 35 per cent and by 2100 to 45 per cent. While more than half of the Japanese population will be over 65 in the year 2100, in China the proportion of senior citizens will triple from the current 10 per cent to 30 per cent in 2050. This would mean that by then there will be only two active workers for every retired person, while today there are ten. The increased dependency ratio and the low workforce may put additional pressure on the tax payers and workforce.

The continent-wise size and growth pattern of population has been given in Table 6.3.

It may be seen from Table 6.3 that the growth of population would continue to be the highest in the continent of Asia, followed by Africa and Latin America. There would be marginal increase in the population of Anglo-America (USA and Canada) and Oceania, while the population of

Table 6.3 *World: Size and Growth of Population (in million)*

World/Continent	*1950*	*1999*	*2025**	*2050**
World	2523.87	5929.83	8039.13	9366.72
Asia	1402.02	3588.87	4784.83	5442.56
Europe	547.31	729.40	701.07	637.58
Africa	223.97	778.48	1453.89	2046.40
Latin America	149.29	462.59	641.40	754.20
Anglo-America	171.61	304.07	369.01	384.05
Oceania	126.12	294.60	40.68	45.69

* Projected figures.

Sources: *World Resource, 1998-99*, World Resource Institute, UNO, New York.
World Guide, 10th edn., 2005-2006.

Europe will decline from about 729 million in 2000 to about 700 million in 2025. The negative growth of population in the continent of Europe is a point of great concern for the Europeans demographers. Most of the European countries have, consequently, adopted the policy to give incentives to the people to increase population.

Table 6.4 *Selected Countries: Size, Growth and Doubling Period of Population*

Rank/Country	*Population (in million) (2005-06)*	*Density (per sq. km.)*	*Doubling period of population (in yrs.) (2000-2025)*
1 China	1322	138	61
2 India	1098	334	35
3 USA	300	31	116
4 Indonesia	225	118	43
5 Brazil	182	21	40
6 Pakistan	161	202	30
7 Bangladesh	152	1060	30
8 Russia	141	8	*
9 Nigeria	130	141	33
10 Japan	128	339	275
11 Germany	83	231	*
12 France	60	110	185
13 UK	59.9	245	280
14 Canada	32	3	180
15 Australia	20	2.6	225
16 New Zealand	3.9	15	250

* Population will not double in infinity.
Assuming current growth rate does not change.

Sources: 1. *The World Guide 2005:200*, 10th edn. Oxford: New Internationalist™ Publications Ltd., pp. 602-3.
2. *Human Development Report*, 2001.
3. *Census of India.*

Each year, the world population grows by almost 90 million—an increase of about three people each second. The world population, other things being remain the same, will continue to grow at least until 2050 A.D.

It may be observed from Table 6.4 that China with a population of 1,322 million (2005-6), having a density of 138 persons per sq. km., has the largest population in the world, followed by India, which has a population of 1,098 million in 2005-6 with a density of 334 persons per sq. km. USA (300 million) and Indonesia (225 million) were the third and fourth largest population countries in the world in 2005-6. Brazil, Pakistan and Bangladesh occupied the subsequent ranks, while Russia with the largest area in the world, has only the eight largest population, being only 141 million.

The doubling period of population of the selected countries between 2000 and 2025 has also been given in Table 6.4. It is interesting to note that the population of Bangladesh and Pakistan will double in 30 years, followed by Nigeria (33 years). The doubling period of population of USA is 116 years while that of Japan and UK is 275 and 280 years respectively. Russia and Germany have entered in a stage in which the population is showing a negative growth rate and their population including several other developed countries of Europe may not double even in infinity.

Future Distribution of Population

The present distribution and density of the world population is highly uneven. The pattern distribution and density during the next few decades is expected to remain similar to that of the present pattern. The main reason for this may be that the international law and world politics do not permit free migration from one country to another. Secondly, the sparsely populated areas have adverse and harsh climate and generally poor in resource potential at the present level of technology.

Prospects of large-scale expansion of population in the cold, arid, mountainous and densely forested areas are dim. The cold climates with their short growing season and long distances from the existing commercial and industrial centres have limited potential for their agricultural or industrial development. Consequently, these areas will remain sparsely populated and that too by the indigenous people that have adjusted to the harsh environment over many centuries or by the outsiders who are engaged in the extraction of minerals, military training and research activities.

In the arid and semi-arid areas, the scarcity of water for daily needs and irrigation purposes is a major constraint in the development of human agglomerations. Most of the deserts, consequently, remain uninhabited or

Figure 6.5 *World Population: Numbers and Projections*

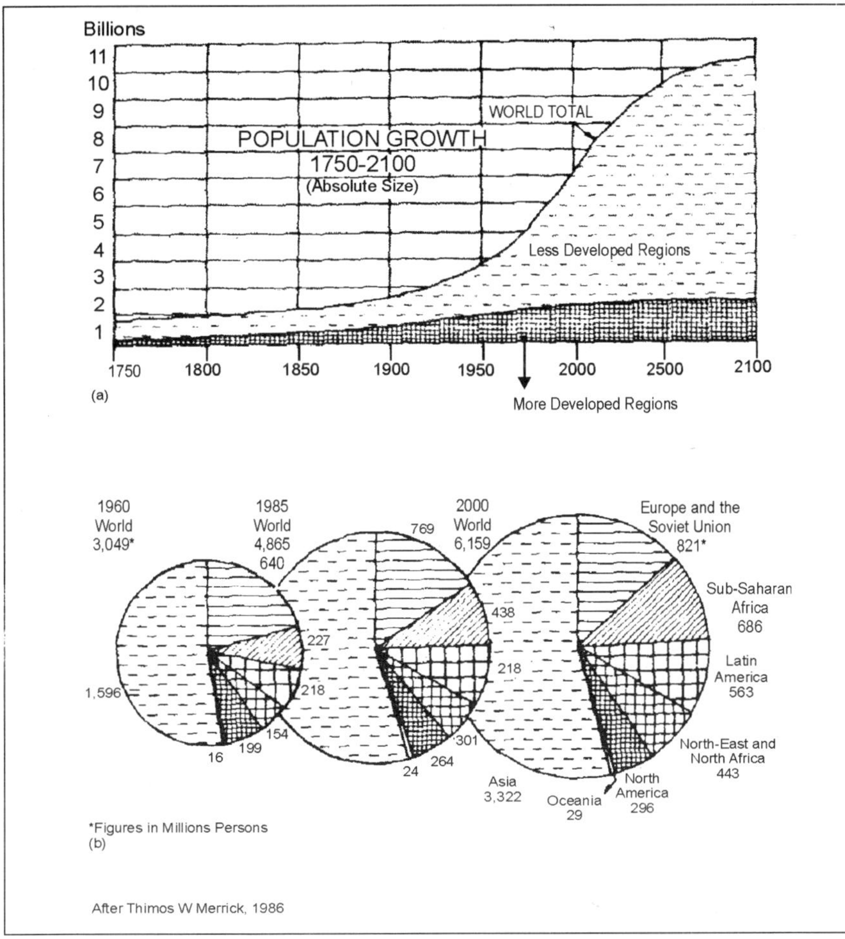

sparsely populated by nomads, hunters, gatherers and indigenous people (Figs. 6.6 & 6.7).

The equatorial rain forests like Selvas of Amazon basin and Congo basin are also not attractive regions for human habitation. Population in these areas may grow only if agriculture is developed by large scale clearing. The felling of trees may, however, be disastrous from the point of view of world climate. In the tropical regions, where the climate at lower altitudes is not conducive, most of the towns, cities and settlements have been developed around 2,000 metres above the sea level.

The rugged and folded mountains are also sparsely populated. Lofty mountains are almost uninhabited. In the Himalayas, Rockies, Andes, Alps and Caucasus, the areas above 2,500 metres are almost uninhabited (Fig. 6.7).

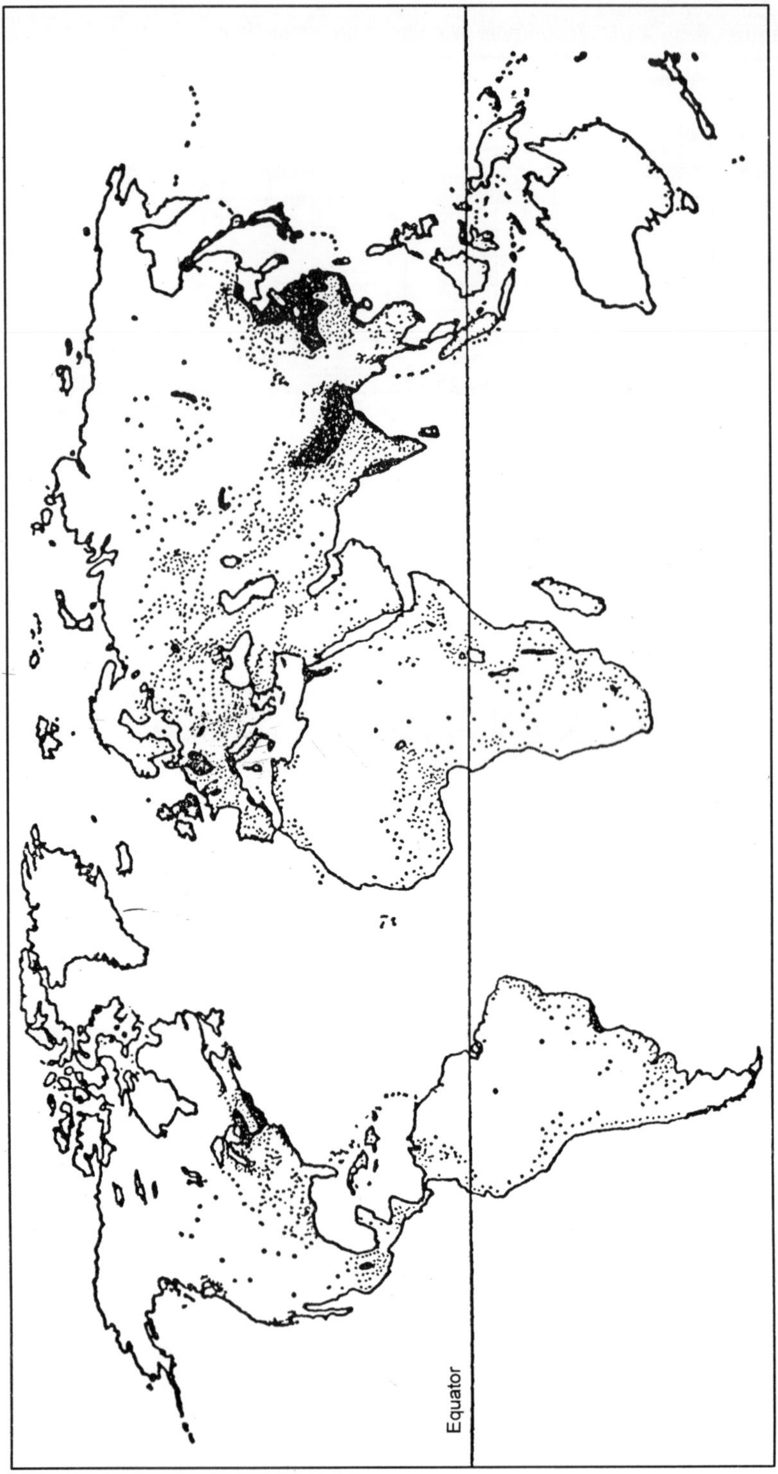

Figure 6.6 *World Distribution of Population*

Figure 6.7 *World: Population Density*

Figure 6.8 *World Population Cartogram*

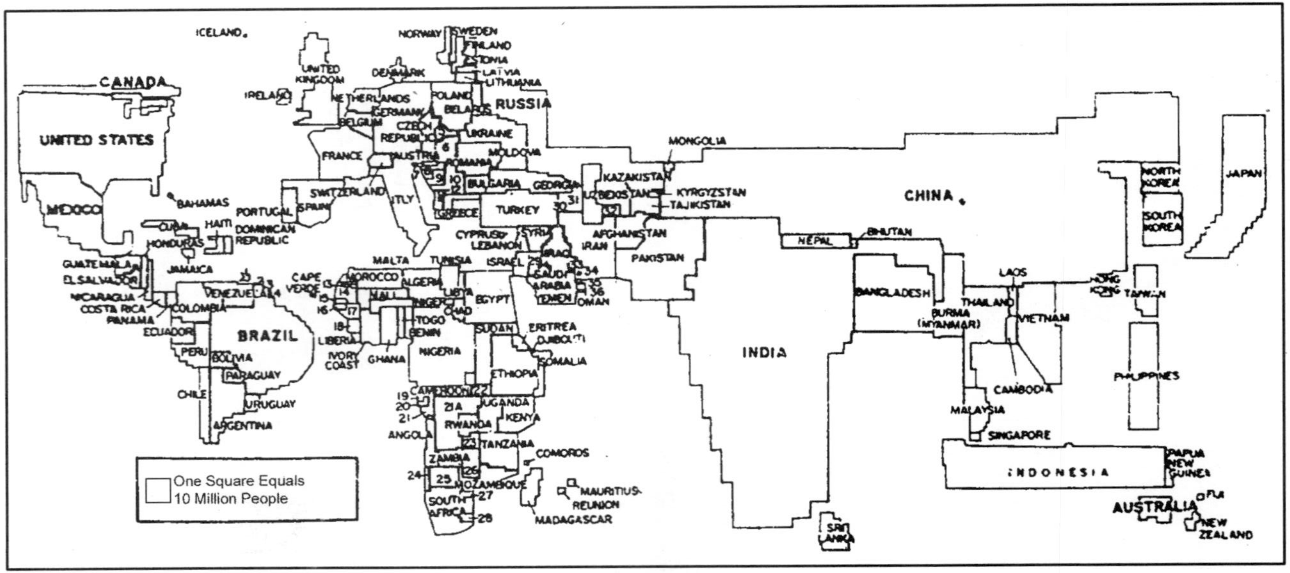

1.	Trinidad and Tobago	9.	Bosnia	17.	Guinea	24.	Namibia	32.	Turkmenistan
2.	Guyana	10.	Yugoslavia	18.	Sierra Leone	25.	Botswana	33.	Kuwait
3.	Suriname	11.	Albania	19.	Equatorial Guinea	26.	Zimbabwe	34.	Bahrain
4.	Fr. Guiana	12.	Macedonia	20.	Gabon	27.	Swaziland	35.	Qatar
5.	Slovakia	13.	Western Sahara	21.	Rep. Congo	28.	Lesotho	36.	United Arab Emirates
6.	Hungary	14.	Mauritania	21.A	Dem. Rep. Congo	29.	Jordan		
7.	Slovenia	15.	Senegal	22.	Central African Rep.	30.	Armenia		
8.	Croatia	16.	Gunea Bissau	23.	Burundi	31.	Azerbaijan		

The world population cartogram (2000 A.D.) has been shown in Figure 6.8. It may be seen from this that China has the largest population, followed by India, USA and Indonesia. The cartogram cannot tell us how the population it reflects is changing. As we see, varying rates of population growth mark the world's regions, some countries' populations are growing rapidly while populations in other countries are actually shrinking. Thus, the cartogram is not a static picture, its pattern changes in space and time.

Population Problems

The growing population in the developing countries and high degree of consumerism in the developed countries have put great pressure on the land, water, forest and mineral resources. The population problems, however, vary in space and time. Since the population problems of the developing countries differ from those of the developed countries, they have been examined separately.

Population Problems of the Developing Countries

About 80 per cent of the total population of the world is living in the developing and underdeveloped countries. China and India together constitute about over 40 per cent of the total population of the world (2005-6). The level of technology in the developing countries is still not adequate to use the natural resources judiciously. There are several countries of the world which are underdeveloped simply because of the non-availability of manpower. Such countries include Brazil, Columbia, Peru, Zaire, Russia (Siberian region), Saudi Arabia, Congo, Kazakhstan, Turkmenistan and Uzbekistan. The resource base in these countries is good to support a large population. Some of the major problems pertaining to population of the developing countries have been briefly discussed in the following:

1. *Rapid Growth of Population*: Birth rate in most of the developing countries is significantly high, while the death rate has been checked appreciably. Family planning in most of the developing countries is not practised seriously. This situation has resulted into a large proportion of young people who are dependent on workforce of the population. For example, in India, Pakistan and Bangladesh, about 33 to 40 per cent per cent of population is below 15 years of age. This large proportion of younger age group puts great pressure on the available medical, educational and other social amenities. The life in developing countries is, therefore, less comfortable and enjoyable.
2. *Unemployment*: In the developing countries, the workforce is largely engaged in primary sector(agriculture, forestry, fishery and mining). The secondary and tertiary sectors (industries and services) are less developed. There are very limited opportunities for the semi-skilled,

unskilled and highly educated people. In rural areas large number of unskilled workers face the problem of unemployment. The educated and skilled technocrats also have limited opportunities of employment. Consequently, both the educated and uneducated, skilled and unskilled workers try to emigrate from rural to urban areas, from their own country to the developed countries of the world or to the countries of South-West Asia in search of better employment and job opportunities. Those who find it difficult, migrate to big towns where it is often more difficult to find employment. Moreover, the towns become overcrowded, making living conditions poorer, resulting into numerous socio-economic and environmental problems.

3. *Poor Standard of Living and Malnutrition*: Poverty, hunger, malnutrition and deprivation are the other major problems of almost all the developing countries. The low standard of living is because of the low per capita income, the poor health is because of poor and undernourishment, and the level of awareness is low because of low literacy and education. The literacy rate is also low, for example, in Niger (16%), Burkina Faso (24%), Mali (26%) and Afghanistan (36%). The female literacy in the developing countries, barring a few, is significantly low. The ignorance and lack of awareness, the inadequacy of medical facilities and lack of financial resources come in the way of improving the health, education, housing and standard of living.
4. *Mismanagement of Natural Resources*: As stated at the outset, the economy of most developing countries is based on primary sector, i.e., agriculture, forestry, fisheries and mining. Agriculture is mostly done by traditional methods, obsolete equipments and rudimentary technology. The farmers have small and fragmented holdings, constrained by financial resources. Owing to lack of funds and finances, the farmers are unable to apply modern inputs (new seeds, chemical fertilizers, insecticides-pesticides, etc.). Consequently, agriculture is subsistence in character, the per unit area and per worker production is low. Many of the tradition-bound farmers, constrained by financial resources, do not accept agricultural innovations. Under these conditions, the farmers are not getting the optimum income from their agricultural resources.
5. *Slow Growth of Industrial Sector*: In most of the Third World countries, the secondary (industrial) sector is generally not strong. There is scarcity of local capital which makes the judicious utilization of resources for the industrial purposes. The workforce, though large in number, is generally unskilled and has no background of industrial development. Though, the large population provides a good market for the finished industrial products, the purchasing power is low. The poor industrial base, lack of capital and poverty of the people create a

vicious circle as a result of which the growth of industries is hampered. The pressure on agricultural resources is increasing and the per capita arable land in many of the developing countries has declined to 0.1 acre.

6. *Orthodoxy*: In the developing countries, people are generally tradition bound and less exposed to the outside world. They are generally staunch religious and respect their cultural values enormously. The tradition-bound people do not accept the new ideas and innovations easily. This is the reason why the family planning and health care programmes are not being implemented seriously in these countries. China, which has gone for one child policy, is an exception. For the removal of such attitudes and for the eradication of blind faith and orthodoxy, awareness through education and mass communication is imperative.
7. *Problem of Underpopulation*: Some of the developing and under-developed counties are underpopulated. In the sparsely populated countries, the growth of population, despite high birth rates, is low. Inmigration from other countries can increase their workforce, but the international law comes in the way. Consequently, the resources of underpopulated countries are not optimally utilized.

   The underpopulated countries, being the areas of isolation or harsh climates, agricultural resource cannot be utilized optimally. Many of the Africa, Latin American and the Central Asian countries fall in this category. A rational immigration policy can help these countries in the proper utilization of their resources.

Population Problems of the Developed Countries

The developed countries of the world are highly industrialized and urbanized. In these countries, the per capita income is high, the literacy and education is almost hundred per cent and the longevity is around 80 years. The standard of living is high and there is a stage of high consumerism. The workforce is mainly engaged in the tertiary sector and only about 10 over cent or less is dependent on primary sector. Despite overall development and high standard of living these countries are also confronted with some of the serious population problems which have been briefly presented in the following:

1. *Long Span of Life*: In the developed countries, the birth and death rates are significantly low. In most of the developed countries one may find the fourth stage of demographic transition in which the population of senior citizens/senile (above 65 years) is at par to the younger population (below 15 years of age). According to the United Nations Population Bureau, the longevity in Japan is 81 years, in Iceland, Sweden and Italy 80 years, in Austria 79 years, and in the

United Kingdom 78.5 years. The proportion of retired population (above 65 years of age) in Italy, Austria, Germany, Norway and Sweden is 20 per cent or above. These retired people are largely dependent on the workforce.

2. *Small Workforce*: As the standard of education improves, children remain for longer period in schools and join the workforce late. This, combined with low birth rate, means that the labour force grows slowly, while employment opportunities in the tertiary and secondary sectors multiply. Despite a high degree of mechanization in most industries, many countries (like Germany, France, Netherlands, Switzerland, Australia, New Zealand, Russia, etc.) are short of workers. The wages in the blue-collar jobs (manual workers and miners) are very high. For example, a coal miner in UK and Germany gets the wages as high as that of a university professor.
3. *Declining Rural Population*: Rural population is increasingly outmigrating towards the urban centres as the educational, medical and recreational facilities in urban agglomeration are more. The fewer, especially the old people, live in rural settlements. Consequently, the rural areas get depopulated. The population of rural areas is decreasing. In countries like Belgium only 3 per cent of the total population is living in rural areas, while 97 per cent live in urban areas.
4. *Urbanization*: Urban places are expanding fast. Many of the million cities have acquired the status of megalopolis and conurbations. As more people move into the cities from the rural areas, the pressure on transport, housing, power, water and health care centres increase. Smoke and chemical effluents from factories result into water, air and soil pollution. Traffic jams and noise pollution are the other problems. Tension created by urban life leads to a far higher incidence of mental illness, heart troubles, depression, loneliness, hyper-tension, madness, diabetic, blood pressure, and breathing problems.

Urban sprawl and slums expansion in some of the developed countries, like USA, are the other major issues, which create many social and environmental problems. The highly productive agricultural land is encroached upon by urban structures, houses, industries, roads and squatters resulting into a decline in arable land.

Thus, it is clear from the above discussion that all the nations of the world—developed, developing, underdeveloped—are concerned about the pattern of population growth and some of their problems are common. Most of the developed countries have areas where agriculture or industry could be improved or where the population is too large and dense. Similarly, the developing countries have large cities where the problems are similar to those of urbanized societies everywhere. It is also important to note the differences between underdeveloped countries. Some have much better resources or smaller population, such as Argentina, Brazil, Mexico and Malaysia which, much more likely, will be able to overcome their problems than those countries with poor resources and large tradition bound, orthodox population, such as Sudan, Somalia, Mali, Niger and Bangladesh.

Population Dilemma of Europe

Although world population has been increasing at an unprecedented rate in most of the countries of Africa and Asia since the Second World War, the population of several of the European countries is either stagnant or declining. At present, the salient features of European population are as under:

1. The population of Europe is older in age than that of any other continent.
2. Population of many of the European countries (Austria, Germany, Norway, Sweden, Finland, Poland, Denmark, Switzerland, Ukraine, Belarus, Romania, etc.) is stagnating or declining.
3. The fertility rate of European countries being 1.2 per cent per annum is below the replacement level. None of the European countries excepting Albania, Bosnia, Serbia, and Latvia is replacing its population at present, through natural increase.
4. Germany and Italy have the oldest population in the world (20 per cent in the senior citizen category) with small proportion of young population (15 per cent).
5. On the whole, the population of Europe has started declining since 1990.
6. During the last three decades, demographers and planners in Europe emphasized on zero population growth—a condition achieved when birth plus immigration equals death plus emigration. An exact equation of birth and death means an increasing proportion of older citizens, fewer young people, and a rise in the median age of the population.

The zero growth rate of population is creating many problems for the society and economy of Europe. Some of such pertinent problems are:

(i) In some of the cities kindergarten, nurseries and schools are closing and universities cut-back in the face of permanently reduced demand.

(ii) It is projected that by 2010 the unemployment will be replaced by a shortage of workers. Germany, Belgium, Finland, Greece, Italy, Spain and Portugal are already facing this problem.

(iii) There will be more burden on the government exchequers as they shall have to provide pensions, medical care and other social amenities to more number of senile population after their retirement.

If the present pattern of population growth, which is almost zero in most of the European countries, is not reversed, the consequences may be serious and catastrophic, which may endanger the very survival of Europeans.

Population Policies

The size of population, its growth rate, age/sex composition and pattern of migration are closely influenced by the population policy of the respective government. Each country has its own population policy. These policies range from encouragement of high fertility to varying degrees of discouragement. Some of the developed countries (Canada, France, Japan, USA, Singapore, Australia and New Zealand) are following the policy of slow growth, also known the policy of *Pronatalism.* In these countries the growth of population is below 1 per cent. Contrary to this, the population policies in some of Afro-Asian and Latin American countries are to reduce the size of population, while Germany, Russia, Romania, Italy, Austria, Hungary, Belarus etc. are adopting the policy of population increase.

In most of the developing countries, the techniques for reducing the birth rate vary from supplying contraceptives and information in some countries to much more direct involvement in others. It is encouraging to note that an overwhelming majority of population of the developing countries prefer a decline in the rate of natural increase in population. In most of these countries, the respective governments have come out with a variety of policies together with mortality control, fertility control, family and individual well-being. It is widely believed that the rate of annual increase can be reduced only if all these measures are combined and the education, income and standard of living improves. Most of the developing countries have fixed the time-bound targets of rate of natural growth by 2010 A.D. For example, Kenya has a target of 2.8 per cent, Egypt 1.5 per cent, Mauritius 1.1 per cent, Sri Lanka 1 per cent and China 0.5 per cent.

The Indonesian government, with the help of United Nations, the World Bank and the World Health Organization (WHO), built more than 3,000 clinics in Java and Bali islands to develop awareness about family planning. Mass media, such as radio, television and newspapers, were used to promote the acceptance of small family and draw people to the clinics.

In some countries, the government's efforts to reduce the birth rate has moved beyond voluntary family planning into various forms of coercion. For example, in Singapore, in which space is extremely limited, there have been heavy financial penalties for parents who have had more than two children. Contrary to this, parents who have been sterilized after the birth of second child have been rewarded with guaranteed free education and employment for their children. As a result of this policy, in combination with a rapid economic development, the total fertility rate in Singapore fell below the replacement level. The government of Singapore is now encouraging the better educated women to have more children.

In India, a voluntary family planning policy has been adopted right from the beginning of 1960s, but a tough sterilization policy adopted in 1976-77 could not bring the desired results. The rigid policy was resisted by the masses. Because India has a democratic form of government and public feelings can be expressed through elections, objections to the tough policy became evident after it was implemented. Most of the political analysts believe the incumbent party (Congress) lost the election of 1977 mainly because of the widespread belief that people were sterilized against their will.

During the last two decades, there had been some dramatic decreases in birth rates in different parts of India, but fertility rates differ from state to state and region to region. The fertility rate in India is still around 1.9 per cent. If this trend continues, according to one estimate, India will be crossing the population size of China by 2025.

Population Policy of China

The demographic characteristics and the population policy of China are very important to examine as this country alone constitutes about 23 per cent of the total population of the world. One of the greatest reductions in the fertility rate through government policy has occurred in China within the last two decades, so it provides an example of the role of government policy in reducing the birth rate substantially.

As stated above, China is the most populous country of the world. According to the UN Population Division, its population in 2000 was about 1.32 billion. The increasing population led to water, food and timber scarcity. The rapid increase in population between 1950 and 1970 made the Chinese government worried. In China, the first family planning campaign was launched in 1972. Up to this year Chinese women had an average of six children. The campaign, conducted under the motto '*no juvenile wedding, more time between pregnancies, and fewer children*' suggested people should wait longer to get married and have children, prolong the time between pregnancies and have smaller families. The first attempt at family planning was positive. The birth control message had reached the masses and birth rates dropped from 5.8 per women in 1970 to 2.7 in 1979.

Faced with extraordinary demographic growth in 1979—at one billion, then almost a quarter of the world's population—the government of the People's Republic of China implemented a severe family planning policy which is consisted of limiting couples to one child. The policy was called *'One Family One Child'*. In order to reduce birth rates, the government imposed fines on parents with more than one child, raised the legal marrying age and demanded the use of family planning methods such as contraceptives, abortion and permanent sterilization. The programme however allowed some exceptions: a second child was allowed in rural areas to help with the chores if the first born was female, and if several years had passed since her birth.

According to the China Internet Information Centre, family planning policies have inspired the people to change their ideas about marriage, reproduction and family, and accept the late marriage and low procreation policy, as well as eugenics. These policies also freed women from frequent pregnancies, heavy family loads and improved internal and children's health. According to one estimate, the one child policy prevented 330 million births in the last three decades.

In 1984, the programme demanded that the legal marrying age be raised from 18 to 20. The officially recommended age has been higher: 23 years for rural and 25 years for urban women.

The programme was effective in reducing overall birth rates. Had the policy not been implemented, it is estimated the country would have 1.6 billion people.

Urban areas such as Beijing and Shanghai, for instance, have enforced the policy. Meanwhile, rural areas average of two children per family.

China is a tradition-bound society. The preference for son is traditional in China. Lineage is handed down through males, who take care of their ageing parents. Women marry and take care of their husband's family. Due to the current one child limit, many couples have only one legal chance to have a child, so they trust ultrasound techniques to identify foetuses and abort females. The programme allows in rural areas—where about two-thirds of China's population lives—that if the first child is a girl, a second girl is often aborted or abandoned until a boy is born.

More than 90 per cent of children in orphanages are abandoned girls between one and four years old, many of them, the second female born to rural families. Many of the peasants consider it a great dishonour to their ancestry not to have a male heir. Couples whose first child is male often sign the one child per family certificate, while those whose first born is female keep trying to have a son at the risk of paying preposterous fines, or, as happened in remote rural areas, the destruction of their home by family planning officials. In Rongshui county, a very poor area peopled by Miao and Dong ethnic groups, parents with more than two children have to pay a $6,000 fine, although the average annual income among peasants barely amount to $60.

The one child policy is anti-female babies. Various means are used to curtail the female birth ratio: deliberate omission of female registration in census and statistical surveys, infanticide, abandonment, criminal negligence and selective abortion.

In China, there is a very simple procedure for sex-selective abortion. China has more than 200,000 clinics and local doctors' offices with modern ultrasound equipment. Although gender-selective abortion is prohibited, the ultrasound test which determines the baby's sex is usually provided through bribes. According to official statistics 98 per cent of abortions are female. Consequently, in some of the provinces, the sex ratio is about 130 females per 1,000 of males.

The 2000 census counted approximately 13 million fewer girls born between 1980 and 2000 than would have been born if China had had normal birth and mortality rates.

In 2006, China had 1,322,273,200 people with an average density of 138 persons per sq. km. The population is unevenly distributed throughout the country. Approximately, 95 per cent of people live in south-eastern half of the country. The central government has tried to encourage migration to the north-west provinces without much success.

The population growth rate has decreased by a yearly average of 1.07 per cent between 1990 and 2000. In 2001 it stood 0.88 per cent. Although China has made transition from high to low birth and death rates, each year the number of births still exceeds the number of deaths by close to 9 million.

The one child policy of China has its own merits and demerits. It has been appreciated as the growth rate of population has come down substantially, and there is continuous improvement in the standard of living. The supporters of the policy also argue that China is having almost one-fourth of the total population of the world and the well-being of future population of the world will be affected by the decision made today about the size of the Chinese families.

The supporters of the policy also insist that it is flexible enough to allow exposition, such as in areas where manual labour is important and in regions of minorities, like Tibet, Sinkiang (Xinjiang), Inner-Mongolia and Uighurs.

Despite all these merits, there are many experts of demography, even within China, who disagree with any policy that will greatly diminish the number of adults in the next generation. According to them, continuous population growth is the key to economic development, which leads to better standard of living, happiness, leisure and increases the life span of the people. In case of low growth of population, the country will suffer from shortage of workers in agriculture, industries and military personnel. The one child policy has also been criticized because of the possible long-range effects of 1-2-4 (age structure one child, two parents, and four

Figure 6.9 *China's One Child Family Model*

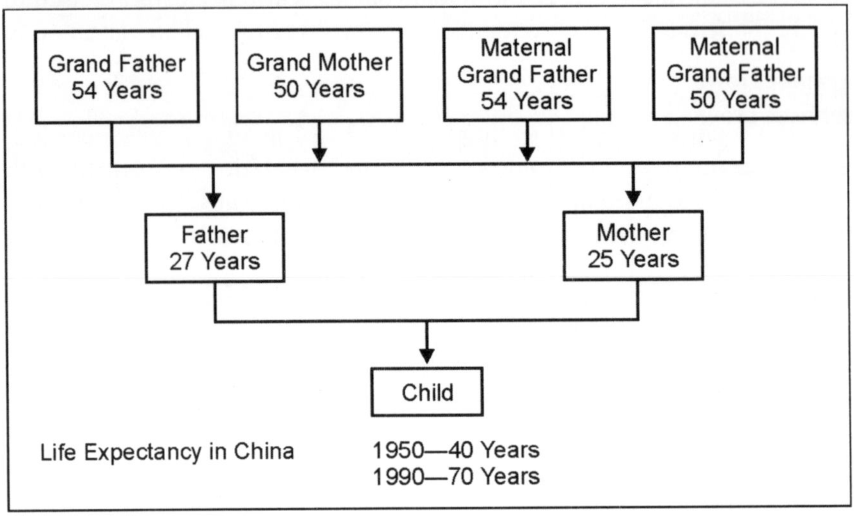

grandparents). In such an age structure there will be more pampering of child. There are complaints about the excessive amount of pampering that single child receives from parents and doting grandparents (Fig. 6.9). Sociologists speculate that these pampered children, when they become adults, will alter the Chinese society unfavourably. Moreover, a continuation of the policy will mean that in future the higher percentage of elderly population will depend on a smaller proportion of working adults. This will lead to high dependency ratio of elderly population which may not be economically productive and biologically reproductive. The pampered children may not be able to bear and sustain this growing population of the senior citizens.

The policy has also been criticized on the ground of individual freedom. Birth statistics by gender, as stated above, indicate that female infanticides have occurred in many areas. These statistics seem to reflect a reluctance by some families to abandon the tradition of having several children, especially sons.

An increase in the proportion of elderly population can affect the future market for various kinds of goods and services, both those sold to young and those sold to elderly people. An ageing population also enlarges the dependency ratio which, in turn, can affect financial systems, such as social security programme. The age and sex pyramid will be distorted, having a short base and heavy top. Such an age and sex pyramid is not sustainable.

Most countries concerned about population growth are attempting to combine economic development with the provision of family planning services. The economic development and social change provide the motivation to have small families. Reducing infant mortality, expanding education systems, broadening the role of women and achieving economic growth—all have been important catalysts in changing traditional attitudes towards the family. Once the motivation is there, family planning services provide the means for bringing the birth rate down.

Policies of Encouraging Population Growth

In the developed countries, the growth of population is either at the level of replacement or showing the trend of negative growth. In several regions of Hungary, Romania, Germany, France, Netherlands, Denmark, Sweden, Norway, Finland, Italy, Austria, Romania, Spain, Portugal, Spain, Russia, etc., the population is declining. This declining trend in population is a matter of great concern for demographers and planners. Usually, "the richer the country, the slower the growth of population". It has also been established that "development is the best contraceptive". Looking at the declining growth rate of population, most of the European countries are taking positive steps to increase the growth rate of their population. France, for example, grants substantial payments and services to parents to help the child rearing. Eastern Europe (Bulgaria, Russia, Hungary, Austria, Romania, etc.) is much more active in encouraging population growth. Three children per family has become the adopted goal in several East European countries. Parental subsidies, housing priorities, and generous maternity leave (with full salary for the mother for one year, and 9 months paternity leave for the father at the 90 per cent salary), and advance increment at the birth of each child are among the inducements for multiple child families. Interestingly enough, in Hungary and Romania, contraceptives are banned and abortions are permitted only on doctor's orders in the hope that birth rates can be increased. In general, despite some short-term success, the Europeans of the East and the West have ignored both blandishment and restrictions. The consequences will only gradually emerge over the next several decades.

The birth rate is Russia has declined substantially after 1991. The low birth rate has resulted in the reduced number of youth entering the labour force and the armed forces. This shortage in labour has created numerous socio-economic and cultural problems. Consequently, a population policy was adopted to encourage large-sized families. In Russia, especially in European Russia, where the rate of natural increase of population in 2005 was only 0.18 per cent, the government is trying to create a favourable environment for high birth rate through incentives and publicity. The Mother Lenin Award was introduced in the Soviet Union as early as 1970s. This award carrying a handsome cash, a furnished accommodation, a car, a certificate used to be given to a mother who gave birth to twelve children. This award has, however, been won only twice by the mothers of Central Asian Republics, and no mother could get it after 1990.

In Germany, where the birth rate has fallen as low as death rate, the government has taken stronger measures to encourage large families. For example, following the birth of second child, mothers are allowed one year's leave from work at 80 per cent of their salaries. Low interest loans for home are available at the time of marriage and when each child is born.

Most of the European countries have no official population policy, other than the pro-natalist policy. In UK it was in 1974 that the contraceptives and abortions received official recognition. Among the Catholics, birth control measures have remained illegal, though late marriage, illegal abortions, rhythm methods etc., have helped these countries in keeping their growth rate low.

Australia and New Zealand have considered themselves as underpopulated countries. The policies of these governments have been pro-natalist and pro-immigration. A policy to have zero growth rate of population has been adopted by these countries.

Japan, Singapore, China, South Korea and Indonesia are the only countries in Asia which have been successful in reducing their birth rates to the level of developed countries, largely by legalizing abortion. The population policy of Japan adopted in 1950 strongly discouraged a family with more than two children.

Most of the South American countries, being Catholics in their faith, do not accept family planning programme, except in the case of medical grounds. Even, in some of the African countries, namely, Congo, Guinea, Ivory Coast, Cameroon, Gabon, Gambia, Mali, Chad, Togo, Guinea Bissau, Niger, Uganda, Somalia, Kenya, Zambia, etc., the respective governments have adopted the policies to increase population. This is based on the belief that continued population growth is the key to economic development. In the opinion of the formulators of this policy, a youthful population is necessary to develop the socio-economic conditions of the country.

Population Policy of India

India has the second largest population (1,096,917 thousands) with a density of 334 per sq km) of the world after China. The Indian policy makers realized the importance of population control as early as 1951-52, but a rigid policy was not adopted to check the fast growth of population. In 1961-71, the population growth rate was 2.25 per cent, which was the highest at any decade after independence. At present, the population growth rate has declined to about 1.9 per cent per annum. During the post-independence period, the death rate has been controlled and medical facilities have been extended to the far-flung villages of the country, yet the explosion of population may be attributed to the numerous economic, social, psychological, cultural and political factors.

When the population policy was designed in the First Five-Year Plan of India, it was realized that the base of population is already very large and the trend of population growth cannot be altered easily and quickly. The plan enunciated that the programme for family limitation and population control should:

(i) present an accurate picture of the factors contributing to the rapid increase of population,
(ii) discover suitable techniques of family planning and devise suitable methods by which knowledge of these techniques could be widely disseminated, and
(iii) give advice on family planning as an integrated part of the service of government hospitals and public agencies. The meager First Plan provision of Rs. 0.65 million for the family planning programme was too little to yield any far-reaching results.

In the Second Five-Year Plan, a voluntary sterilization population policy was introduced. The family planning programme was provided an amount of Rs. 50 million. This amount was utilized mainly to establish 1,650 family planning centres in different parts of the country.

In the Third Five-Year Plan, logistics were provided for family planning, which motivated about one million people to accept sterilization.

More emphasis was laid on the family planning programme in the Fourth Five-Year Plan. The most distinctive feature of the Fourth Plan was that it set a time-bound target reducing the birth rate from 3.9 per cent to 2.3 per cent by 1979. The outlay for this plan was raised to Rs. 2,860 million. Consequently, by the end of this plan, about 9 million couples were covered under sterilization and about 6 million by other family planning methods. Thus, about 7 million births were estimated to have been averted during this plan period.

In the Fifth Five-Year Plan Rs. 5,000 million were provided for the family planning programme. The programme sought to integrate most of the basic social services, including education and public health. A more rigid policy with an element of compulsion, monetary incentives, penalties and legalization of abortion during the Fifth Plan made the Indian population policy more effective.

High priority was given to the family planning programme in the Sixth Five-Year Plan. The strategy during the plan was to integrate health, family welfare and nutrition services at all levels. Monetary incentives and full rebate in income tax for specified donations for welfare purposes were given by the government. The strict population policy, followed by the Indira Gandhi government, was, however, opposed by the masses.

In the Seventh and Eights Plans, a more pragmatic policy was adopted. There was more emphasis on persuasion, publicity and family and individual well-being.

Despite all these measures and policies, the population of India is growing at a faster pace and taking the shape of an explosion. The economic development and rising standard of living of some of the people are not adequate to bring down the population growth rate. The time factor is so pressing, and the population growth so formidable, that we have to get out of the vicious circle through a direct assault on the problem.

Indian National Population Policy, 2000

The immediate objective of the National Population Policy, 2000 was to address the unmet needs for contraception, health care infrastructure, and health personnel, and to provide integrated services for basic reproductive and child health care. The medium-term objective is to bring the total fertility rate to replacement. The long-term objective is to achieve a stable population by 2045, at a level consistent with the requirements of sustainable economic growth, social development and environmental protection.

In pursuance of these objectives, the following national socio-demographic goals to be achieved in each case by 2010 are formulated:

1. Address the unmet needs for basic reproductive and child health services, supplies and infrastructure.
2. Make school education up to age 14 free and compulsory and reduce dropouts at primary and secondary school levels to below 20 per cent for both boys and girls.
3. Reduce infant mortality rate to below 30 per thousand live births.
4. Reduce maternal mortality rate below 100 per 100,000 live births.
5. Achieve universal immunization of children against all vaccine prevalent diseases.
6. Promote delayed marriage for girls not earlier than the age of 18 and preferably 20 years of age.
7. Achieve 80 per cent institutional deliveries and 100 per cent deliveries by trained persons.
8. Achieve universal access to information/counselling, and services for fertility regulation and contraception with a wide basket of choices.
9. Achieve 100 per cent registration of births, deaths, marriages and pregnancies.
10. Contain the spread of AIDS and promote greater integration between the management of reproductive tract infection (RTIs) and sexually transmitted infections (STIs) and the National Aids Control Organization.

11. Prevent the control of communicable diseases.
12. Integrate Indian System of Medicine (ISM) in the provision of reproductive and child health services.
13. Promote vigorously the small family norm to achieve replacement level of total fertility rate (TFR).
14. Bring about convergence in implementation of related social sector programmes so that welfare becomes a people-oriented programme.

Population and Human Well-being

Although the world population is continuously growing, it is growing at a slower rate that demographers had projected only a few decades ago. Recent major gains in average life expectancy, reduced rates of child and infant mortality, and the increasing proportion of children now attending school provide ground for optimism about human well-being. Many of these gains have been made possible by unprecedented rates of economic growth in many of the developing and developed countries. Some 4 to 5 billion people are expected to experience substantial improvement in their standard of living by the end of 2015.

Yet, these global successes mask urgent, sometimes worsening, problems at the local or regional level, especially among developing countries. More than one-fourth of world's population has not shared in the economic and social progress experienced by the majority and still lives in poverty. Hunger, disease, illiteracy, and restricted freedom of choice or action are the persistent problems in many of the least developed countries of the Sub-Saharan Africa, South Asia and in Central America. The pressure of population can contribute to human deprivation, especially in poor rural areas, where competition for land and water can strain the capacity of local environments. Rapid population growth is also fuelling problems in many cities, especially the mega cities, where it can overwhelm the capacity of municipal authorities to provide even elementary services.

The interaction between population growth and human well-being is complex and not a matter of number alone. The capacity of countries to support growing population is enhanced when those countries achieve a sufficient, equitable distribution of wealth, technological development, effective government, strong institutions and social stability.

What are the implications of approximately doubled population? Population growth concerns most where countries are least able to deal with its consequences. Key issues in providing for increased populations will include the achievement of adequate income levels, food security, employment, and provision of basic social services. Also critical is the sound management of natural resources which, in many developing countries, still directly supports the livelihoods of majority of inhabitants.

A significant number of world's people already face critical shortfalls of the essential items needed for a healthy life. Some 1.5 billion live in absolute poverty, 840 million are undernourished, roughly 1.5 billion lack safe drinking water, and about 1 billion are illiterate. In many areas, population growth is accelerating the rate of degradation of forests, fisheries and productive soils. Populations at direct risk from environmental degradation are concentrated in the least developed countries of the Sub-Saharan Africa, South Asia and parts of Latin America. If the future generations in the developed and developing countries continue to use resources at the present rate, the consequences in terms of global climatic change, loss of vital renewable resources and toxic pollution will be severe.

Migration

Migration, together with fertility and mortality, is a fundamental element determining population growth and population structure in a country, region or area. It is, however, difficult to provide a scientific criterion for classification of migration. In fact, migration may be international, intra-national, inter-regional, inter-urban, intra-urban, rural to urban and urban to rural.

Migration is a characteristic of homo sapiens like that of tool making and culture building? Man is the most widely distributed social animal on the earth surface. From their probable origin in East Africa, human groups had spread out to occupy the major continents of the earth excepting Antarctica about 20,000 years before present (B.P.), long before the beginning of agriculture in 8000 B.C. and written history. Thus, migration of human beings is a geographical phenomenon that seems to be a human necessity in every age. Since man has a tendency to leave the areas in which life is difficult, he migrates to the areas where life may be easy and better.

Migration has been defined differently by different experts. In its most general sense, migration is ordinarily defined as the permanent or semi-permanent change of residence of an individual or group of people over a significant distance. Migration may be permanent or semi-permanent. Recently, geographers have been increasingly concentrating and probing on the issues of population dynamics and the problems associated with migration.

On the basis of time criterion, migration may be temporary or permanent. If we take into consideration the distance, the migration may be long or short. On the basis of number, migration may be individual or mass; it may be politically sponsored or voluntary. On the basis of social organization, migration may be that of family, community, clan or individual. On the basis of causes, migration may be economic, social, cultural, or political. Migration may be stepwise or direct.

Causes of Migration

The causes of migration may be physical (climate, droughts, floods, earthquakes, volcanic eruptions, epidemics), or socio-economic, cultural and political. The rapid growth of population and increasing pressure on resources may be the cause of permanent or temporary, long or short distance migration. Many a time the differences between groups in the level of technology, economic opportunities and cultural clash may cause large-scale migration. Some of the important causes of migration are briefly described below:

1. *Technology*: People with more sophisticated technology invade and conquer new areas. Contrary to this, technologically inferior groups may be attracted to the areas of greater opportunities provided by a more developed society. For example, ancient Romans conquered vast areas of Europe, North Africa and South-West Asia. During this period many people migrated to Rome, which provided better economic and employment opportunities. During the medieval period, the Arabs were quite advance in education and technology. Their rule was from Iberian Peninsula (Spain and Portugal) to Central Asia, Mongolia and China. After the Industrial Revolution in 1779, the Europeans emigrated in large numbers to North America, Australia, New Zealand, Africa, West Indies and South America. The large-scale emigration from the European countries continued up to the first part of the 20th century. The Europeans colonized the countries world over to exploit their resources.
2. *Economic Causes*: One of the prime motives of emigration seems to be economic. Man's need to have virgin land to till has inspired him to migrate to distant areas. The slaves were exported from the eastern and western coasts of Africa to the different European colonies in the world (Fig. 6.10). It was because of this reason that the slaves (African Negroes) were transported to the transplantations in tropical North America, Latin America and West Indies (Figs. 6.11 and 6.12). The Indians and Sri Lankans labourers were taken to Malaysia, Indonesia, Mauritius, Fiji and Bali islands. These workers subsequently got settled in the respective countries. Lust for virgin land also motivated the Europeans to emigrate from UK, Ireland, France, Germany, Italy, Poland, Holland, Spain and Portugal to settle in the Prairies of Canada and USA. The temptation to acquire virgin land in America and Australia and New Zealand became a great magnetic force for the Europeans to settle in the different parts of the world.

   Non-availability of jobs and increasing unemployment are also the factors responsible for the outmigration of job seekers, both from the developing and developed countries.

Figure 6.10 *Sources of African Slaves*

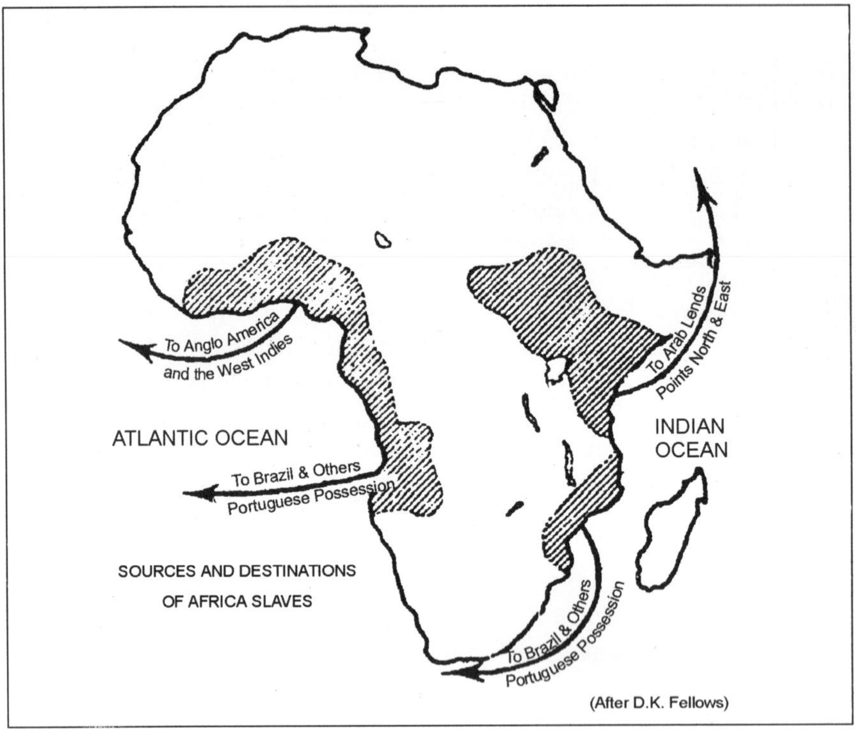

The agricultural labourers, if unemployed, also leave their native places. Throughout the history, the workers emigrated to neighbouring and distant fertile tracts. For example, in about 2000 B.C., Abraham migrated from Ur (Mesopotamia) to Cannan—present Lebanon, Palestine and Israel. The Greeks migrated to Scythia, the Jews migrated to the Nile valley and Mesopotamia. At present, in India, labourers and semi-skilled workers migrate to Punjab and Haryana to work in agriculture fields and industries, while professionals and skilled people migrate to USA, Canada, European countries, West Indies, Middle East, Japan, Australia and New Zealand.

3. *Overpopulation*: Rapid growth of population in relatively poor resource regions may create a condition of overpopulation. Overpopulation may exist at the local, regional or national level. In the past as well as at present people migrated from the overpopulated regions in search of employment and better job opportunities. The migration from the Third World countries is mainly because of the push factor of overpopulation. The emigrants include domestic servants, agricultural labourers, unskilled and semi-skilled workers, technicians, engineers, doctors, academicians, scientists, and experts of

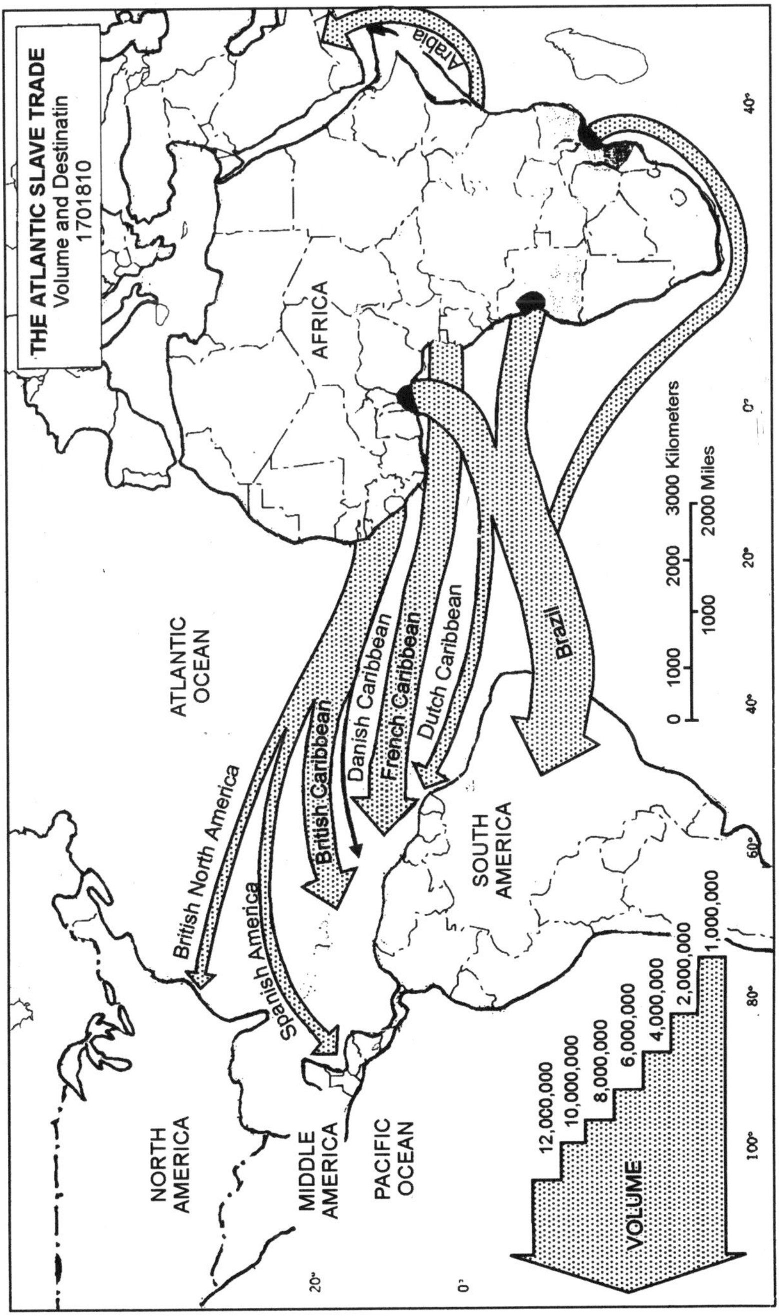

Figure 6.11 *Slave Trade*

Figure 6.12 *Triangular Slave Trade*

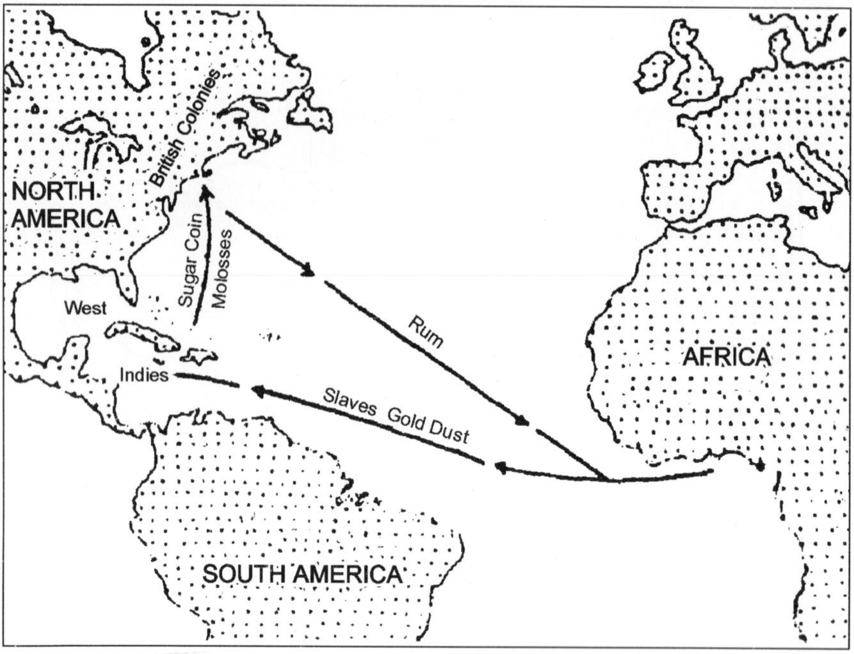

information technology and management. Each year, from India alone, about 3 lakh skilled and semi-skilled people emigrate.

4. *Socio-cultural and Religious Causes*: The human desire to stay, work and enjoy with the people of his ethnic and cultural groups is also an important cause of human migration. Throughout the human history, socio-cultural factors led to large-scale emigration. In the early part of the 20th century, there had been expulsion of Jews from Germany, Poland, Austria, Spain and Russia. A large-scale emigration of Muslims took place from Bosnia and Serbia (erstwhile Yugoslavia) in the 1990s. Muslims are moving out of Myanmar. The refugees from Afghanistan migrated to Iran, Pakistan, Tajikistan and Uzbekistan. This process is ongoing and with the passage of time becoming as the single most important factor for human migration (Fig. 6.13).
5. *Political Causes*: After the Second World War, political factors have become as one of the main causes of human migration. Political refugees is a world-wide phenomenon. One can cite numerous examples of political migration. These included, Turkish, Armenians, and white Russians in the early part of the 20th century; European Jews after the First and the Second World Wars; and Palestinians, Chinese, Hungarians, Cubans, Indians, Pakistanis, Bangladeshis, Kashmiris, Tamils, Vietnamese, Afghans, Iranians, Iraqis, Somalians, Kurds, Serbians, Bosnians, and Albanians. All these are the examples of forced migration induced by political factors.

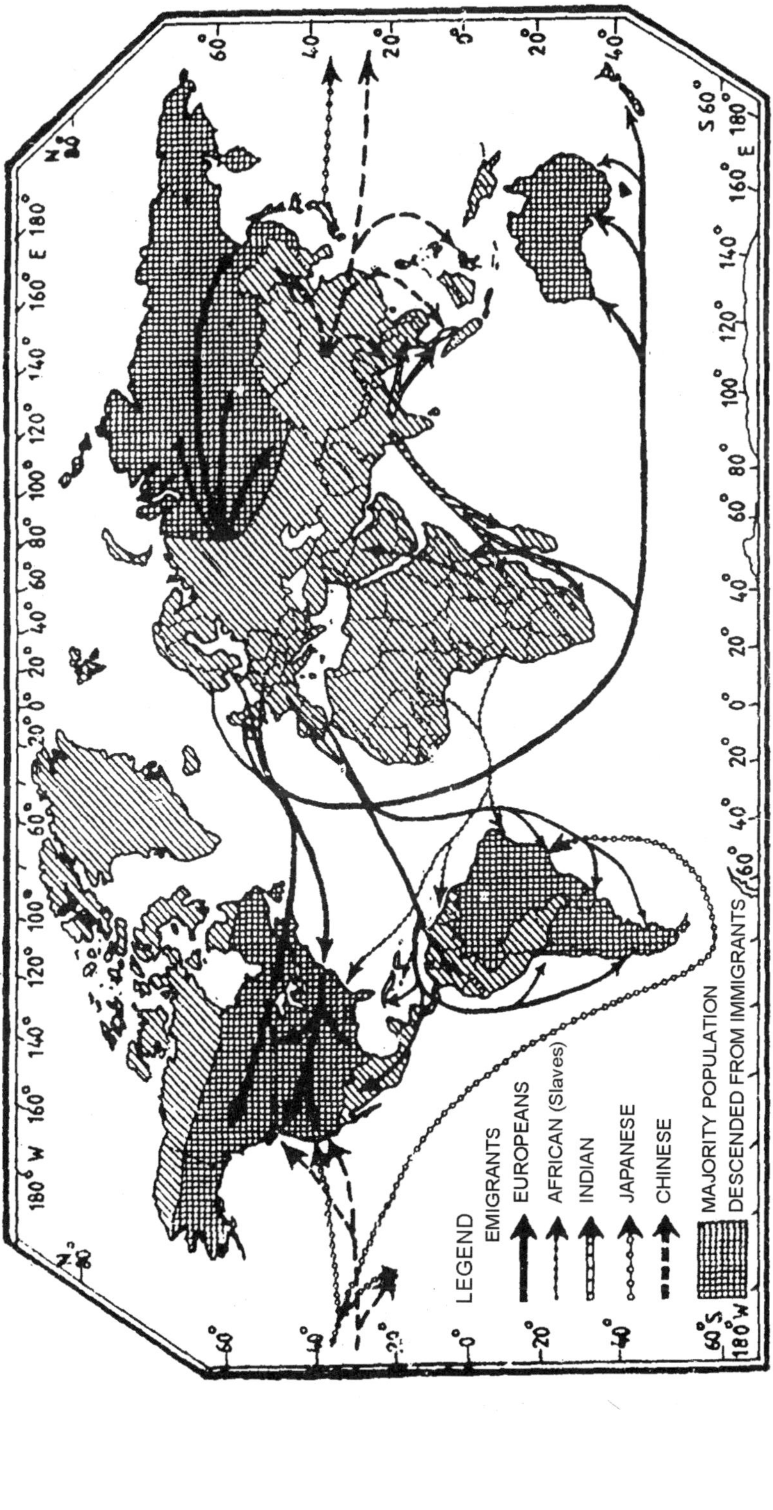

Figure 6.13 *Principal Migrations of Recent Centuries*

Another form of involuntary migration is the expulsion or exchange of minorities by nations. For example, the Germans repatriated from Czechoslovakia after the Second World War and Hindus and Muslims got exchanged when the sub-continent was partitioned in 1947 (Fig. 6.14).

Figure 6.14 *Forced Migration in South Asia, 1947*

The forced and compulsory migration is always bound up with tragic highlights in human history. During the medieval period, Negroes used to be sold in the urban markets of (Lisbon) Portugal, Madrid (Spain), Marseille (France) and Rome(Italy). These slaves used to be exported to USA, Canada and Latin American countries.

6. *Demographic Causes*: A number of demographic factors also play a vital role in the migration of human population. For example, age has been recognized as one of the important demographic factors controlling the degree of desire to move among the potential migrants. It is not surprising that adults are more migratory than other age groups. It is the rate of growth of population that determines the extent of population pressure in an area. The great historic movement of the Europeans across the Atlantic Ocean was an expression of increasing pressure of population on the resource base of Europe. Similarly, in India, the large-scale emigration from the densely populated parts of Orissa, West Bengal, Bihar and Uttar Pradesh is largely due to poor population-resource ratio.
7. *Diffusion of Information*: The availability of information through mass communication, education, cultural contacts and spatial interactions also increase the chances of population migration. The communities characterized by orthodoxy and illiteracy and strong communal ties are relatively less mobile than those who are socially more awakened, technologically and educationally more advanced. The information

network and cultural contact increase the horizon of job opportunities. Thus migration generates more migration, which signifies the role of information technology in the stimulation of migration. In India, the Sikhs, Gujaratis and people of Kerala are most mobile being adventurous and educated. People from these states may be found all over the world including the countries in which language and cultural constraints are serious impediments.

8. *General Rise in the Level of Aspirations*: With the advancement in science and technology, new items of comfort and luxury appear in the market and the level of aspirations of both the educated and uneducated people goes up. Everybody is tempted to enjoy the better standard of living. In all the developing countries, including India Pakistan and Bangladesh the young men who were better-off than their fathers were nonetheless satisfied, and many sought to better themselves overseas. It is mainly because of this reason that Indian engineers and doctors are emigrating to USA, Canada, Australia and South-West Asian countries.
9. *Wars*: Wars have been one of the important causes of human migration. Wars have always involved upheavals, particularly in the regions where these have been fought. The First World War (1914-19) involved the displacement of about 6 million people, and the Second World War (1939-45) involved the involuntary displacement of about 60 million people. Some people were forced to move out to avoid political and religious persecution even long before the war. Forced labour movements and deportations accounted for several millions, and subsequent resettlement involved still further movement. Nearly a million Jews and Poles were deported by Germany during the war, another million by Russia. The population of Europe was in constant flux during the war, from Volga to the far eastern part of Siberia. All these migrations during the war resulted into the changes of human resources, in land use, customs, traditions and human and cultural values. The Indo-Pakistan wars in 1965 and 1971 also resulted into the migration of millions of refugees from one part to another.
10. *Government Policy*: Each government has a foreign policy. The government policy of a particular country also favourably or adversely affects the pattern of population migration. There is enough scope in Russia, Australia, Canada, Brazil and Gulf countries to provide job to millions of people, but the respective governments are selective and do not permit any person to enter their country in search of job and employment opportunities.

Consequences of Migration

The process of migration has a direct bearing on the demography, society, economy and ecology of the place of origin and destination of the migrants. It has been pointed out by the experts of population migration that each migrant by nature seeks to create something of the original milieu in the midst of new environment and consequently, enriches the culture and civilization. Some of the main consequences of migration are presented as under:

1. *Reallocation of Resources*: The place of origin and place of destination, the individuals and the families of the migrants undergo a qualitative and quantitative change in their population and demographic structures. In general, people from the crowded and overpopulated regions emigrate to the areas of scanty population with better resource base. Thus, it leads to reallocation of human resources, which helps in maintaining a balance between human resource and natural resources.
2. *Change in Demographic Characteristics*: The migration of population from one place to another brings a tangible transformation in the demographic characteristics. The absolute number of population, the density of population, sex ratio, fertility rate, age composition, literacy rate and occupation structures are transformed at both the places, i.e., place of origin and destination of the migration. For example, at the place of immigration, an increase in density of population brings more pressure on resources. The new arrivals are generally young, mostly unmarried as a result of which the fertility rate goes up at the place of destination and declines at the place of origin, where the elderly people, females and children remain staying. Moreover, the outmigration of the educated and skilled people brings a quantitative and qualitative change in the human resources of the two places.
3. *Change in Sex Ratio*: The sex ratio at the place of destination drops as the male members are added, while the sex ratio at the place of origin increases as male youth generally migrate. One of the high sex ratios in Uttranchal and Himachal Pradesh is mainly because of the male-selective emigration for several decades from these areas.
4. *Economic Gains*: There is more judicious utilization of natural resources at the place of destination as a result of which the production in the agricultural and industrial sectors goes up and the economy prospers. The emigrants sent money back home to their families that brings prosperity to the population of place of origin also.
5. *Transformation of Ethnic Characteristics*: The physical and marital contacts of people belonging to two different ethnic groups may

change the biological characteristics of the migrants and the host population. Greater interactions between the Negroes and the Whites and the Red Indian with the European have resulted into the new ethnic groups like Mulatto and Mestizo. The Anglo-Indians and Africo-American can also be cited as such examples. In many cases more interaction between the different ethnic groups is the cause of some fatal diseases like AIDS and plague. As a matter of fact, the interaction of people of different ethnic and cultural backgrounds when stay together leads to the enrichment of civilization. This may be said as the greatest benefit to human society from migration.

6. *Transformation of Cultural Values*: When large-scale migration takes place, as it happened during the 18th and 19th centuries from Europe to Americas, the cultural values of the groups of people underwent substantial transformation. The immigrants bring with them their language, religious and cultural values. Integration and amalgamation of people causes serious problems, particularly when the people having different languages stay together. For example, the French and English in Canada, the Dutch and English in South Africa, the Indians, Pakistanis and Bangladeshis in South-West Asian countries, and the Chinese in Indonesia are confronted with such problems. The people belonging to different religious groups, whenever move and start living with the people professing other religious faiths, it may be either mean a healthy spread of religions in areas or may cause people of different religious faiths to flee for fear of suppression and persecution by the other. The immigration of Catholic, French and British protestants in Canada and the multitude of Asian religions to UK are the examples where a healthy spread of different religious faiths has emerged. Contrary to this, the Jews and Arabs in West Asia provide an example of the latter category where followers of one religion have been conspicuous of their persecution and suppression.

   The dietary habits of both the ends are also significantly influenced. For example, the Indians and Chinese, who migrated to UK, USA and other European countries, opened the restaurants and hotels in which they serve the dishes of their countries. It is because of this reason that Indian dishes like *Idli, Dosa, Biryani, Pulao, Chapati, Kaba, Tandoori, Samosa, Rasgulla, Barfi*, etc. and Chinese dishes are increasingly becoming popular in European countries. Contrary to this, the continental breakfast (bread, butter, toast and egg) is becoming more popular in urban India, Pakistan and Bangladesh, which provides an example of the influence of English rulers on the upper and middle classes of their erstwhile colonies.

7. *Psychological Consequences*: People, who migrate from rural to urban areas, especially the big industrial and commercial cities, suffer from

lack of proper residential accommodation. The highly polluted atmosphere of the new place of work, toxic fumes, and pollutants affect their health adversely.

They also have to adapt to new dietary habits and timings of food. There are evidences to prove that the incidences of respiratory and water-borne diseases, cardiac problems, psychological disorders and loneliness are common among the migrants.

References

Bennett, D. Gordon, 1984, *World Population Problems,* Campaign, III Park Press.

Census of India, 2001.

Chandna, R.C. 1990, *A Population Geography,* New Delhi: Kalyani Press.

Clark, W.A.V., 1975, *Migration and Development,* The Hauge: Mouton.

Clarke, J.I., 1972, *Population Geography,* Oxford: Pergamon Press.

— (ed.), 1984, *Geography and Population: Approaches and Applications,* New York: Pergamon Press.

Encyclopaedia Britannica, 1991, London.

Encyclopaedia of Social Sciences, 1995, London.

Fellman, J. et al. (eds.), 1985, *Human Geography*, Dubuque: W.C. Brown.

Findlay, A. et al., 1987, *Population and Development in the Third World,* London: Methuen.

Grigg, D.B., 1982, 'Migration and Overpopulation in Historical Perspective' in Paul White and Robert Woods (eds.), *Geographical Impact of Migration,* London: Longman.

Haupt, Arthur and Kane, T. Thomas, 1978, *The Population Reference Bureau's Population Handbook*, Washington, D.C.

Jackson, J.D. (ed.), 1969, *Migration,* Cambridge University Press.

Jacobson, J.L., 1987, Planning the Global Family, *World Watch Paper 18,* Washington D.C.: World Watch Institute.

Johnston, R.J. et al., 1990, *The Dictionary of Human Geography*, Oxford University Press.

Jone, H.R., 1981, *A Population Geography*, London: Harper and Row.

Lewis, G.J., 1982, *Human Migration,* London.

McNeill, W.H. and R.J. Adans, 1978, *Human Migration, Patterns and Policies,* Ind. and London: Indiana University Press.

Malthus, R.T., 1798 (reprint 1926), *An Essay on the Principles of Population,* London.

Merrick, T.W., 1986, World Population in Transition, *Population Bulletin,* 41, No. 2, Washington D.C.: Population Conference Bureau.

National Academy of Sciences, 1986, *Population Growth and Economic Development: Policy Question*, Washington D.C.: National Academy Press.

Newland, K. 1979, *International Migration: Search for Work, World Watch*, Paper 33, 1979.

Newman, A.J. et. al. 1984, *Population Patterns: Dynamics and Prospects*, Dubuque: Iowa, 1979.

Paters, G.L. and R.P. Larkin, 1979, *Population Geography: Problems, Concepts and Prospects*, Dubuque: Iowa.

Repetto, R., 1987, Population, Resources, Environment: An Uncertain Future, *Population Bulletin*, 42, No. 2, Washington, D.C.

Rotberg, R. and T. Rabb (eds.), 1984, *Hunger and History: The Impact of Changing Food Production and Consumption Patterns on Society*, New York: Cambridge University Press.

Scientific American, 1974, *The Human Population*, San Francisco: W.H. Freeman.

Stamp, L.D., 1964, *The Geography of Life and Death*, Ithaca: Cornell University Press.

The United Nations Development Programme, *World Resources, 1998-99.*

The World Guide—*See the World from a Different Angle*, 2005, Tenth Edition, Oxford: New Internationalist™ Publications Ltd.

Trewartha, Glenn, 1969, *A Geography of Population: World Pattern*, New York: John Wiley & Sons.

United Nations Population Division, 1988, *World Population Prospects*, New York.

Wrighly, E.A., 1969, *Population and History,* New York.

Zelinsky, W., 1966, *A Prologue to Population Geography*, Englewood Cliff: Prentice Hall.

Zelinsky, W. et al., 1970, *Geography and a Crowding World*, New York: Oxford University Press.

7

Climatic Change

According to the experts of climatology and meteorology the planet earth is getting hotter day by day. The academic community has reached a strong consensus regarding the science of global climatic change. The causes of climatic change are both physical and anthropological. After the Industrial Revolution, the global warming is, however, largely the result of emission of carbon dioxide and other greenhouse gases from human activity including industrial processes, fossil fuel combustion, and change in land use, such as deforestation. The last decade 1995-2005 was the warmest since records began to be kept. The year 1998 was the warmest year of the 20th century. The scientists believe that future years will be even more warmer. Thus, the world is undoubtedly warming. According to a large majority of the international scientific community, rising temperatures will lead to great change in global ecosystems, human health, activities and efficiency, and distribution of human population.

Climate is a summary of mean weather conditions over a time period, usually based on 30 years of records of temperature, precipitation, pressure, and other climatic phenomena. Climates are largely determined by location with respect to land and sea, air masses, to large scale patterns in the general circulation of the atmosphere, latitude, altitude, distance from the sea, ocean currents, and local geographical features.

In fact, global climatic changes have been great and diverse during the earth's approximate 4.6 billion years' history. The global climate has undergone more or less cyclical variations. Some epochs have seen a warm climate, some cold, and, in many instances, there was an abrupt change from one to another. The early climatic history of the world is, however, not well understood but it is known that during the last one million years there has been alternating of glacial and inter-glacial episodes.

The trend of change in surface temperature for the past 18,000 years has been shown in Figure 7.1. It may be seen that in about 18,000 years

Figure 7.1 *Trend of change in surface temperature for the past 18,000 years*

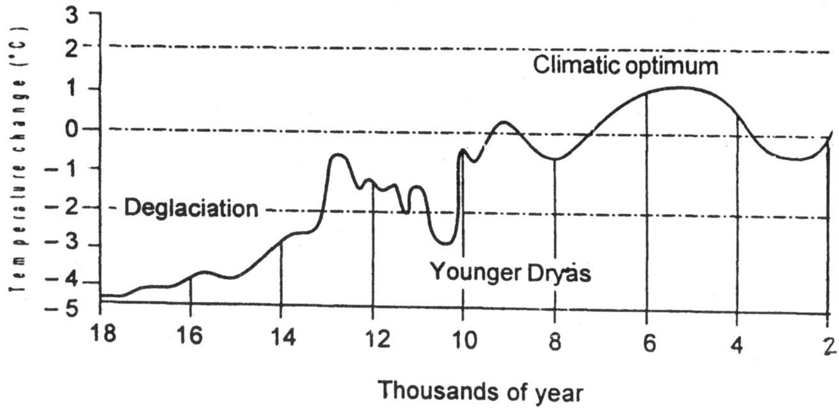

(After J.E. Oliver and J.J. Hidore, 2003)

before present, the air temperature of the earth was around –5°C, which showed a fluctuating trend, crossing the 0°C around 9,000 years before present. Since then, there is a steady increase in the surface temperature of the earth being around 15°C at present (Fig. 7.2).

Since the beginning of the Industrial Revolution in 1779, atmospheric concentrations of carbon dioxide have increased nearly 30 per cent, methane concentrations have more than doubled, and nitrous oxide concentrations have risen by about 15 per cent. These increases

Figure 7.2 *Trend of change in surface temperature for the past 50 years (1950-2000 A.D.)*

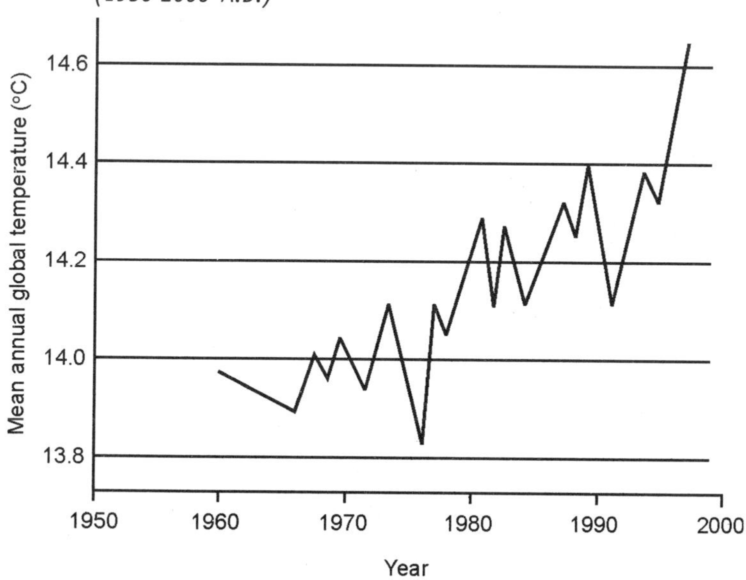

Figure 7.3 *Anthropogenic Release of Gases into Atmosphere*

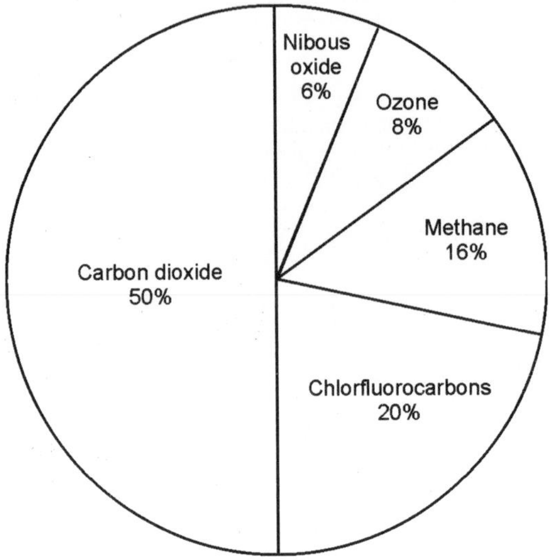

have enhanced the temperature trapping capability of the earth's atmosphere (Fig. 7.3).

Causes of Climatic Change

In support of the gradual rise in air temperature and climatic change, the following evidences are advocated by the experts of climatology:

1. Temperature records.
2. Melting of mountains and valley glaciers.
3. Warming of ocean water at global level.
4. Rise in sea level.
5. Upward shifting of snow lines of the tropical and sub-tropical mountains.
6. Thawing of permafrost areas.
7. Shifting in seasonal weather phenomena and change in precipitation patterns.
8. Spread of tropical diseases towards the temperate and polar regions.
9. Appearance of grass on some slopes of the Antarctica.
10. Occurrence of snowfall in the Gulf countries (Ras-al-Khaima, 2005).

Increase in Air Temperature

According to the experts of climatology, the increase in the air temperature in the 20th century has been about 0.5°C to 0.7°C. The last two decades of the 20th century and the first six years of the 21st century (2001-2006) have been the warmest years since the systematic recording of temperature began in 1878 (Figs. 7.2-7.4).

Figure 7.4 *Trend of Change in Surface Temperature since 1880*

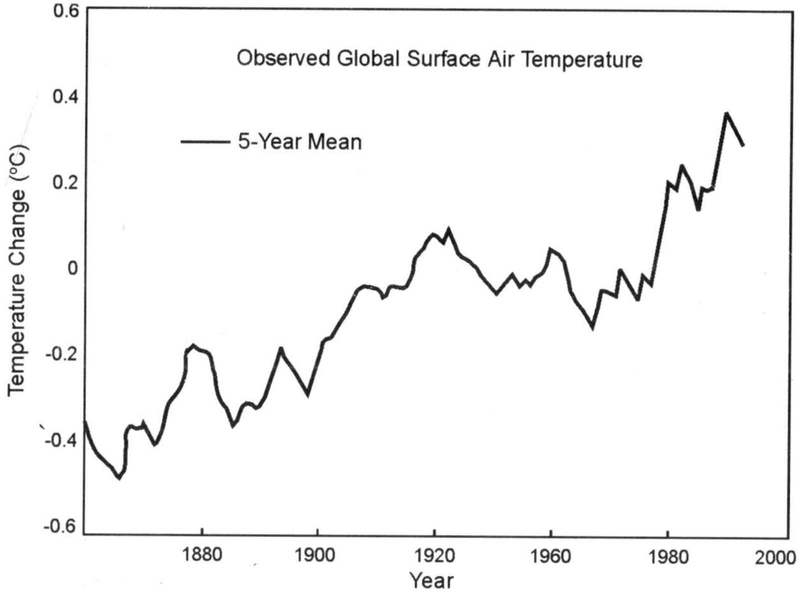

Source: H. Wilson & J. Hansen

Melting Glaciers and Ice Sheets

Perhaps, the most pervasive climatic effect of increased warming would be the rapid escalation of ice-melt. The additional water, especially from continental ice masses (Antarctica and Greenland), would raise sea level worldwide.

The recent evidences have shown that the ice sheets of the Antarctica, Greenland, Baffin Island, etc., are breaking and melting. The mountain glaciers are also shrinking in both depth and length. The regular monitoring of the Antarctic ice sheets has shown about 100 metres annual rate of their shrinking. A rise in temperature of about 4°C has been reported in the western Antarctic peninsula since 1950.

There are ample evidences to show the melting and retreat of the glaciers of Himalayas, Andes, Rockies, Alps, Caucasus and other young folded mountains. According to the latest survey of the Himalayas by Chinese scientists, there is reduction in the height of Himalayas since 1966. The measured height of Mt. Everest was 8,850 metres in 1965, which dropped to 8,848 in 2000 A.D.

Thus, there is conclusive evidence for a worldwide recession of mountain glaciers. This is among the clearest and best evidences for change in energy balance at the earth's surface since the end of the 19th century.

Sea-level Rise

In consequence of the melting of continental and mountain glaciers, the sea level is rising (Fig. 7.5). It may be observed that about 18 thousand years before present the sea level was about 82 metres below the present sea level. Since then it is continuously rising. The more significant rise in the sea level occurred about 6,000 years before present.

The possible sea-level rise must be expressed as a range of values that are under constant reassessment. Estimates of future sea-level rise range from a low of 50-110 centimetres to a high of 6 metres (about 20 feet). These levels could be further augmented by melting of alpine and mountain glaciers, which have experienced about 30 per cent decrease in overall ice mass during the century.

The consequences of global warming are bound to be serious in the form of sea-level rise that will gradually inundate coastal areas, changes in precipitation patterns, increased risk of droughts and floods, reduction in biodiversity and a number of potential challenges for public health.

Apart from the above, there are also some other indicators of climatic change. For example, the gradual spread of a few tropical diseases in the

Figure 7.5 *Fluctuations in Sea Level with Reference to Present Sea Level (preset datum line at 0 metre)*

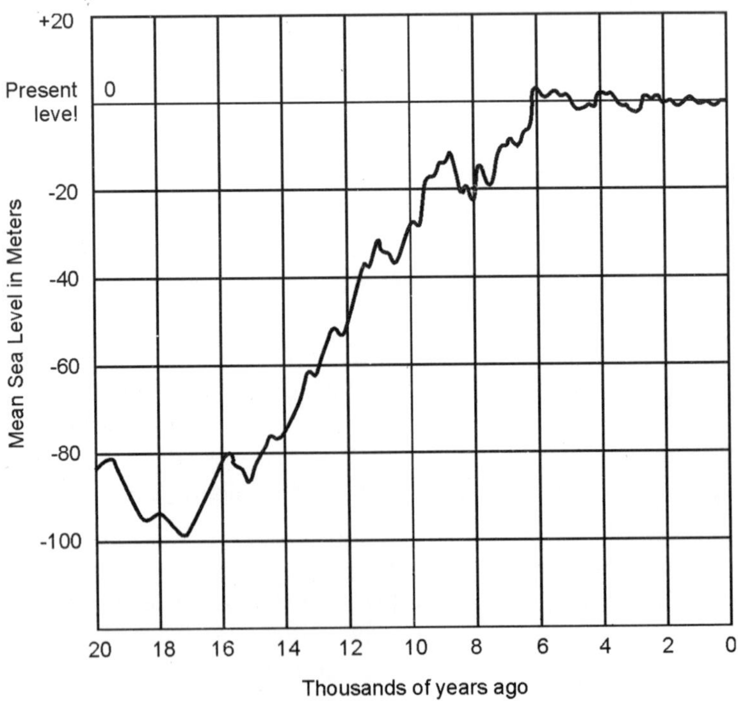

(After Fairbridge, in H.J. Critchfield, 2002)

temperate latitudes, such as malaria, cholera, dengue fever, chikengunia fever, yellow fever, Egyptian fever, etc., is also an important indicator of climatic change. The insects, bacteria and viruses of such diseases are the characteristics of hot and humid climatic conditions.

Decrease in the Antarctic penguins population by about 40 per cent during the last three decades, and large-scale destruction of corals due to coral bleaching also suggest an increase in the temperature of oceans, which also supports the hypothesis of global warming and climatic change. The continued thawing of permafrost areas in Alaska and Siberia is another testimony of global warming. About 10,000 Alaskan lakes have dried up in last 52 years.

Africa's two highest peaks—Mount Kilimanjaro and Mount Kenya—will lose their ice cover within 25 to 50 years if deforestation and industrial pollution are not stopped. Kilimanjaro has already lost 82 per cent of its ice cover over the last 80 years, while Mount Kenya has lost about 92 per cent of its ice cover over the last 100 years, according to Fredrick Njau of the Kenyan Green Belt Movement.

Theories of Climatic Change

Change is the law of nature. Climate is not an exception to this. There are geological, paleontological and geographical evidences which prove that changes have occurred in the climate of the earth throughout its history. The experts of climate have put forward a number of theories and concepts about the climatic change. Some of them have been presented briefly in the following paragraphs:

Astronomical or Orbital Theory

Sun is the primary source of heat and light to the earth. The eccentricity of the earth's orbit around the sun changes. The earth's orbit undergoes a change of shape during a cycle of 90,000 to 100,000 years. Sometimes, the orbit forms a longer ellipse and then it returns to more circular shape. At the time of greater eccentricity of the earth's orbit, the amount of heat received at the earth's surface at perihelion may be 20 to 30 per cent greater than that of aphelion. Variation of this magnitude in the solar radiation would lead to change in the solar constant, thereby affecting the temperature and climate of the earth.

As a result of variation in the angle between the earth's axis and the elliptic, the position of equator, Tropic of Cancer, and Tropic of Capricorn will change. At present, the earth's axis is inclined at an angle of 23.5 degrees. But, according to the experts, this angle undergoes a change. The inclination of the earth varies from 22 degrees to 24.5 degrees during the cycle of 41,000 years. These changes in the axis of the earth may lead to gradual and more warm phases on the earth's surface.

Theories of Continental Drift

The continental drift theory of Wegener, the Sea Floor Spreading by H.H. Hess, and the Plate Tectonics by Morgan reveal that the continents and ocean basins have drifted apart. The drift was started about 300 million years back. Consequently, the positions of equator and poles have undergone significant changes and that affected the horizontal distribution of continents and oceans. The geologists and geo-physicists opine that during the geological past the continental drift must have brought many climatic changes which are reflected in the distribution of coal in Greenland, Ruhr basin, Donetz basin, Kuznet basin, Pechora basin, Siberia, Greenland and Antarctica. The changes in the position of continents and oceans might have affected the place flow and direction of the ocean currents. Resultantly, the distributional patterns of temperature and precipitation through the ocean currents might have been affected.

Theories about the Changes in the Atmospheric Composition

The contents of carbon dioxide, nitrous oxide, methane, water vapour, etc., are not constant in the atmosphere. The proportion of carbon dioxide changed in the past because of the volcanic eruptions. It is changing more fast at present because of the heavy consumption of fossil fuel by man.

The carbon dioxide theory about the climatic change was put forward by T.C. Chamberlin. According to this theory variations in the carbon dioxide content of the atmosphere play an important role in causing the worldwide climatic change. This gas is transparent to the incoming solar radiation but absorbs outgoing long-wave terrestrial radiation. This leads to greenhouse effect. Thus, any change in carbon content in the atmosphere would bring about changes in the temperature of the atmosphere. The high rate of industrialization and urbanization has been directly adding more carbon dioxide to the atmosphere, which has been attributed as the main cause of climatic change.

According to one estimate, the present carbon dioxide content of the atmosphere is 410 ppm (parts per million). It is likely to reach 500 ppm by the end of the 21st century (Table 7.1). If the present rate of increase in CO_2 continues, this atmospheric gas may double by the year 2050. Consequently, the global temperature may rise by about 1.5° to 2° C by 2050.

The CO_2, CH4, CFCs and NO_2 (nitrous oxide) are continuously increasing in the atmosphere. It has been established that between 1980 and 2001 the surface temperature of the oceans has raised at the rate of 0.11°C per annum. This increase in temperature will also lead to sea-level rise.

The CFCs are thought to contribute about 25 per cent of the global warming. The CFC gases deplete the ozone layer and thereby enhance the greenhouse effect.

Table 7.1 *Lower Atmospheric Concentration of CO_2*

Year	*CO_2 concentration*	*PPM percentage*
1825	0.021	210
1885	0.028	
1985	0.035	350
2000	0.036	360
2050 (projected)	0.045	450

Source: World Resource Report, 2000-2001.

The Volcanic Dust Theory

The volcanic dust deflects light of short wave lengths coming from the sun. Contrary to this, long wave terrestrial radiation can easily pass through volcanic dust without any loss. Thus, volcanic dust may lower down the earth's temperature to a certain extent. The volcanic dust is considered as responsible for 'Little Ice Age'. In the geological past the frequent volcanic eruptions initiated the process of Ice Age.

Geographic Theories (Sunspots Theory)

There is a close relationship between the sunspots and the heat and energy received at the earth surface. The number of sunspots increases and decreases in a cyclic manner. The sunspot cycle averages eleven years in length which has been called as 11-year cycle. The variation in solar constant is thus also responsible for the climatic change.

Atmospheric greenhouse gases (water vapour, carbon dioxide, methane, and other gases) trap some of the outgoing energy, retaining heat somewhat like the glass panels of a greenhouse. Without this natural 'greenhouse effect' temperature would be much lower than it is now, and life as known today would not be possible. Instead, thanks to the greenhouse gases, the earth's average temperature is a more hospitable about 15°C (60° F). However, problems may arise when the atmospheric concentration of greenhouse gases increases.

Recent researches suggest that human agency is altering the climate because of the emission of carbon dioxide, methane and CFC gases leading to greenhouse effect. The current scientific community strongly agrees that during the last 100 years, as a result of the growing concentrations of gases, such as carbon dioxide, methane, nitrous oxides and chlorofluorocarbons, there occurred a substantial change in the composition of gases in the troposphere.

The CO_2 has been increasing steadily in the atmosphere after the 1779 Industrial Revolution. At present, the rate of increase in CO_2 in the atmosphere is about 8 to 10 billion metric tons per annum. The increase in the quantity of CO_2 alone is thought to be responsible for about 60 per cent of the global warming in the troposphere.

Climate and Development

Climate change is no longer an environmental problem but is now also a development issue, since it threatens to increase poverty, hunger, malnutrition and disease, and affects national, regional and international security. It is well known that industrialized countries are the main culprits when it comes to greenhouse emissions, while developing countries are the least protected.

For example, there are 1.3 billion people worldwide without access to clean water, 2 billion without sanitation, and as many without electricity, and about 1 billion are malnourished. Climate change could compound these problems, since it would diminish the quality and quantity of water, worsen droughts, floods and insect transmitted disease, reduce the agricultural yields and increase human displacement.

Social problems derived from an increase in global temperature will also lead to more conflicts, since poor countries have fewer means of adapting to climate change. Developed countries can modify their infrastructures in order to resist, but the relatively poor and the poorest countries can neither undertake these changes, nor buy solutions.

The Greenhouse Effect

It is the warming of the atmosphere as some of its gases absorb the heat given out by the earth. The energy, the earth receives from the sun, has to be balanced by the radiation emitted from the earth's surface. This is known as the earth's effective radiating temperature. For example, without an atmosphere, surface temperature would be approximately –18°C. In fact, earth's surface temperature is approximately 15°C.

The reason behind this is that the ultraviolet, the short-wave radiations, can pass through the atmosphere easily and gets absorbed by the matter present on the earth. The infra-red radiations are emitted out from the matter but being of longer wave length they are weak carrying lesser energy. Hence, infra-red radiations get trapped into the atmosphere and its components. This results in an increase of temperature.

Short-wave radiation from the sun warms the earth during daylight hours, but this heat is balanced by outgoing long-wave radiation over the entire 24-hour period. Much of this radiation is absorbed by atmospheric gases, most notably water vapour, carbon dioxide, ozone, methane and chlorofluorocarbon. All these may be called greenhouse gases. Without this absorption, which is also known as counter-radiation, the temperature of the atmosphere would fall by 30°–40°C.

Through human actions, such as the clearance of rain forests, increased area under paddy, and the increased rearing of livestock, the

Figure 7.6 *The rise in atmospheric carbon dioxide since 1750*

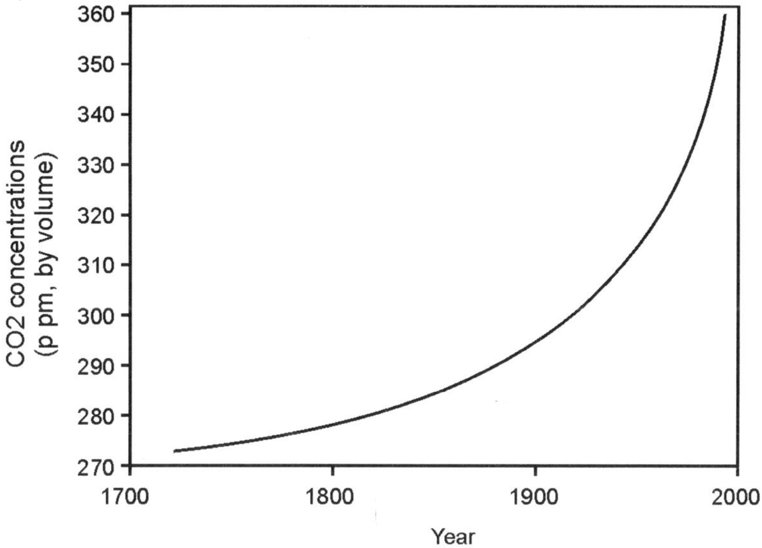

(After Watson et al. 1990)

concentration of greenhouse gases in the atmosphere is increasing. Measurements taken at Mauna Loa (Hawaii) show that the concentration of atmospheric carbon dioxide (CO_2), for example, increased by 8 per cent between 1960 and 1985, mostly because of the increased use of fossil fuels (Fig. 7.6). It would follow, therefore, that increased concentration of such greenhouse gases would lead to a rise in global temperatures, and, indeed, global mean temperature has increased over the last century, but the cause of temperature rise has not been unequivocally put down to the increase in greenhouse gases. It may be that uptake of CO_2 by the oceans actually increases with higher temperatures. Others argue that increased concentration of CO_2 foster improved rates of photosynthesis in plants, so that faster-growing trees, for example, might partially offset increased concentration of CO_2. Thus, general models of effect of growing greenhouse gases level do not give unequivocal predictions of future trends in climates.

One of the many threats to life-sustaining systems comes directly from an increase in the use of resources. Carbon dioxide is released through the burning of fossil fuels and the clearing and burning of forests. The accumulation of this gas, along with others, traps solar radiation close to the earth's surface, causing global warming. One of the results of the greenhouse effect is maintaining a concentration of water vapour in the lower atmosphere, much higher than would be possible in the low temperatures that would prevail if the phenomenon did not exist.

Effects of Global Warming

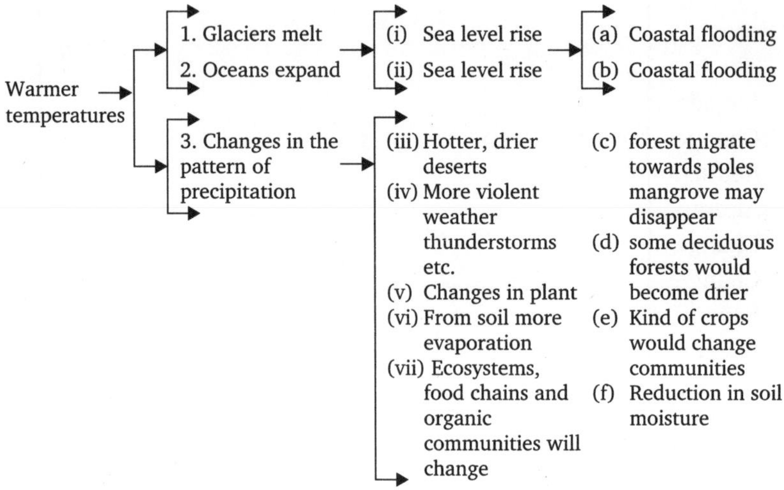

Consequences of Climatic Change

It is difficult to forecast how this situation will affect us since global climate is a very complex system. If a key aspect is altered, such as average global temperature, ramifications from this change will have far-reaching consequences. Wind and rain patterns prevalent during thousands and millions of years (and on which the lives of millions of people depend) could change; sea level could rise and threaten islands, sea beaches and low-lying coastal regions. This might result into famines, floods, landslides, avalanches, hurricanes, natural hazards and catastrophes unprecedented in their severity and also large-scale migration of human population and plants and animal communities.

As stated above, rising global temperatures are expected to raise sea level, change in precipitation and other local climatic conditions. Changing regional climate could alter forests, crop yields and water supplies. It could also affect human health, animals and many types of ecosystems. Deserts may expend into existing arable lands, and features of some of our national parks may be altered permanently.

Some of the effects of climatic change are already being felt. For example, the Arctic polar cap has lost 42 per cent of its thickness and 27 per cent of coral reefs in the world disappeared. While scientists are relatively confident that climate change will lead to higher temperatures, they are less sure of how it will affect precipitation. The following are some of the possible consequences of climatic change:

1. Sea level change.
2. Atmospheric circulation, changes in the patterns of permanent winds, ocean currents and groundwater.
3. The Inter-Tropical Convergence Zone (ITCZ) may move northward.
4. Increase in tropical cyclones, cloud cover and storms.
5. Change in the patterns of precipitation.
6. Reduction in soil moisture.
7. Alteration in natural vegetation and soil belts.
8. Change in cropping patterns and crop yields.
9. Change in the availability of water supplies.
10. Expansion of deserts and more desertification within the deserts.
11. Effect on food supply.
12. Effect on organism and human health.
13. National parks and natural reserves may be altered.
14. Impact on international trade.
15. Change in the geo-politics of the world.

These predictions of climatic change are described as under:

1. *Sea Level Change*: As a consequence of global warming and climatic change, the sea level may rise by about two to three metres by the end of the 21st century. The sea beaches, seaports, urban centres, numerous coral and volcanic islands will submerge under water. At present, nearly 50 per cent of the urban population of the world is residing along the coastal areas that will be adversely affected. The seaports have to be pushed back and the harbours including the infrastructures of the coastal urban places will be under water. It will be a great economic loss to the humanity.

Moreover, the rise in sea level will inundate farm land and making coastal ground water saline, which may become useless from the agricultural point of view. At stake are the river deltas, lowland coasting farming valleys, and low-lying mainland areas. In the worst possible scenario, the rise of 6 metres (20 feet) would flood 20 per cent the pampas of Argentina, submerge Bahamas, Netherlands, the Maldives, coastal Bangladesh, India, while 20 per cent of Florida state will be under water.

As the sea level rise exceeds half a metre, the area affected in the Asia-Pacific region rises to over half a million sq km, affecting hundreds of millions of people. The areas likely to be adversely affected are Bangladesh, India, Vietnam, China, rivers deltas, and Pacific islands. The Commonwealth Scientific and Industrial Research Organization (CSIRO) found global warming in the Asia-Pacific region could cause sea level to rise by up to 16 cm by 2030 and by 50 cm by 2070. Consequently, there will be increased number of environmental refugees fleeing their flooded homelands, citing growing migration from some south Pacific island states already suffering from rising sea levels.

The rising sea levels and increased rainfall would spread infectious diseases in regions, leaving millions more at risk of dengue fever, chickengunia fever and malaria. Moreover, the regional economies would be hard hit by chronic food and water scarcity.

2. *Atmospheric Circulation and Changes in the Direction of Ocean Currents*: The ocean and the atmosphere are extensions of each other, and human activity has changed the atmosphere as it has changed the ocean. Pollutants injected into the air can have global consequences for the ocean and for all the earth's inhabitants. Depending on the change in the temperature and salinity of the oceans, the direction, velocity and dimensions of the ocean currents will also change.

According to forecasts by some of the experts of weather and climate, there would be greater snowfall and rainfall in the higher latitudes during winter due to possible interruption or decrease of the Gulf of Stream, which is responsible for maintaining high median temperatures in the North Europe. As per one estimate the warm ocean currents are an important agent of transfer of heat and energy from the lower latitudes to the temperate region. The greatest transfer of heat by ocean currents occurs in the mid-latitudes, where 10^{15} (10 million billion calories) are transferred each second—more than a million times the power consumed by the population of the world.

3. *Northward Movement of the Inter-Tropical Convergence Zone (ITCZ)*: In the lower latitudes, the Inter-Tropical Convergence Zone may move northward, bringing more rainfall in the desert areas. The monsoon mechanism may become even more complicated. The ecosystems and the plants and animals of the tropical deserts will change. The desert areas may also move northward in the Northern Hemisphere and southward in the Southern Hemisphere. This may lead to more desertification within the deserts.

4. *Increase in the Number of Tropical Cyclones*: Since the area of oceans will increase, there will be more evaporation. The moisture and latent heat reaching the troposphere will create conducive conditions for the origin and development of tropical cyclones, hurricanes, typhoons and tornadoes. There will be intense rainfall with thunder and lightning in short duration. The rainfall intervals and rainfall variability will increase. This will accelerate the soil erosion. The countries of the south (tropics and sub-tropics) would experience an intensification of droughts and an expansion in areas today affected by desertification such a Sahel, North Africa, Southwest Asia, Southeast Asia, South Asia, Central America and the Mediterranean region (Spain, Italy, Portugal, Balkan Peninsula, Greece and Coastal Turkey), where it is predicted that the problem will achieve significant proportion.

5. *Change in the Patterns of Precipitation*: The precipitation belts tend to shift pole-ward, because the average temperatures are expected to

increase more near poles than near the equator. If there is increased precipitation in the middle and higher latitudes (temperate zone), it will not be in the form of more rainy days, but rather a decrease in the number of rainy days and an increase in the intensity of rain. In fact, the weather will be stormy even in the temperate latitudes. This will cause floods, landslides, avalanches, mudslides, land creep and soil erosions.

The mean sea level will rise due to thermal expansion of water and the melting of mountain glaciers. The increase is calculated to be between 25 to 50 cm by 2025 and up to a metre by 2050. This rise would mean the contamination of water reservoirs with marine water, receding on a scale which is thought could do away with 15 per cent of Egypt's, and 14 per cent of Bangladesh's fertile lands.

6. *Reduction in Soil Moisture*: Because of high temperature, the rate of evaporation will increase, and consequently, the soil moisture will diminish. Many of the crops which need more moisture may not be grown in the areas where they are grown at present.

7. *Alteration in Vegetation and Soil Belts*: The climatic belts and vegetation belts overlap each other. With the change in temperature and precipitation regimes, the vegetation belts will also change. The coniferous forest may push northward by about 200 km within the 21st century and same may be the case with the temperate and tropical grasslands, the deciduous and equatorial rain forests. The beech tree vegetation of North America will be pushed more west and northward (Fig. 7.7).

Figure 7.7 *Impact of Global Warming on Vegetation in North America*

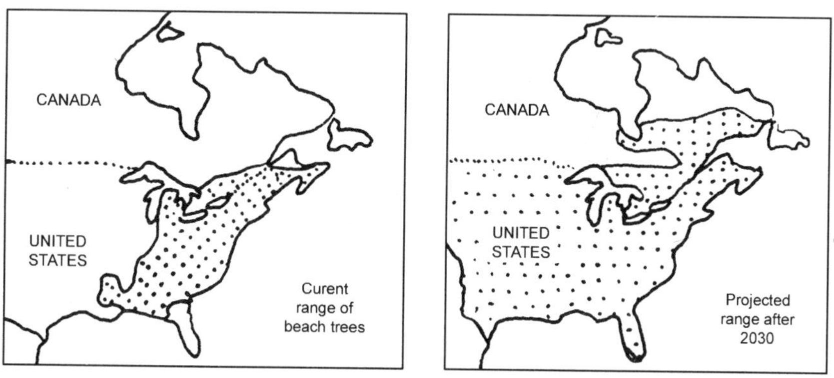

8. *Change in Arable Land, Cropping Patterns and Crop Yields*: All crops need a minimum and a maximum temperature and precipitation under which they give optimum returns per unit area. Increase in the concentration of CO_2 may boost crop productivity. In principle, higher levels of CO_2 should stimulate photosynthesis in certain plants, especially wheat, rice and soybean, by about 30 to 100 per cent.

The copping patterns in Siberia, Canada, USA, Japan, Korea and Europe will change. The permafrost soil may become under plough after an increase in the temperature. The limits of arable land will shift northward. The climate of these regions will be conducive for crops like barley, oats, wheat, sugar-beet and fodder (Figs. 7.8 and 7.9).

9. *Change in the Availability of Water Supply*: Though there will be stormy weather in the lower latitudes, resulting into torrential rainfall, but the recharge of the underground water will not be adequate. Consequently, there may be a scarcity of water in the areas where adequate underground water is available at present. Water available in the oasis and mountainous springs may also diminish.

10. *Expansion of Deserts*: Under the condition of high temperature, there would be more evaporation from the land and soil also, resulting into extension of deserts and more desertification within the deserts. According to one estimate, the deserts and arid lands cover about 34 per cent of the total land area of the earth, which may grow to 36 per cent within the 21st century.

11. *Effect on Food Supply*: In the developed countries of north-western Europe and USA extensive agriculture is practised in which only one crop is harvested in a year. Modern single crop agriculture is more delicate and susceptible to changes in temperature, water demand and irrigation needs, and soil chemistry than traditional multiple cropping.

The southern and central grain-producing areas of North America are forecasted to experience hot and dry weather by the middle of the next century as a result of higher temperatures. Today, the United States and Canada sell wheat to China and Commonwealth of Independent States (former USSR). If the foreign grain areas become wetter and gain a longer frost-free period, grain sales could shift in the other directions. Specifically, if US production decreases, the longer growing seasons, increased temperatures and thermal energy patterns, and possible increased precipitation at higher latitudes, could benefit Canadian cropping patterns.

12. *Effect on Human Health*: According to Huntington, there is a positive correlation between the prevailing temperature, humidity and precipitation and the health and efficiency of human population. The temperate climatic regions may record higher temperature and less precipitation. Under the new weather and climatic conditions, the efficiency of the people may go down unless artificial environment is created at the place of work.

The land-based animals will have to adapt to changing patterns of available forage. Increased temperatures are already causing thermal stress to some embryos as they reach their thermal limits. Warming of

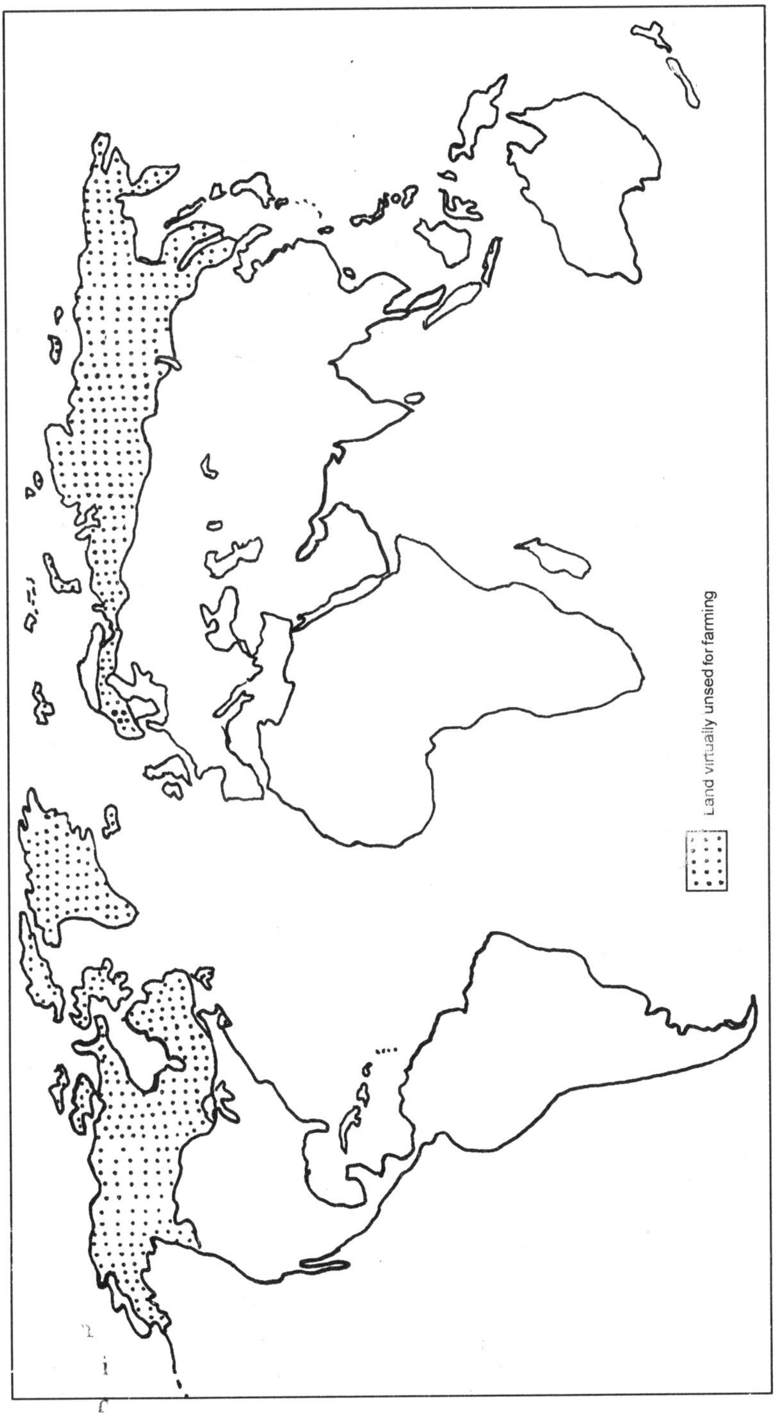

Figure 7.8 *Limits of Arable Land in 2001*

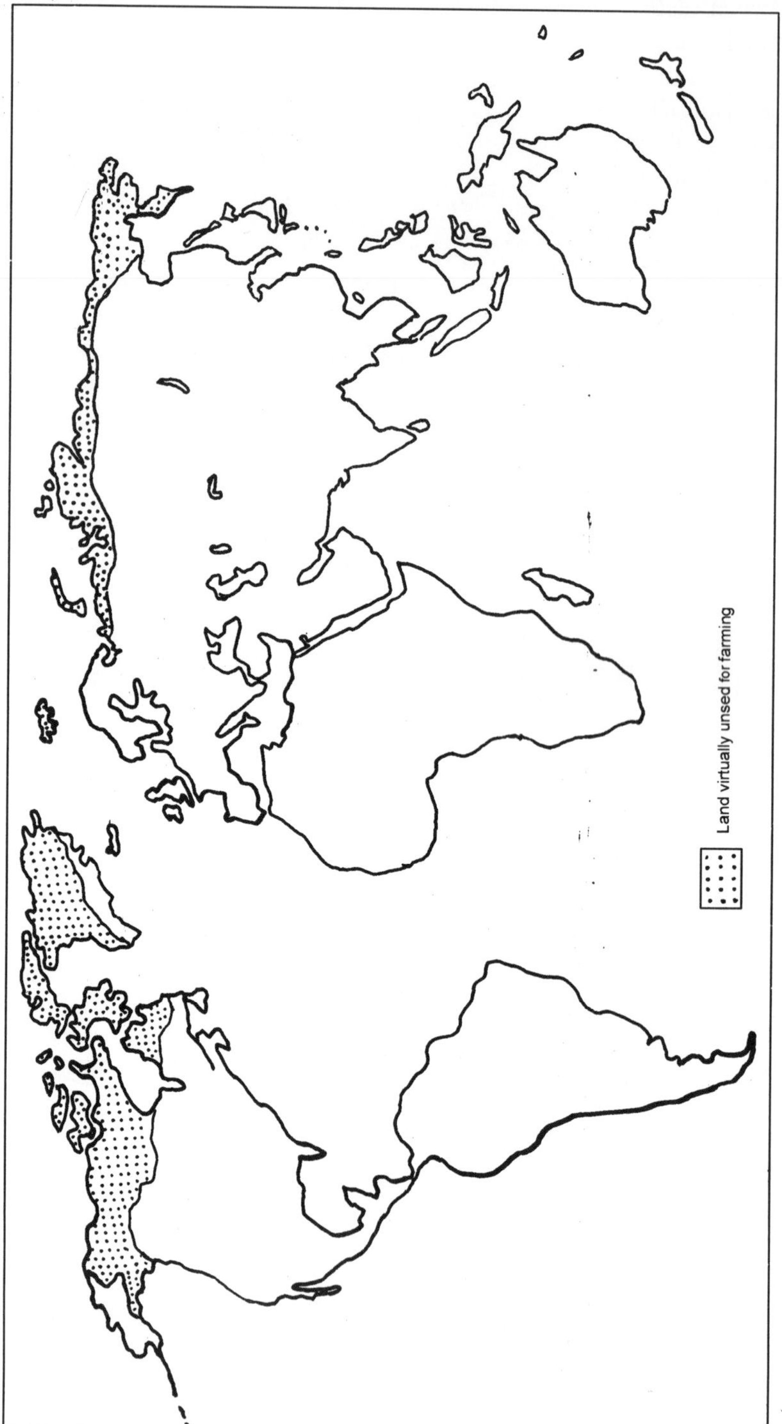

Figure 7.9 *Limits of Arable Land in 2025*

ocean water is endangering corals worldwide and appear to be deteriorating at record levels.

Another effect of climatic change will be that forest will recede to the interior of continents. The lack of water and food in some places will cause serious food problems, which in turn could result into malnutrition, undernourishment, and increase in diseases like malaria, dengue or Nile-fever in the tropical regions.

13. *Effect on the Location of National Parks and Bio-Reserves*: The existing bio-reserves and national parks will also move north and southward in both the hemispheres if the temperature and precipitation regimes change.

14. *Impact on International Trade*: Generally, the surplus countries supply the agricultural products to the food-deficit countries/regions of the world. If the Siberian region started growing more crops and higher production, Russia may become a surplus country in the production of cereals and dairy products. Same may be the case with northern Canada, Japan, Finland, Norway, Sweden, Argentina and Chile. Consequent upon the trade relations of the international community the trade pacts will change.

15. *Ozone Depletion*: Ozone is a molecule formed of three atoms of oxygen. Ozone forms naturally when lightning strikes through air, larger quantities are generated spontaneously in the stratosphere. A diffuse layer of ozone mixed with other gases—the ozone layer—surrounds the world at a height of about 20 to 40 kilometres (12 to 25 miles).

Harmless synthetic chemicals released into the atmosphere—primarily chlorofluorocarbons (CFCs) used as cleaning agents, refrigerants, fire-extinguishing fluids, spray-can propellants, and insulating foams—are converted by the energy of sunlight into compounds that attack and partially deplete the earth's atmospheric ozone. Ozone levels in the stratosphere have decreased by 50 per cent at the South Pole, 40 per cent at the North Pole, 4 per cent over Australia and New Zealand, Chile and Argentina, and 3 per cent over most of the United States, since 1969. The amount of depletion varies with latitude (and with the seasons) because of variations in the intensity of sunlight. The ozone hole has been shown in Figure 7.10. It may be observed from this that at the polar areas (middle part of Fig. 7.10) the ozone layer is thinning. A reduction in the thickness has already started affecting adversely the health in the higher latitudes of the Southern Hemisphere.

The decline in ozone alarms scientists because stratospheric ozone intercepts some of the high-energy ultraviolet radiation coming from the sun. Ultraviolet radiation injures living things by breaking strands of DNA and unfolding protein molecules. Species normally exposed to sunlight have evolved defenses against average amounts of ultraviolet radiation, but increased amounts could overwhelm those defenses. Land plants, such

Figure 7.10 *Ozone Depletion*

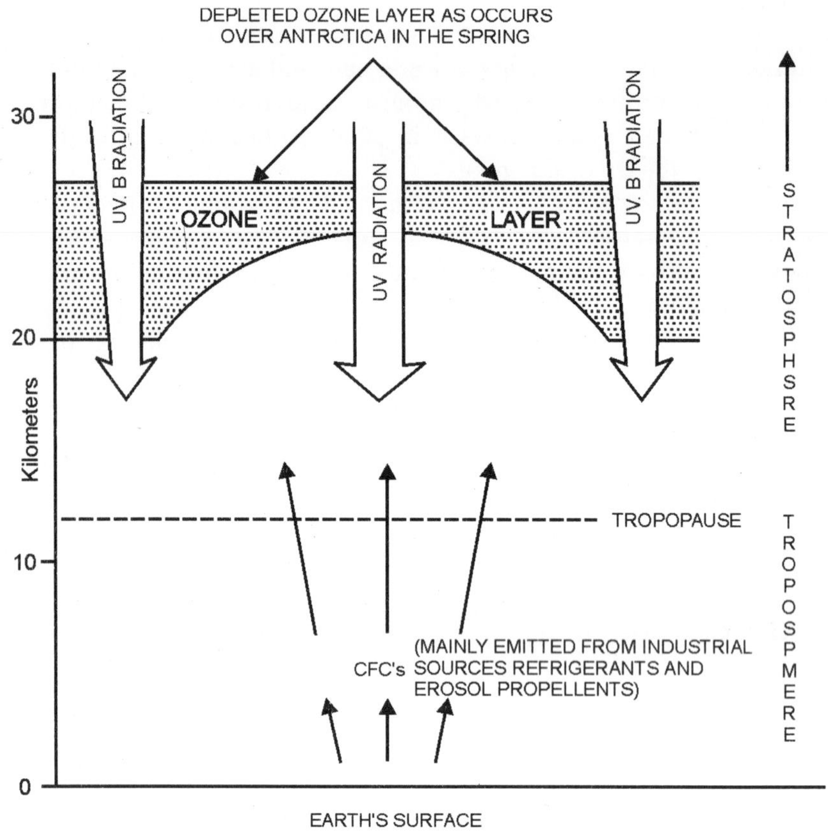

as soybeans and rice, would be subjected to sunburn that decreases their yields. Even plankton in the uppermost two metres of ocean would be affected. Our own species would not escape: A 1 per cent decrease in atmospheric ozone would probably be accompanied by a 5 to 7 per cent skin cancer. According to medical researchers, the 4 per cent ozone depletion over Australia and New Zealand will cause at least a 20 per cent increase in human skin cancers over the next two decades. Strong ultraviolet light can also suppress the immune system and cause eye cataracts and cancer.

16. *Acidification*: Acidification is yet another major problem of environmental degradation. Acidification has created potential danger to our environment since 1960. The main gases involved in the production of acid rains are nitrogen oxides, especially sulphur dioxide (SO_2) and nitrous oxide (NO). The major source of these gases is the burning of fossil fuel and industrial processes. The processes involved in the formation and deposition of acid pollution have been shown in Figure 7.11.

Figure 7.11 *Acid Pollution*

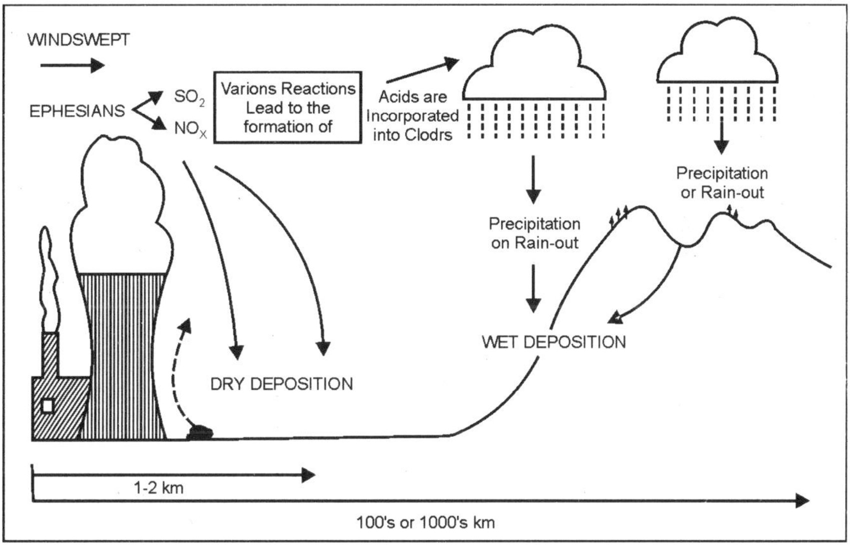

The acidification creates low pH values in clouds. The acid rain water may be deposited hundreds of kilometres from the source of pollution. Consequently, the forest, fish and marine life are adversely affected.

17. *Changes in the International Relations and Geo-Politics of the World*: The climate of country has a direct bearing on the foreign policy of the country on which depends the international relations and the geo-politics.

Despite all these reasonable conjectures, the real impact of global climatic change is difficult to assess.

Climatic Change in India

As per the opinion of the experts of climatology, the global average temperature has gone up by about 0.5°C. This was enough to cause an apparent surge in heavy rain and heat waves. According to one calculation, global average temperature can rise up to 5.8°C by the end of this century. How many extremes we are going to see in the future? What disaster do we face is unpredictable?

The average general picture about climatic change conceals important regional issues. The impact of rising temperature and greenhouse gases in the atmosphere may be more serious at the regional levels. The scientists of the Indian Institute of Tropical Meteorology (IITM) found that temperatures would increase by about 5°C in many parts of India by the end of the century, and 3-4°C in southern India. The incidences of violent weather and their intensity have increased by about 50 per cent.

This will lead to more heat waves, more torrential-concentrated rainfall and more prolonged dry spells. Such extreme events are bound to change life, economy, society and ecology, but we do not know precisely how much because we cannot predict where and when they will occur. This result is in consistent with the recent observations in London-based journal *Nature* that storms over the Pacific and Atlantic Ocean have increased in both intensity and number by 50 per cent since 1970. Year 2005 was the worst ever in the world in term of losses from violent storms.

Rising temperature can affect every aspect of life. It can throw the monsoon out of gear, reduce crop yield, create severe water shortages, spread disease and play havoc with entire ecosystems. According to the scientists of the Indian Institute of Technology, New Delhi, the water in our rivers could decrease by about 20 per cent in another five decades. According to the experts of Indian Agricultural Research Institute (IARI), the yield of some crops could decline significantly (15 per cent or more) as temperature rises. Imagine, what a 15 per cent decline in agricultural output could do to the economy and society.

Increasing evaporation and reduced flow in rivers will decrease fresh water availability at a time when 50 per cent increase in demand is expected in five decades. Under these conditions rain water harvesting needs to be done. Moreover, investment in water management on a massive scale, and specifically on diversions and storage technologies, is the need of the hour.

The scientists of the Centre for Ecological Sciences at the Indian Institute of Science (IIS) in Bangalore found that our forests would change in nature and shift in geographic belts.

According to the Indian Institute of Tropical Meteorology (IITM) the total number of rainfall in India will increase, but the number of rainy days will decrease. So rains will come in torrents and then stay away for long periods. Figure 7.12 shows how the number of rainy days will decrease and Figure 7.13 indicates the amount of rainfall per day increase.

The consequences of climatic change on the economy have not yet been studied thoroughly in India. According to the experts of environment and climate, a warming earth intensifies the hydrological cycle. More water will get into the atmosphere and then back to the ocean as rain. Increase in rain does not translate into increase in available fresh water, as this rain would come in concentrated spells. The IITM experiment gave a hint of trouble—rains will increase in flood-prone areas and decrease in drought-prone areas. The unseasonably hot and cold spells, heavy rains would not be good for agriculture. Change in climate can have a devastating effect on crops.

Figure 7.12 *The Future Rainfall Intensity in India*

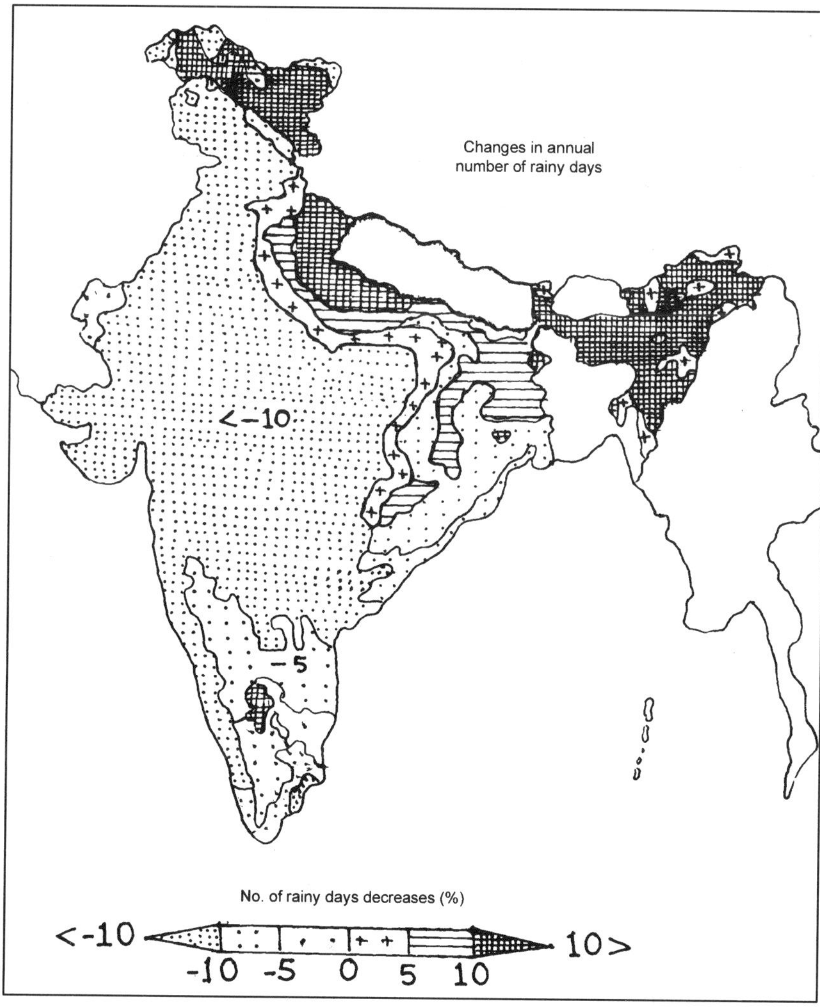

Source: Indian Institute of Tropical Meteorology.

Contradictory Opinion

Although there is a strong scientific consensus about the causes and consequences of climatic change, a group of scientists from the US believes that such theories are only hypotheses that have not been proved. These researches suggest that emissions which produce the greenhouse effect may not be the main cause of global warming.

The future climate is not only unknown but impossible to ascertain within the relevant limits for the implementation of rules; the models used do not adequately describe clouds, vapours, aerosols, ocean currents and solar effects.

Figure 7.13 *The Future Rainfall Intensity in India*

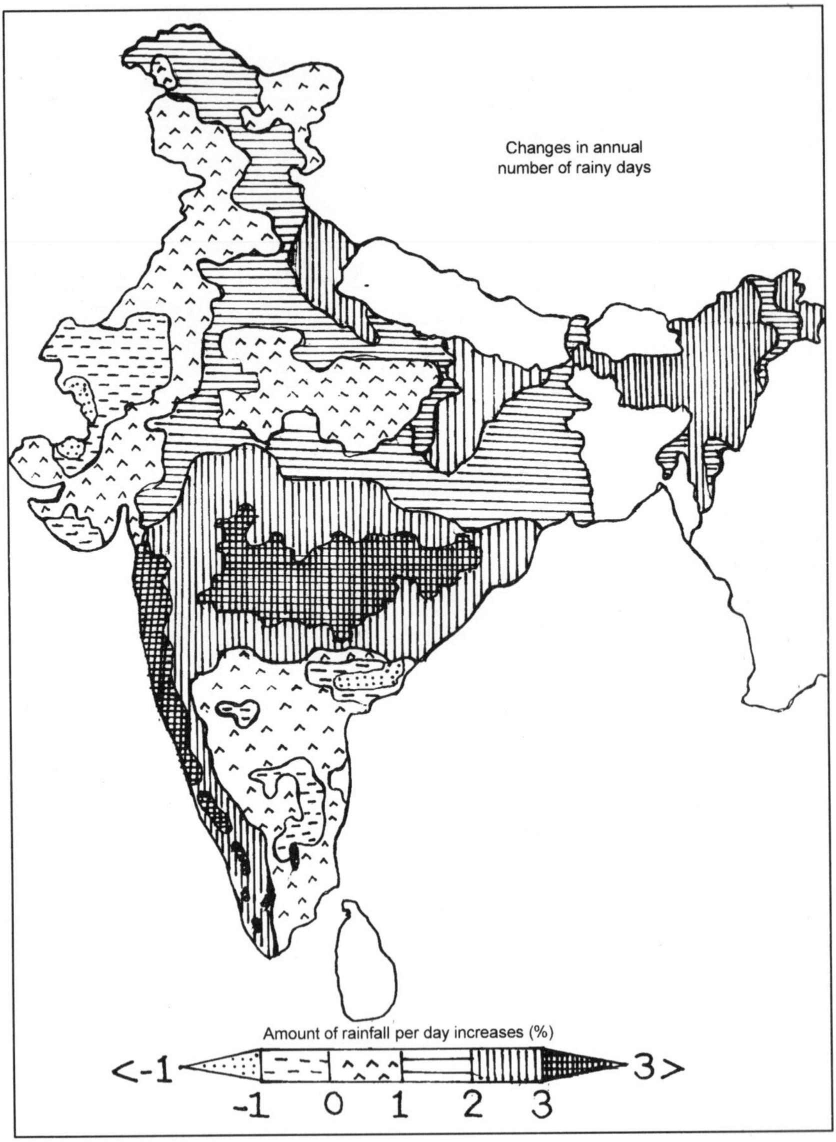

Source: Indian Institute of Tropical Meteorology.

At the same time, these scientists also state that temperature simulations made by the UN Panel on Global Climate Change look more like an arbitrary model of curves than a genuine demonstration of humanity's influence on the global climate.

Contrary to the hypotheses of climatic change, some of the Dutch scientists opine that in near future there will be global cooling, leading to Ice Age.

Global Cooling

In the 1980s and early 1990s, several volcanic eruptions also affected global temperatures. The 1982 El Chichon eruption in southcentral Mexico sent a cloud of sulfuric acid and ash 25 km into the atmosphere. Within 21 days the cloud had circled the globe, increasing the atmospheric albedo and lowering temperatures by 0.5°C. Some of the finer particles remained in the atmosphere through 1985.

The largest eruption so far in the century began in June 1991. Mount Pinatubo in the Philippines caused significant global temperature effects by injecting a blanket of acid mist and ash into the atmosphere. This increased atmospheric albedo, which decreased insolation reaching the earth's surface. The volcanic aerosols increased short-wave reflectance substantially. This change in net radiation lowered average temperature in the Northern Hemisphere from 0.5°C to 1°C, and about half of that amount in the Southern Hemisphere.

Another potential cooling effect was postulated by Paul Crutzen (a Dutch atmospheric scientist) and John Birks in "The Atmosphere After a Nuclear War: Twilight at Noon". Their study, published in the Royal Swedish Academy of Sciences publication *Ambio* in June 1982, launched an avalanche of scientific analyses, assessments, and general confirmations.

The Nuclear Winter Hypothesis, as it has come to be known, now encompasses a whole range of ecological, biological and climatic impacts, associated with detonation of a relatively small number of nuclear warheads within the biosphere. Resulting urban fires and fire storms would produce such a pall of insolation-obscuring smoke that surface heating would be drastically reduced. Temperatures could drop 25°C and perhaps more in the mid-latitudes. This potential global impact of modern warfare affects literally all aspects of earth systems.

Climatic Real Battle

Although these warnings on the phenomenon of global climate change have been around for more than four decades, they had never been accepted by institutions such as Pentagon and the World Bank. But, in 2003, both institutions warned of the terrible geopolitical consequences climate change could have on community.

A secret report by the Pentagon published by UK's *The Observer* in late 2003 warns that major European cities would be submerged by rising seas, while by 2020 Great Britain's climate would be similar to that of Siberia. Major famines, generalized uprisings and nuclear conflicts over the control of resources could take place worldwide.

The document warns that abrupt climate change caused by polluting industrial gas emissions could take the planet to the brink of anarchy while countries develop their nuclear capacity to defend and secure the provision of food, water and energy. According to the Pentagon report, the threat to global stability is greater than that of terrorism. The Pentagon goes on to say that "disturbances and conflicts will be endemic features of life".

The report adds that as soon as 2007, climate change will generate violent coastal storms, causing some areas in the Netherlands, for instance, to become uninhabitable. On the other hand, the report flatly dismisses the idea that the planet's resources could sustain all of humanity as long as there was a fair distribution of wealth. In other words, the relationship of sustainability between the planet and its inhabitants will become seriously unbalanced.

Another report, this time by the World Bank, admits that the planet is in grave danger and that the main challenge for the future is not terrorism but rather human dependency on fossil fuels. Ninety-four per cent of World Bank energy investments are in fossil fuels and only 6 per cent in renewable resources of energy. The report suggests that investments in renewable energy should increase by a minimum of 20 per cent. If current levels of fossil fuels consumption continue, global warming and polar ice cap meltdown could multiply the rise in sea level in ten years in relation to 2003 levels.

The report recommends immediate action in order to counteract the effects of global warming.

Solution

Some of the leading experts of climate opine that society has adapted to many problems in the past. Capital can be raised and technology harnessed to block the flood and guard the coastlines, deltas and estuaries. The real solution of the climatic change, however, lies in the reduction of greenhouse gases, which shall reduce the severity of the hazard. In addition, individual conservation and a demand for energy efficiency remain cheap ways to stretch present energy resources, reduce imports, cut carbon dioxide emission, and save money.

One of the significant agreements that emerged from the 1992 "Earth Summit" was the U.N. *Framework Convention on Climate Change*—a legally binding agreement signed by over 160 nations that is a first ever attempt to evaluate and address global warming on an international scale.

In the opinion of Fredrick Njau of the Kenyan Green Belt Movement, the industrial and developed nations need to step up support to help poor nations to adapt to global warming with drought and heat resistant crops and alternative energy sources so that people do not cut down trees for fuel. African forests, he added, are soaking up pollution from industrialized nations for free and should reap some kind of reward or benefit for that (*Times of India*, October 14, 2006, p.33).

The Kyoto Protocol, 1997

The Kyoto Protocol, which was agreed in Japan in 1997, is the only international instrument which includes obligations for industrialized countries: a 5.2 per cent reduction in greenhouse emissions by industrialized countries during 2008-2012, based on 1990 figures. To be enforced, the treaty needs to be ratified by 55 industrialized member-nations, amounting to at least 55 per cent of global greenhouse emissions.

If the US were to take part in this agreement—having withdrawn from it in 2001 after a decision by President George W. Bush—that proportion of emissions would rise to 80 per cent. If Russia ratifies it, as recommended by the Putin Cabinet in September 2004, the proportion of committed emissions would rise to 61 per cent. If both countries took part, the proportion would rise to 97 per cent. However, due to its modest goals, the implementation of the Kyoto Protocol is, in fact, only a small step. Rather than 5.2 per cent, a 60 to 80 per cent reduction in greenhouse emissions would be needed to stabilize carbon dioxide concentration in the atmosphere to normal level (between 0.03 and 0.04).

It seems ironic that while the US leads a war to defend the world from terrorism, at the same time, it punishes that same world with carbon dioxide emissions, causing 36 per cent of the damage related to global warming. This can only be classified as terrorist attack on a global scale.

Also, among industrialized countries, the Netherlands, Finland, France, Greece, Spain, Austria, Denmark, Belgium and Ireland are well below the level of controls they should apply in order to reduce greenhouse emissions. This is mainly due to the indiscriminate use of transportation by land and their huge industries. Only the UK and Sweden will meet the objectives of the accord within the established targeted time-frame.

References

Ahmad, A., The Scary Story of Climate Change, *Business World*, September 18, 2006, pp. 46-54.

Houghton, J.T., G.J. Jenkins, and J.J. Ephraums (eds.), *Climate Change 1992—The Supplementary Report to the IPCC Scientific Assessment,* World

Meteorological Organisation/United Nations Environment Programme, New York: Cambridge University Press.

Houghton, J.T., Richard A. and George M. Woodwell, Global Climate Change, *Scientific American, 260*, No.4 (April 1989): 36-44.

Husain, M., 2002, *Human Geography*, Jaipur: Rawat Publications.

Kumar, K.K., Climate Change in India, Tracking Climate, *Business World,* pp. 52-54.

Mather, J.R., 1974, *Climatology: Fundamentals and Applications*, New York: McGraw Hill.

Oliver, J.E. and Rhodes W.F. (eds.) 1987, *The Encyclopedia of Climatology*, New York.

Turco, R.P., et al., 1983, Nuclear Winter: Global Consequences of Multiple Nuclear Explosions, *Science*, 222, No. 4630, pp. 1283-92.

Glossary

Aborigine: A member of an indigenous people existing in a land before invasion or colonization by another race. This term is especially used for the original inhabitants of Australia (Australoids).

Absolute Zero: The lowest temperature theoretically obtainable: -273.15°C

Accessibility: The ease of approach to one location from other locations. This may be measured in terms of the distance traveled, the cost of travel, or time taken. In network analysis accessibility may be expressed by measures of connectivity.

Acculturation: The adaptation to, and adoption of, a new culture. This may occur simultaneously as two cultures meet but more often occurs as an immigrant group takes to the behaviour patterns and standards of the receiving group. A major example is the acceptance of American norms by the millions of European immigrants who arrived in the USA in the first half of the twentieth century.

Acropolis: Literally, the highest part of the city. The best known example is the Acropolis in Athens.

Action Space: The area in which an individual moves and makes decisions about her or his life, including, for example, shopping, studying, or working: the set of place an individual is aware of (see mental map).

Activity Index: A measure of the extent to which a local authority develops policies to attract industry. Such policies may range from the provision of sites to the construction of factories.

Activity Rate: The percentage of people of working age who are actually employed. This may be calculated for a region or a nation. A low activity rate indicates a high level of unemployment.

Activity Space: The spatial separation of the sexes during the working day (hours), that part of action space with which an individual interacts on a daily basis. There seems to be a hierarchy of activity spaces for most people, increasing in spatial extent from family space, to neighbourhood, to economic space, and then to urban space. With movement up the hierarchy, the individual's knowledge of the space becomes less comprehensive (see mental map).

Adaptation: Any change in the structure or functioning of an organism that makes it better suited to its environment.

Additional Worker Hypothesis: The view that a rise in unemployment leads to rise in the working population. When the major wage earner becomes unemployed, other members of the family who were not in paid work now, seek employment in order to sustain the household. This might be valid if part-time work is recognized.

Administrative Principle: The principle advanced by Christaller (1933: trans: C. W. Baskin 1966) which proposes that, in a region with a highly developed system of central administration, settlement is so arranged that one major centre administers six centres of lesser rank, each of which, in turn, oversees a further six centres. The number of settlements at progressively lower levels, starting with the highest in rank, thus follows the sequence 1, 7, 49, 343 ...This hierarchy is known as k=7.

Advanced Economy: A synonym for a more developed country, generally defined as having per capita GNP in excess of $10,000 per annum (1995) and an agricultural workforce of less than 6 per cent of the working population.

Adventitious Population: People living in the countryside by choice and not because they are employed in rural occupation.

Aesthetic Landscape: The landscape in terms of taste and beauty.

Affinity Environment: A spatially bounded social environment based on voluntary preference among residents for salient attributes such as ethnicity, lifestyle, occupation, income, family status, faith and religion.

Afforestation: The planting of trees. Tree-planting can stabilize soils by increasing, intercepting and reducing run-off, reduce flooding through the reduction of silting, improve soil fertility (in agro-forestry schemes), provide timber and firewood, and counteract possible warming through the absorption of the increasing levels of atmospheric carbon dioxide.

Age Dependency: The dependency on the contributions of those in full employment of some members of the population too young to be employed full-time. In India below 14 or above 60 years are reckoned to be age dependent.

Age Structure: The composition of a nation by age groups. The type of age structure has a profound effect on the future of the nation, with old age dependency at one extreme and explosive population growth at the other.

Age-Sex Pyramid: A set of two histograms set on a vertical axis and back to back which depicts the numbers of two sexes in different age groups. Males are usually on the left with females on the right. The youngest group is at the base; the oldest group at the apex. Actual figures or percentages of the age groups may be used.

Ageism: Prejudice towards and/or discrimination against people because of their age.

Agency: Another term for action, as in human agency.

Agglomeration: The concentration of activities, usually industries, near to each other, for example in a specialized industrial region, such as Okhla Industrial region, Hooghly, Mumbai, Coimbatore, Gajraula (U.P.), and Ahmedabad industrial regions.

Agrarian: Describing an agricultural system which combines horticulture with the rearing of animals.

Agribusiness: Large agricultural operations which are run like an industry. A single business can concerned with the whole of agricultural output: the ownership of land, agricultural processes, the manufacture of agricultural machinery, the processing of the product and its shipment.

Agricultural Density: The density of the agricultural population per unit area of agricultural land. It is however, rather difficult to define and isolate the agricultural population, as many rural inhabitants are engaged in essentially urban occupations (see adventitious population).

Agricultural Geography: The study of spatial patterns in agricultural activity.

Agricultural Revolution: The relatively abrupt change, which first occurred as estimated 10,000 years ago, involving the domestication of plants and animals. Sometimes the term *agricultural revolution* is also applied to later periods of rapid changes in agricultural methods and production.

Agricultural Location Theory: An attempt to explain the pattern of agricultural land use in terms of costs, distance and prices.

One explanation is concerned with the effect of the city on rural land use. It predicts that circular zones of different agricultural production will surround the city. Zones may also occur beside a coastal strip, or along both sides of a transport artery. A pattern of agricultural land use zoning may occur from a unit as small as a single farm to one as large as a subcontinent. The classic theory of agricultural location was put forward by von Thunen.

Agricultural System: Any method of farming may be seen as part of an agricultural system. 'Inputs' include seeds, water, fertilizers, pesticides, herbicides, and livestock which are introduced to the 'plant': the buildings, machinery, and land. The 'output' is the produce of the farm.

Agro-forestry: Any agricultural system which incorporates the planting or encouragement of trees on land where animals are raised or crops grown. Thus agro-forestry includes 'slash and burn' or the use of shade trees.

The planting of trees may aid farmers since tree roots can bind soil and limit soil erosion, deep rooted trees can tap new nutrient sources. Leguminous trees can fix atmospheric nitrogen and improve soil fertility, leaf litter can add organic matter, firewood, and timber.

Agro-town: A town of up to 20,000 people with agriculture as the main economic activity. As such they are not truly urban settlements, although they may be the only significant settlements in the region. Agro-towns occur in parts of Mediterranean Europe and may find their counterparts in the Yoruba region of West Africa.

Alonso Model: An explanation of urban land use and land values, developed by William Alonso (1964). It is grounded on the concept of bid rents whereby the urban land user seeks central location, but is willing to accept a location further from the city centre if the rents are lower in compensation. The use that can extract the greatest return from a site will be the successful bidder. To this basis, Alonso, in a study of housing, added the quantity of land required, and variations in the amount of disposable income used on land and transport costs on one hand, and on all goods and services on the other. If the amount of goods and services is held constant, the price of land should decrease with increasing distance from the centre. The well-off will choose to live at lower densities at the edge of the city; the poor remain in high density occupancy near the city centre. The quantity of land that may be bought should increase with distance from the centre, but commuting costs will rise with distance from the centre so that the quantity of wealth available for land will decrease. Each household represents a balance between land, goods, and accessibility to the workplace.

Alonso also explained that high-income groups, who are less constrained than lower-income groups in their choice of residential location, may prefer to accessibility to the CBD offered by the inner city to space, quiet, and cheaper land of the suburbs, so that gentrification may result. The assumptions on which this theory rests range from all land being of equal quality to lack of planning constraints. This means that the theory is a long way from reality, although it does reflect some aspects of urban morphology.

Analogue Model: A model which explains a phenomenon by reference to some other occurrence; for example, H.T. Odum likened an ecosystem to an electric circuit. Energy input was shown as batteries, food flows as electric currents, and energy dissipation by amperage and voltage chambers.

Animism: Belief that inanimate objects, such as plants and stones or natural phenomena such as thunderstorms, floods, droughts, and earthquakes have discrete spirit and conscious life.

Anthropogenic: Brought about by human agency (see anthropo-geomorphology).

Anthropogeomorphology: The study of human effects on the physical landscape. This effect may be direct, through construction, excavation, or hydrological interference (damming, dredging, canal construction and so on) or indirect, when human agency has unlooked for consequences. Examples for the latter include accelerated soil erosion, the generation of earthquakes through the construction of reservoirs.

Anti-natalist: Concerned with limiting population growth, the people's Republic of China has pursued anti-natalist policies, notably the one child strategy, for about two decade. While anti-natalist government policies may be instrumental in lowering birth rate, state coercion may have unexpected and damaging results; reports in 1995 suggested that abortion of female children had become common in China, so that male: female sex ratio at birth had become grostesquely imbalanced (rising in one province to 140:100), and so that female infanticide and the abandonment of girl babies is common. Research in population growth suggests that improving opportunities for women is the best contraceptive (see birth rate).

Apartheid: The system of social segregation first promulgated by the largely Afrikaner National Party of South Africa in 1948. Petty apartheid meant the separation of facilities such as lavatories, transport, parks, and theatres into two groups: white and non-white. With the election of South Africa's first democratic government in 1994, the last vestiges of apartheid were officially removed, but the policy will have left its mark on the South African landscape and its society for many years to come; for example, under the Nationalist Party regime, housing was zoned for racial groups so that predominantly African, coloured Asians, White areas still exist, as do the homelands, to which every African was assigned, according to each one's major tribal group.

Aquaculture: The use of waters, other than the sea, for agricultural production, usually the production of fish (see fish farming).

Arable Land: Originally meaning fit for cultivation, as opposed to pasture or woodland, the term is now applied to agricultural land used for growing crops.

Areal Differentiation: A recognition of the different regions of the earth's surface, the study of the way human and physical phenomena vary over the earth's surface which is aimed at interpreting the variations in the character of the settled world. As it may be seen as chorology, or traditional regional geography. As systematic geography developed, using the quantitative approach, many geographers abandoned regional geography from the mid-1960s and onwards.

In 1980 areal differentiation was again seen as a central theme in geography. The major topics in this field emphasized by humanistic geographers include the social construction of space, the sense of place, and the iconography of landscape, while other geographers, notably Marxist geographers, are concerned with those spatial variations in the quality of life and in economic activity which are referred to as uneven development. Another approach that contextual geography considers the interaction of place and human agency in producing geographical sameness and difference.

Systematic themes have by no means been abandoned, but it is generally recognized that every process will be modified according to the unique nature of each separate environment.

Arithmetic Density: The total number of people divided by total area.

Artifacts: The material evidence of human subsistence, shelter and production.

Assimilation: Also known as acculturation or integration, this is the integration of an immigrant, outsider, or subordinate group into the dominant, host community. Initially the migrant group is segregated from the host culture but then there is often a blurring of cultural lines, and the concept of assimilation does imply that the minority group eventually takes on the values of the host, or charter group. This contrasts with the view that new groups can affect the values of dominant group, or live alongside it in a multicultural society.

The rate of assimilation depends on the race, religion, customs, occupations, and cultures of the migrants and the dominant group.

Balance of Payment: A comparison between the payments made by one country to other nation of the world and the revenue it receives from them. If receipts exceed outgoings, the balance is positive. The capital account records payments made in settlement of old debts or establishment of new ones; the current account shows payments made on goods and services, including interest payments. The balance of trade is a similar record, but registers only visible exports and imports.

Balance Growth: A strategy of growth with an equal emphasis on agriculture and industry. Agricultural developments provides the food required and releases labour from the land engage in industry. Industrial wealth stimulates markets for agricultural growth—or such is the theory. Unbalanced growth denotes a strategy which focuses on agriculture or industry alone.

Basin Cultivation: A form of tropical cultivation where low earth ridges enclose small fields-basins. The ridges are to check run-off from heavy tropical rains, conserving soil moisture and limiting soil erosion.

Behavioural Environment: The perceived environment; the impression people have of their environment, which is the basis for decision making, as individual organize the facts of phenomenal world into patterns and give them values and meanings in accordance with their cultural context.

Behavioural Geography: The view of geography counters the simplistic views of geographical determinism and neo-classical economics and suggests that, far from being an economic man, an individual is a complex being whose perception of the environment may not correspond with objective reality. A distinction is made between the objectively observed environment—things as they are—and the perceived environment—things as they are seen by individuals. Individuals react to their perceptions, rather than to the phenomenal (real) environment. Furthermore their decisions may not be rational or optimizing, but may depend greatly on chance (see satisficer).

Behavioural geography is concerned with understanding the flow of events which produce, reproduce, or transform a system; an analysis of processes rather than outcomes. It is concerned with the selectivity abstracted structures (mental maps) which are used as part of the decision making process, whether by individuals or corporations. The value of such decisions depends upon the perceptions of the decision maker and his or her ability to respond to that perception.

Behavioural Model: A model which takes into account the vagaries of human nature rather than depending on the concept of economic man.

Behavioralism: The view that the actions of an individual occur as a response to stimuli. Through constant repetition, the individual learns to make the same, 'correct' response to a given stimulus.

Behaviouralism has been widely rejected by social scientists who note that it over-simplifies human behaviour and takes no account of the mental processes involved in the perception of, and response to, a stimulus; it neglects all the aspects of human behaviour which cannot easily be observed.

Binary Pattern: Used in the analysis of settlement-size frequencies to describe the situation where the upper end of the settlement hierarchy is dominated by a number of settlements of similar size.

Birth Rate: The number of births in a year per 1,000 of total population taken at the mid-year mark. This is also known as the crude birth rate.

Boot Strap Strategy: An approach to urban renewal which involves the conservation and upgrading of existing properties without displacing the occupants. It is based on the belief that everything is improvable and worth improvement.

Break of Bulk: A point or location where a cargo must be transferred from one carrier to another, typically at ports where goods are transferred from water to land transport.

Buffer State: A generally neutral state which lies between two powerful and potentially belligerent neighbours. Invasion by its more powerful neighbours is often the lot of a buffer state. Poland, as a buffer between Germany and Russia, has suffered particularly badly in this respect.

Bush-fallowing: A type of subsistence agriculture where land is cultivated for a period of time and then left for some years to regain its fertility.

Capitalism: An economy system for the general production of goods and services based on private enterprise in which most of the resources/means of production and the power to exploit them are in the hands of relatively few private individuals and are subject to largely uncontrolled economic competition.

Carrying Capacity: The maximum potential number of inhabitants which can be supported in a given area.

Cartogram: Broadly defined as a map using statistical symbols, a more specialist usage defines a cartogram as a type of map transformation based on a scale other than a true scale.

Cash Cropping: The growing of crop to produce goods for sale or for barter rather than for the subsistence of the farmer and family.

Caste: One of the classes or distinct hereditary orders into which the Hindus are divided according to religious law.

Catch Crop: A fast-growing plant which is intercropped between the rows of the main crop. It is often used as green manure.

Central Business District (CBD): The heart of an urban area, usually located at the meeting point of the city's transport systems, which contains the highest percentage of shops and offices. Land values are high because of high accessibility, therefore land use is at its most intense in order to offset rent costs. In consequence, in many countries development is upwards rather than sideways. Within the CBD, specialist areas, such as a jewellery or garment-making quarters, may arise in order to benefit from external economies. Vertical land use zoning is also common, so that retail outlets may be on the ground floor, with commercial users above them and residential users high up.

The CBD is increasingly under threat due to a combination of traffic congestion, which has led to parking restrictions, and to the growth of out-of-flow developments, including superstores.

Central Place: A settlement or a nodal point which, by its functions, reserves an area round about it for goods and services.

Central Place Theory: A theory advanced by W. Christaller (1933, trans. 1966) and later, Losch 1954, concerned with the way that settlements evolve and are spaced out. Christaller envisaged an isotropic plain with an even distribution of purchasing power. Travel costs were the same in any direction and all parts of the plain were served by the central place, so that the spheres of influence of the central places completely filled the plain. Central goods and services were to be purchased from the nearest central place and no excess profit was to be made by any central place. Christaller contended that each central place should have a hexagonal market area since this polygon represents most effective packing of the plain and is most nearly circular.

To ensure that goods and services are freely available, central places emerge at the centre of a hexagon containing six lower-order places. One higher order place will serve a total of two lower-order neighbours. This may mean that two distinct lower order places are served or that the central place will serve one-third of each of the six lower order places surrounding it. This will bring to two total of lower order places served, and, with the addition of the central place itself, three places are served. This method of serving the market is known as k=3 system.

A different system of hexagon would evolve if transport costs are to be minimized. The hexagon is rotated to that the settlements are located evenly at the mid-point of the hexagon's sides. Now the central place serves a half share in the surrounding six settlements: a total of three places plus the central area. Therefore, k=3+1 = 4; k=4 system.

The most efficient pattern for the administration of settlements sees all six lower-order centres inside the hexagonal area of the central place.

All these places fit into a hierarchy. Higher order places stand out from the hexagonal pattern of lower-order centres, but are themselves packed in hexagonal around an even higher-order central place. Christaller envisaged this hierarchy as going all the way up to major regional centres. It should be said that hexagonal patterns are very rarely found in real life.

Losch, who had worked independently of Christaller, extended these ideas. He plotted the ten smallest market areas, each with a different 'k' value. Each network surrounded a common central place. Tracing of each network were laid over each other and the tracings were positioned so as to produce the largest number of places occurring for each 'k' value. The result was a central place with city-rich and city-poor areas spread out in wedge around the major central place. Such a pattern is found around Indianapolis (USA).

Common sense tells us that the basic postulate of these models do not exist but they still give insight into the nature of town development and distribution.

Chain Migration: A migration process which depends on a small number of pioneers, who make the first moves to set up a new home in a new place. They send information back-home, and this encourages further migration from the originating area. A British example comes from young men from the Indian subcontinent, who, after finding work and a place to stay in the UK, would send for other members of their families over the years. Chain migration is particularly apparent in international migration, where it is less easy for prospective migrants to get good information about their destinations, so that the importance of contact with family and friends is much greater. One result is that, in British cities for example, there is often a distinct segregation of migrant groups into different neighbourhoods according to the place of origin, as the original links are maintained.

Child-woman Ratio: The ratio of the number of children below five to the number of women of child-bearing years, which may be expressed as:

PO-4/Pf 15-44

Where P refers to a population, the numbers refer to their ages, and f denotes women.

In the absence of universal registration of births, the child-woman ratio is a relatively good indicator of fertility.

Chorography, Chorology: The establishment and description of geographical regions. This is the oldest practice of Western geography; Greek geographers sought to describe the parts of the earth. It has been seen as a synonym for 'areal differentiation'. R. Hartshorne (1939) was probably the most forceful advocate of geography as a chorographic science, and chorography is often seen as being in opposition to spatial science (Systematic Geography).

City: City is defined as a large urban centre functioning as a central place which can provide specialized goods and services. There is no world-wide, or even European agreement over limiting figures of population size or areal extent for a city.

City-filling activity: Activity which provides goods and services for use within the city. A synonym for non-basic activity.

City Region: An area that surrounds an urban settlement which functions as the regional centre. This area is tied to the regional centre of higher-order functions although oriented to local towns for lower order ones. It is an area considerably larger than the urban settlement area. Also known as complementary region, hinterland, sphere of influence, umland, urban-field and tributary area.

City Size Distribution: The frequency with which the settlements of a country or region occur in certain, arbitrary defined, population size groups, such as under 10,000, 10,000-25,000 and so on. In general large settlements are less frequent than small ones, but when size is graphed against frequency, several patterns may be recognized: the most common being the binary, log-normal, and primate pattern.

Club of Rome: A society, formed in 1968, of one hundred members ranging from academics such as scientists, social scientists, economists, and teachers, to managers and civil servants, with the aim of understanding the workings of the world as a finite system and suggesting alternative options for meeting critical needs. Concerns have included inequality, on global and regional scales, pollution, unemployment, and inflation. Its publication limits to growth (1972) argued the necessity of slowing down population growth in order to ward off Malthusian checks.

Colonialism: The acquisition and colonization by a nation of other territories and their peoples. In this respect, colonialism is as old as society.

Common Market: A group of two or more countries that agree (a) to remove tariffs and other restrictions on trade and on movement of capital and labour between members; and (b) to establish a *common* external tariff on goods imported from outside the single *market* area.

Commuting: A daily journey from home to place of work and back again.

Concentric Zone Theory: The theory, proposed by E.W. Burgess (1926), that urban land use may be classified as a series of concentric zones. Zone I, the CBD, lies at the centre of the city. Zone II is in transition. It is the crowded, multi-occupied zone of the city first invaded by migrants. Within this zone are the ghetto areas. In zone III are the working men's houses, the area of second generation immigrants, one step up from zone II. Zone IV and V are residential; zone IV for the better-off and zone V for the commuters. The model assumes uniformly flat, and available land, and ignores the importance of transport routes, but relies on the theory that city growth results from distinct waves of in-migrant.

Contextual Theory: A theory which argues that the contexts in which human activity takes place—the time, the space, and the place in the sequence of events—are crucial to the nature of that activity. T. Hagerstrand (1984) who gave birth to the idea, explained that processes are constrained and shaped by the terrestrial space and time in which they take place: very action is situated in, and shaped by, a particular space and time.

This way of thinking strengthens the re-emergence of regional geography as a discipline, but space is not just the context for human activity, it is also consequence of it.

Contour Ploughing: A method of ploughing parallel to the contours rather than up and down. It is used to check soil erosion and the formation of gullies.

Conurbation: A term coined by Patrick Geddes to describe the extensive built-up area formed by the coalescence of one separate, and mainly urban settlements of million cities.

Core Area: The heartland of a nation, usually more advanced than the rest of the nation, with an intense feeling of native culture and nationality. In the opinion of some experts, the core area acts as the crucible from which a state grows, and while this is true of some European states, many African states, for example, were created quite arbitrarily, and have alter forged a nation identity. It has also been suggested that if the core fails, the state will languish.

Core-Periphery Model: J. Friedmann (1966) maintained that the world can be divided into four types of regions. Core regions are centres, usually metropolitan, with a high potential for innovation and growth, such as Sao-Paulo in Brazil. Beyond the core area are the upward transition region areas of growth spread over small centres rather than at a core. Development corridors are upward transition zones which link two core cities such as Delhi and Faridabad, and Kanpur and Lucknow.

The model has been criticized on a number of counts. Most notably, it has been argued that uneven development is not the inevitable consequence of development, but of the particular mode of production used to bring about that development, and that Friedman's model represents the effects of the capitalist mode.

Cosmopolitan: Literally, 'a citizen of the world'; one free from local and national prejudice. In geography, the word tends to be used to denote an ethnically and culturally mixed population, e.g. London, Paris, Beirut, Dubai and Sharjah.

Counter-urbanization: The movement of population and economic activity away from urban areas.

Critical Isodapane: In Weber's theory of industrial location (1929 trans. 1971) the cost of goods increases with transport costs from the point of production. The critical isodapane is isoclines where this increase is exactly offset by the saving made by decreased costs brought about by cheaper labour.

Crop Combination Analysis: A technique evolved by Weaver to delimit agricultural regions which, he argued, are not regions of simple monoculture as suggested by the names Corn Belt, Cotton Belt or Spring Wheat Belt, but are areas of combinations of crops.

Crop Rotation: The practice of planning a succession of crops in a field over a period of years. Rotation of crops can maintain soil fertility since different crops use different soil nutrients, so excessive demands are not made of one nutrient. In certain rotations, plants (like legumes pulses, peas and beans, etc.) are grown to restore fertility.

Cultural Distance: A gap between the cultures of two different groups such as that between the culture of rural societies and that of cities.

Cultural Ecology: This term, coined by J. Stewart (1955), describes the study of the relationship between nature and culture in human societies. Cultural ecology is the study of interactions of societies with one another and with the natural environment, and as such is a branch of cultural anthropology. More recent perspectives have stressed that societies are composed of individual persons acting within the given structures, such as societal constraints.

Cultural Evolution: The successive stages of societies, involving the diversification of cultures as they adapt to particular environments (specific evolution) and the progression of all cultures from hunting and gathering to industrial societies (general evolution).

Cultural Geography: The study of the impact of human culture on the landscape. Themes which have been explored include the effects of plants and animals domestication, fire, hydrological techniques, farming methods and settlements.

Cultural Hearth: The location in which a particular culture has evolved. C.O. Sauer (1952) chose Mesopotamia as the cultural hearth of agriculture where wheat and rice were domesticated first.

Cultural Landscape: The landscape which results from many generations of human occupancy. Many features of present landscapes were fashioned by past societies who effected more or less permanent changes. The cultural landscape is evolved from the natural landscape by a cultural group. The cultural landscape is the man-made landscape as an expression of the response of a culture to its natural surroundings.

Cultural Region: A region characterized by a common culture, in other words, an area that is distinguished from other areas on the basis of ethnicity, beliefs, religion, and customs (i.e., type of uniform region)

Culture: 1. The phenomenon that binds people together by shared beliefs and values and by a common image of themselves and the world. 2. The learned behaviour which is socially transmitted, such as faith, belief (religion), language, customs, morals, technology, and art; everything in society which is socially, rather than biologically transmitted. 3. In Geographic Information Systems, human, as opposed to physical features.

Dairying: The production of milk from cows, buffaloes, goats and sheep. It is an agricultural system with very high labour and plant costs but using relatively little land. Income is earned throughout the year.

Death Rate: The number of deaths in a year per 1,000 of population as measured at mid-year. This crude death rate may be expressed as:

$$D/P \times 1000$$

where D=the number of deaths, P=the mid-year population.

Decolonization: The process by which a colonial power withdraws from an overseas territory and hands over political control to an independent state.

Demographic Regulation: The notion that a population will restrict its fertility when a reduction of mortality causes population to grow beyond the ability of the environment to sustain it. This is not a simple response and is accompanied by changes in society in an attempt to maintain an equilibrium.

Demographic Relaxation Theory: The view that overpopulation leads to war. A major example is the professed desire for *lebensraum*—space in which to expand—shown by the German government in the late 1930s and which led to the invasion of Czechoslovakia and Poland. The resulting wars destroy populations. In this way, war is seen as relieving overpopulation.

Demographic Transition: An account, but not a complete explanation of changing rates of fertility and mortality, and natural increase. These changes are held to occur as a nation progresses from a rural, agrarian, and illiterate state to a predominantly urban, industrial, literate one. Four stages have been recognized:

1. The High Stationary Stage: Birth and death rates are high and the death rate fluctuates from year to year.

2. The Early Expanding Stage: Fertility remains high, but improved conditions mean falling death rates. Population therefore increases.
3. The Late Expanding Stage: Death rates are low and fertility is declining but population is still increasing.
4. The Low Stationary Stage: Birth and death rates are low and the birth rate fluctuates.

Demography: The observed statistical and mathematical study of human population, concerned with the size, distribution, and composition of such populations. The main components of this study are fertility, mortality and migration.

Density gradient: The change in density in an urban area from the centre to the periphery.

Dependency Ratio: The ratio between the number of people in a population between the age of 15 and 60 in India and 15 and 65 in developed countries and the dependent population: children (0-14) and elderly people (60/65 and over).

Depopulation: The decline, in absolute terms, of the total population of an area, more often brought about by out-migration than by fall in fertility or excessive mortality.

Determinism: The view that human actions are stimulated and governed by some outside agency (environment or the economy). Individuals have no choice in regulating their actions, which may be predicted from the external stimuli which triggered them. This view is recently rejected.

Developed Country: A country at a relatively advanced stage of development; sometimes called a more developed country.

Developing Country: A country at a relatively early stage of development; sometimes called the less developed country.

Development: The use of resources to improve the standard of living of a nation; the means by which a traditional, illiterate, low technology society is changed into a modern high-technology society, with a corresponding increase in income.

Development Indicators: Development indicators as used by the World Bank include details of birth and death rates, fertility, life expectancy, health, urbanization, industrialization, production, consumption, investment, capital, income, education, energy, consumption, and trade.

Development Stages of Growth: A theory of growth which suggests that all economic development passes through the same sequence. Initially, there is subsistence economy, but with the better transport, regional specialization develops and village industries emerge. Agriculture becomes more intensive and industrialization takes place. Over time, processing gives way to service industries (see Rostow's Model).

Diffusion: The widespread dispersal of an innovation from a centre or centres. This innovation may be anything from an epidemic disease to a political belief, or even the wearing of reversed baseball caps. Tortsen Hagerstrand's model of diffusion (1968) implies the existence of a mean information field which regulates the flow of information around a regional system. The flows are moderated by barriers which can obstruct the evolution of information into innovation, and thus mould the diffusion wave; the ripple of innovation which spreads from one location to another.

Diminishing Return, Law of: The principle that further inputs into a system produce ever lower increase in output. Any extra input will not produce an equal or worthwhile return. Thus, while early application of fertilizers may increase yields, further application will not see a corresponding rise in output, and even further application may actually damage the crop, an excessive fertilizer can burn plant tissue.

Dispersed Settlement: In comparison to nucleated settlement, a settlement pattern characterized by scattered, isolated dwellings. In a dispersed settlement most of the people live in scattered farmhouses and cottages, rather than in settlements. Dispersed settlement is characterized by the wide spacing of individual homestead. This lower-density pattern is a characteristic of the lower Himalayas, Wester Ghats, hilly tracts of North East India, the khadar lands, and eastern slope of Aravallis. Highlands, poor soils, and ubiquitous water supply help to create dispersed settlements as do cultural factors.

Distance Decay: The lessening in force of a phenomenon or interaction with increasing distance from the location of maximum intensity; the inverse distance effect. For example, the intensity of land use declines with distance from the market.

Domestication: The bringing under human control of a plant or animal.

Dormitory Town: A settlement made up largely of daily commuters who are employed elsewhere in a larger centre. These commuters have displaced the original residents or live in new housing at the edge of the town or village.

Downtown: A term used in USA to denote the heart of the city, the CBD.

Dry Farming: Farming without irrigation, using techniques which conserve water for the crop. Strategies include mulching, frequent ploughing and fallowing, working the soil to a fine tilth, and frequent hoeing and weeding.

Dry Point Settlement: In a marshy, damp or frequent flooded area, raised sites such as low mounds or gravel terraces attract settlement.

Dual Economy: An economy comprising two very different systems, and found in many developing countries where an advanced economy co-exists with a traditional economy and the two have little contact with each other. For example, in developing countries one may find the subsistence agriculture with plantation.

Dynametropolis: A metropolis which exhibits continuous growth.

Ecological Fallacy: The danger of making an analysis at one level apply at other levels, for example, of interfering individual characteristics from group characteristics. The most famous expose of this type of reasoning was made by W.S. Robinson (1950), who showed that there was a correlation between the number of black people in American states and literacy levels. He then showed that the assumption leading from this that black people were inclined to illiteracy—was by no means certain.

Ecological Perspective: An emphasis on the interrelationships between humans and their environment.

Economic Geography: The analysis of the spatial distribution of the transportation and consumption of resources, goods and services, and their effects on the landscape.

Economic Growth: The growth in wealth of nation, as measured by an increase in gross national product or in national income.

Economic-man: A theoretical being who has perfect knowledge of an economy and has the ability to act in his own interest to maximize profits. The idea of economic-man has proved useful too in neoclassical economics but other writers suggest that the concept of the satisficer is more realistic.

Economic Rent: Economic rent is synonymous with profit, since built into the concept is the notion of opportunity cost; the real cost of choosing one alternative goods or service in terms of sacrifice of the next best alternative.

Ecosphere: the ecosystem that includes the entire earth.

Ecosystem: The complex relationships that function within an environmental setting.

Ecumene: The portion of the earth's surface occupied by permanent human settlement.

Ecumenopolis: A term for the futuristic city that will eventually cover the ecumene as a continuous system forming a universal settlement.

Emigrant: A person who leaves a homeland to settle in another country.

Entrepot: A centre to which goods in transit are brought for temporary storage and re-export. The term is also used as an adjective, as an entrepot port (e.g. Singapore, Hong-Kong, Rotterdam).

Entropy: A measure of the disorder within a system.

Environmental Perception: The way a person or group becomes aware of and comprehend the world.

Ethnic Group: A group of people united by a common race and ancestor.

Exceptionalism: The study of the unique and particular; the idiographic. In the 19th century it was argued that geography was exceptionalist, since it was concerned with the differences between regions, and in the 1960s regional geography was attacked for its perceived idiographic nature. Model building became the paradigm. Recent thinking has combined the search for 'laws' about spatial patterns with an understanding of the importance of the regional context (see contextual theory)

Extended Family: A family unit which consists of relatives by blood and by marriage as well as two parents and their children.

Extensive Agriculture: Farming with low inputs of capital and labour and generally with low yields per acre. It is associated with regions of cheap available land where high revenues are important. Certain forms of commercial agriculture may be extensive, such as cattle farming in the Australian ranches.

Favela: Shantitown on the outskirts or within an urban area in Brazil.

Factory Farming: A system of livestock farming in which animals are kept indoors throughout the greater part of their lives in conditions of very restricted mobility. Pigs, laying hens, broiler chickens, and veal calves are the animals most often kept under these conditions. Factory farming leads to a standardized product raised (see agribusiness).

Feminist Geography: A geography which emphasizes the oppression of women and the gender inequality between men and women.

Fertility: The level of child bearing in an individual, but more often in a society or nation. General fertility rate, or fertility ratio, which shows the number of births in a year per 1,000 women of reproductive age, generally given as being between 15 and 45 years.

Fish Farming: The rearing of fish in man-made pools, tanks or fields.

Footloose Industry: An industry whose location is not influenced strongly by access either of raw materials or market, and which can therefore, operate within a very wide range of locations. Any form of 'direct' business, operated almost entirely through telephone and fax lines, would be an example.

Formal Imperialism: The extension of control by a state over other territories and peoples through political and military power.

Formal Sector: All those type of employments which offer regular wages and hours, which carry them employment rights, and on which income tax is paid.

Frontier: That part of the country which lies on the limit of the settled area. It differs from the boundary because the term frontier indicates outward expansion into an area previously unsettled by a particular state. Some frontiers have occurred where two nations advance from different directions, leading to boundary disputes.

Frontier Region: A sparsely populated, little developed, peripheral region, often with hostile physical environment, such as Sahara desert, Arctic Siberia. These are the resource frontier regions.

Functional Classification of Towns: A categorization of cities according to the functions they discharge: administration, defense, industry, culture, commercial, provisions of goods and services. In reality, any one city may fulfill a number of functions.

Functional Region: A type of region characterized by its function such as a city-region or drainage basin.

Gap-town: A town located in a pass to an upland area which benefits from being a focus of routes.

GDP (Gross Domestic Product): The total value of the production of goods and services in a nation measured over a year.

Gross National Product: The total value of goods and services produced (and recorded by monetary transactions) in a country during a year.

Generative City: A city that acts as a centre of change and stimulates economic growth.

Geography: The discipline has many interpretations, which might best be understood if they are taken chronologically. Whereas the Greeks were concerned with what we now call 'chorography' or areal differentiation' throughout most of the human history, geography was concerned with exploration, in order to increase a knowledge of the earth.

In the 1950s a new practice developed. F. Schaefer (1953) initiated a move to seek 'laws' which would explain geographical phenomena, particularly within the field of human geography, initiating the shift to logical positivism and the concept of geography as a science of spatial distribution. This movement to locational analysis was seen as a revolution in geography, and was accompanied by extensive quantification. In time, positivism came under fire, for it was felt that this approach redified human beings. In consequence, there has been an exploration of the role of perception, at the heart of cultural geography, and of the deep structure, most notably capitalism, which underlie human actions. This is the field of structuration, proposed by Giddens, but modified by geographers, who stressed that the results of the reactions between social structure and human agency depend upon location. According to some experts, geography is the study of the landscape: a view that goes far uniting the many concerns of the geographers.

Geopolitics: The view that location and the physical environment are the important factors in the global power structure; the state may be seen as a realm in space. Early proponents of the study were Halford Mackinder, and Friedrich Ratzel, and the 1970 domino theory may be seen as a branch of geopolitics.

Geopolitik: A view of geopolitics developed in Germany in the 1920s. Individuals are subordinate to the state which must expand with population growth, claiming more territory – lebensraum – to fulfill its destiny. These ideas are not synonymous with national socialism, but were used as a quasi-science by the Nazis to justify their territorial demands.

Geosophy: The study of geographical knowledge. One aspect may be the development of geography as a branch of knowledge; the other relates to human beings as a whole. Thus all human beings have some geographical knowledge, although the standpoint of this knowledge may be different from that of the professional geographer, and, even, though may be false. It may be acted on in any case. Alternatively, the mind may erect barriers to certain places, even though there are now no unexplored parts of the earth (see mental map).

Ghetto: A part of city, not necessarily a slum area, occupied by a minority group. The term was first used for the enforced concentration of Jews into specific residential areas in European cities from the Middle Ages, but has now spread to include other ethnic groups in unofficial ghettos, especially black minorities in USA. Lifestyle within the ghetto differ distinctly from those of the 'host' population and the prejudices of the host confine the sub-group to particular location. Although ghettos are characterized by social disadvantage, most ghettos display a spread of socio-economic groups and better-off may move to the affluence of the 'gilded ghetto'.

Ghost Town: A term used to describe a once flourishing settlement, now completely abandoned or inhabited by few people. Many former mining settlements, particularly those associated with the great gold rushes of 19th century North America, South America and Australia.

Gravity Model: A model of the interaction between two population centres based on Newton's Law of Universal Gravitation: two bodies in the universe attract each other in proportion to the product of their masses and inversely as the square distance between them.

Habitat: The environment of an organism or group of organism.

Hamlet: A small clustered settlement, sometimes consisting only of several farmsteads, but in other instances somewhat larger and with a few services. This is the smallest in the hierarchy of settlements, less than a village. In India, *Purva,* and *Majra* are the examples of hamlet.

Heartland: A term suggested by Halford Mackinder (1904) to indicate the interior of Eurasia. Mackinder maintained that whoever controlled the heartland would eventually control the world as political units become larger and larger. He did not live to see the rise of nationalism in the late twentieth century.

Heat-Island: The phenomenon whereby temperatures within a city or large town often significantly higher than those of surrounding rural areas. There are various factors involved, such as (i) the release of heat by domestic and industrial fuel consumption, (ii) the high capacity for heat absorption of many urban surfaces such as roads, walls, roofs and bricks, and (iii) the reduced consumption of heat by process.

Heavy Industry: Manufacturing industry which often needs large quantities of bulky materials. These are usually transported by water or rail as in iron-ore smelting or ship-building. These industries have a high material index. Productivity per worker is generally low, and heavy industries are often dirty and noisy.

Hinterland: The hinterland is the area serving and being served by a settlement. The term was originally applied to ports, and one port may share part of its hinterland with another, but has now been extended to refer to the sphere of influence of settlement. Christaller's Central Place Theory was based on nested, hexagonal hinterland.

Historical Geography: The study of past human geographies; of past landscapes.

Horticulture: Originally garden cultivation, this now refers to the intensive production of fruit, vegetables, and ornamental plants.

Human Geography: A generalized term for those areas of geography not dealing exclusively with the physical landscape or with technical matters such as cartography and remote sensing. It is concerned with the relationships between man's activities and physical environment.

Humanistic Geography: A view of human geography centred on human perception, capability, creativity, experience, and values.

Iconography: In geography, the study of the way in which images of the landscape reveal symbolic meaning. A human landscape is not only shaped by a society and its culture, it also helps to shape the society.

Idiographic: Concerned with establishing the uniqueness of a phenomenon: for example, an individual, a place, or a region

Industrial Complex: A large concentration of manufacturing industry in a relatively small area.

Industrial Geography: The study of the spatial arrangement of manufacturing industry.

Industrial Location Theory: Theories of the forces leading to the location of industrial activity. One choice might be the least-cost location. Another is the locational interdependence approach which stresses the influence of other enterprises especially under conditions of imperfect competition. A third is the profit maximization approach, although it is by no means certain that firms do maximize profits.

Industrialization: The process by which manufacturing industries develop from within a predominantly agrarian society. Characteristic features of industrialization include the application of scientific methods to solving problems, mechanization and factory system, the division of labour, the growth of the money economy, and the increased mobility of labour force—both geographically and socially.

Informal Imperialism: The extension of economic and political influence by a state over other territories and peoples, but without formal rule, also called neo-imperialism or neo-colonialism.

Inner City: That part of the built up area close to the city centre, considerable areas of which are characterized by old and obsolescent housing, by multiple family occupation of dwelling, by poor commercial and social services, and by the presence of industrial and commercial enterprises which contribute to the urban blight typical to such area.

Intensive Agriculture: Agriculture with high level of inputs—capital and labour—and high yields.

Isochrone: A line joining places of an equal journey time to the same location.

Isodapane: A line joining up places of equal total transport costs for industrial production and delivery between the points where the raw materials are located and the markets (see Weber's theory).

Isotim: A line drawn about a source of raw materials or a market where transport costs are equal.

K-Value: Used in the Central Place Theory to denote the number of central places of a given order dependent on a central place of the next highest order in the central place hierarchy.

Kibbutz: A type of agricultural system, first established by Jewish settlers in Palestine and then supported by the Israeli state. In this system all the agricultural land and resources are communally owned and all farming strategies are decided jointly. The domestic arrangements, including child care, are shared by all. Later Kibbutzim began to use waged labour. Even today, nearly 3 per cent of the population of Israel have been part of the movement.

Land Use Classification: The analysis of land according to its use: agricultural, industrial, recreational, and residential.

Land Use Survey: In U. K. the first land use survey was carried out between 1931 and 1939. Showing seven categories of land, the original maps are kept at the London School of Economics where they may still be studied. The maps were also published at a scale of six inches to mile.

The second land use survey was carried out between 1961 and 1969 using base maps of the same scale but having 256 categories. These maps can be consulted at the Land Use Research Unit at King's College, London.

Land-locked State: A nation with no access to sea. There are 40 landlocked states ranging in size from the tiny Vatican City to Mongolia.

Landscape: An area, the appearance of an area, or the gathering of objects which produce that appearance. Carl Sauer first used the term in geography in 1925, stressing the concept of the landscape as the expression of interaction between humans and their environment (see iconography).

Lebensraum: Literally, 'living space', the room needed for a nation's expansion. The concept was used by the Germans to justify their territorial growth. The term was introduced after 1870 and became a central concept in the propaganda literature of the Nazis.

Life Expectancy: The average number of years which an individual can expect to live in a given society, normally derived from a national life table.

Life Space: The limited time and space which an individual has in which to pursue a necessarily limited range of opportunities. Life space is the interaction of individual with her or his behaviour setting.

Light Industry: The manufacture of relatively small articles (toasters as opposed to girders), using small amount of raw materials. Such industries are more footloose than heavy industries.

Linear Village: An elongated ribbon of settlement usually formed along a routeway such as a road or canal.

Locational Triangle: A model devised by Weber to establish a least-cost location. In this model, S_1 and S_2 are the sources of the required raw materials, located at two corners of the triangle, and M is the market, at the third corner. Assuming equal transport costs in all directions, the least cost location, P, located somewhere within the triangle, is derived from the amounts of 'pull' from each of the three corners.

Losch Model: A model of central place developed by A. Losch (1954) which is less narrow than that of Christaller, in that it treats the range, threshold, and hexagonal hinterland of each function separately. The resulting pattern of central places is much more complex than Christaller, and yields a continuous, rather than stepped, distribution of population sizes.

Low Order Goods and Services: Goods and services with low range and with low threshold of population like daily newspaper, bread, and hairdressing. The goods are often convenience goods (see Central Place Theory).

Lumbering: The extraction of timber from forests.

Market Economy: An economy in which the major parts of production, distribution, and exchange are carried out by private individuals or companies rather than by the government. A market economy is characterized of capitalism.

Market Principle: One of the three principles underlying Christaller's central place theory (see central place theory).

Marxist Geography: Marxist geography attempts to explain the world and also to change it. Marxism sees human beings gradually transforming themselves from stage to stage until they reach social perfection, and this transformation is seen as an aim towards which societies should be moving. This change is to brought about by 'dialectical' processes-conflict between opposing forces—bringing forth a new synthesis which again is contradicted, and so on. The forces shaping society are seen as entities, which include capital, labour, capitalism and other modes of production, the state, class, society, and the market.

Marxist geography highlights the dialectical relationships between social processes and the natural environment and spatial relationships. It is concerned with the modes of production which underlie the superstructure of society. It sees spatial and environmental problems, such as the destruction of habitats or even development, originating deep within the social formations of capitalism. It aimed at changing the fundamental operations of social processes by changing the workings of production. Marxist geography is the study of the inherent contradictions of capitalism as they appear in the landscape and as they relate to each other.

Megalopolis: A Greek word applied by Gottman (1964) to the almost continuous extent of urban settlements stretching from Boston to

Washington D.C. The term is used for the merger of several million cities.

Mentifacts: The world view, values, and beliefs, such as those expressed in religion and art, that hold society together.

Metropolis: A term applied loosely to any large city (metropolitan area), but specifically to that city in a country which is the seat of government, or of commercial activity.

Mental Map: A map of environment within the mind of an individual which reflects the individual's perception of, and preferences for, different places and is the result of the way in which an individual acquires, classifies, stores, retrieve, and decodes information about locations and places.

Migration: The movement of people from one place to another. The terms in-migration and out-migration are used for internal migration, where no national boundaries are crossed, and the simplest classification separates this from international migration.

Migration Chain: The process by which early movers encourage others to move by providing information and communal assistance.

Milieu: The sphere in which each individual lives and which he or she is affected by. This sphere includes the tangible objects and people, the social and cultural phenomena, and the images which influence human behaviour (see phenomenal environment).

Model: A representation of some phenomenon of the real world made in order to facilitate an understanding of its working. A model is a simplified and generalized version of real events, from which the incidental details, or 'noise' has been removed. An iconic model represents reality on a smaller scale, an analogue model shows reality in maps and diagrams, and a symbolic model uses mathematical expression to portray reality. Probabilistic models take into account the fact that human behaviour cannot be predicted with absolute certainty, while simulation models use mathematical laws of probability to stimulate the consequences of human behaviour. Finally is the economic model, the variables are defined in cash terms.

In geography, models were at their most popular in 1960s, since that time, few new models have been created, and many classic models, such as those of von Thunen or Hagerstrand have been reworked.

Modern Society: A society whose members have high per capita income, high standard of living, high longevity, low birth rate, low death rate, and access to high quality educational and medical services.

Monotheism: The doctrine or belief of existence of one God only.

Multiple Nuclei Model: A model of towns growth advanced by C.D. Harris and E.L. Ullman, based on the fact that many towns and all large cities grow about many nuclei rather than around a simple CBD. Some of these nuclei are pre-existing settlements, others arise from urbanization and external economies.

Nation State: A political unit with defined boundaries that encompass a people who share common cultural traits and a sense of identity.

Natural Environment: The physical elements of earth, such as landforms, climate, vegetation and soils.

Necropolis: Literally the city of dead but the term is sometimes applied to cemeteries.

Nodal Region: An area connected to a common place (centre or node) by lines of communication.

Normative Theory: A theory of what ought to be, rather than what actually occurs.

Nucleated Settlement: A term used for the compact rural settlements

Nuptiality: The frequency of marriage within a population, usually expressed as a marriage rate. Nuptiality varies with the age of first marriage common in any society and with the age structure of the population; obviously it will be lower in an ageing population than in a very young one.

Occupancy Rate: The number of people dwelling in a house per habitable room (kitchen and bathrooms are not counted). A rate of one person per room is taken as acceptable; more than one person per room represents overcrowding.

Occupational Mobility: The ability of the individual to change jobs after the acquisition of a new skill.

Old Age Index: The number of people over retirement age in the population, as percentage of the total adult population.

Open-field: Large farm areas that are divided into unenclosed strips for the use of individual farmers.

Optimum City Size: Some urban geographers argue that there are benefits to be gained as a city grows—in terms, for example, of cheaper transport and the more economical provisions of services—up to and until the optimum size is reached. A figure of around 5,000,000 has been suggested as an optimum size. Beyond this point diseconomies operate: pollution and congestion set in, and the city is a less efficient and less pleasant place to live in.

Optimum Population: A theoretically perfect situation where the population of an area can develop its resources to the greatest extent, and achieve maximum output while enjoying the highest possible standards of living.

Overpopulation: 1. A vague but popular term that usually relates the number of people to the resources of an area. 2. Too great a population for a given area to support. This may be because population growth has outstripped resources or because of the exhaustion of resources. The synonym of such a situation include high unemployment, low incomes, low standard of living, and possibly malnutrition and famine.

Over-urbanization: Refers to the fast growth of urban places in the developing countries in which the social amenities are less leading to numerous problems for the urban dwellers.

Parasite City: A city, that dominates the economic, social, and political activities of an area to the extent that it has a dampening effect on economic growth.

Paradigm: The prevailing pattern of thought in a discipline or part of a discipline. The paradigm provides rules about the type of problem which faces investigators and the way they should go about solving them. For geographers, for example, the paradigm would be referred to when questions such as 'what is geography?'; what are the legitimate areas of investigation for geographers?; How should geographers go about their investigation?; are asked. Perhaps the most powerful paradigm for western geographers has been the 'scientific method'.

Perceived Environment: The portion of the total environment that is consciously sensed and which serves as a basis for decision making.

Periphery: When applied to a comparison of national economies, the term has historically referred to all countries other than those identified as belonging to the 'core.'

Periodic Market: A trading market held on one or more days of each week and on the same days of the week.

Physiological Density: The number of people per unit area of land suitable for agriculture.

Planning: As practiced by local or national government, the direction of development.

Plantation: An agricultural system, generally a mono-culture, for the production of tropical and sub-tropical crops, especially, tea, coffee, banana, cotton, palm-oil, rubber, sisal, and sugarcane.

Political Geography: One of the three divisions of human geography, the others are economic and social geography. This is the geographical analysis of political studies. It is concerned, among other things, with the spatial expression of political ideals, the consequences of decision-making by a political entity, and with those geographical factors which influence political activities and problems.

Popular Culture: Those aspects of culture which are widely promoted and maintained through mass media.

Population Density: The ratio of a population to a given unit of area. Crude density is simply the number of people living per unit area and can be very misleading. India and Japan almost similar crude densities—but very different living standards. Accordingly, densities may be plotted using different criteria.

Population Geography: The study of human population, their composition, growth, distribution, and migration movements with an emphasis on the last two. It is concerned with the study of demographic processes which affect the environment, but differs from demography in that it is concerned with the spatial expression of such processes (see demography).

Population Persistence: The degree to which a population continues to reside at one population place over period of time. The persistence is usually expressed as the percentage of remaining population to the total population, and then adjusted for births and deaths.

Population Potential: The accessibility people from a given point: that is, a measurement of how near people are to a point. The population potential at one place is the sum of the ratios of population at all other points to the distances from the place in question to those points.

Population Problems: Between about 9000 years before present until about 1800 AD it is estimated that world population grew from 5 million to 800 million, with an average growth rate of 0.1 per cent per annum. Between 1800 and 2000, world population grew to over 6 billion. World population is set to double every 35–45 years. If this rate of growth is sustained, there would be standing room only in the next century. The problem is more acute in the Third World because a large proportion of the population is very young. The eventual reproductive capacity of such regions could result in an expansion of population which would end in Mathusian disasters.

Population Projection: The prediction of future populations based on the present age-sex structure, and with the present rates of fertility, mortality and migration.

Port: Defined as a place where ships may anchor to load and unload cargo, a port may be classified by its function. Terminal ports are final destination of cargo-carrying ships. Some specialized ports handle predominantly one particular type of traffic and include container ports, ferry ports, fishing ports, and naval ports. Ports of call lie between terminal ports and may handle part of cargo of a vessel.

Possibilism: A view of the environment as a range of opportunities from which the individual may choose. This choice is based on the individual's needs and norms. It grants that the range of choice may be limited by the environment, but allows choices to be made, rather than thinking on deterministic lines.

Post-Industrial City: A city exhibiting the characteristics of a post-industrial society. Service industries dominate with a strongly developed quaternary sector and footloose industries abound, often on pleasant open space at the edge of the city. Post industrial cities are also characterized by large areas of office blocks and buildings for local government administration. These cities often exhibit marked inequality of income distribution because of the contrasts between those who are appropriately skilled—professionals, managers, administrators, and those in high technology service industries—and the poorly paid service workers who look after their needs, together with the unemployed. The former can afford high house prices, and, in fact, contribute to them; the latter cannot.

Post-Industrial Society: A post-industrial society has five primary characteristics: the dominance of service, rather than manufacturing industry, the pre-eminence of professional and technical classes, the central place of theoretical knowledge as a source of innovations, the dominating influence of technology, and levels of urbanization higher than anywhere else in the world. At present the development of a post-industrial society is linked only with very advanced economies. If it exists at all.

Postmodernism: 1. An architectural style which is composite of past styles, characterized by a variety of colours, stylistic details from many periods, and what is claimed to be return to a vernacular style.

2. A philosophical stance which claims that it is impossible to make grand statements—meta-narratives—about the structures of society or about historic causation because everything we perceive, express, and interpret is influenced by our gender, class, and culture. No one interpretation is superior to another. It has brought to geographers a recognition that space, place, and scale are social constructs, not external givens. Of particular interest to geographers is the way that time and space have been *compressed* by modern transport and communication systems, especially by jumbo jets, internets and mobiles: nowhere is very far away any more. As a result, cultures are transformed.

Some geographers claim that post modernism challenges the dominance of time and history in social theories and instead stresses the significance of geography and spatiality.

Postmodernism has also been linked with late capitalism, and with post-Fordism, but these links are, of course, meta-narratives; the very interpretations which postmodernism reject.

The postmodern tradition also stress and, indeed, champions differences, and this is a strand which has been welcomed by feminist geographers, who would claim that geography has been speaking in an authoritarian, masculine voice for too long.

Primary Industry (Primary Activity): Economic activity such as agriculture, forestry, fishing, mining and quarrying, concerned with the extraction of natural resources. All such activities fall into the primary sector.

Primary Urbanization: Urbanization which results from forces arising within the country as a spatial expression of a region's culture, as in the wave of urbanization in U.K. which resulted from the industrial revolution.

Primate City: The concept of primate city was developed by Mark Jefferson. The largest city within a nation which dominates the country not solely in size—being more than twice as large as the second city, but also in terms of influence. The development of primacy is not fully understood but some researchers have suggested that the importance of the primate city tends to decline as the economy grows and that, therefore, primate city may be encouraged by colonialism, as it occurs often at the major port. Capital cities of the past empires tend to be over-large.

Principle of Least Effort: The thinking behind movement minimization procedures.

Probabalism: Possibilism sees individuals or groups making choices within the scope of environment. Probabalism suggests that some choices are a good deal more likely than others.

Psychic Income: The enjoyment, which cannot be measured in financial terms, that people drive from location in a particular place. The example usually cited is an entrepreneur who chooses his factory site near a favourite golf-course in order to receive enjoyment as well as gaining profit from his industry. Although no manufacturer will readily operate at a loss, many industrialists will locate away from the optimum point in order to benefit from psychic income.

Pronatalism: A policy that encourages births.

Public Goods: Goods freely available, either naturally, like air or from the state, like education in most developed countries. Pure public goods, like defense, are provided for all. However, pure public goods may be distributed impurely, as when an area with high crime rate has a higher level of policing. Impure public goods, like, libraries, are provided at particular locations, so they are more accessible to some than others.

Public Sector: That part of national economy owned and controlled by the state. It includes nationalized industries, national and local government services, and public corporations.

Quality of Life: The degree of well-being felt by an individual about his or her lifestyle. Preference vary, but most assessments of the quality of life consider amenity, together with social benefits such as health, welfare and education, and economic aspects such as income and taxation.

Quaternary Activity: Economic activities which specialize in information technology, solicitors, accountants, management, banking, education, research and media.

Radical Geography: A description of the geographical writing which began to appear in the 1970s, and which was based on Marxist geographical analysis. Major topics of concern are health, hunger, poverty, malnutrition, unemployment, discrimination, and crime, and the aim is not simply to analyse what is happening, but also to advocate change (Marxist Geography).

Raison d'eter: The 'reason for being' or justification for the existence of a nation.

Ranching: Large scale and extensive cattle rearing, best developed on temperature grasslands, like the pampas of Argentina and Uruguay.

Rank Size Rule: The rule formulated by Zipf (1949) states that if all the settlements of a country or region are ranked according to their population size, the population (Pn) of the nth settlement in the ranking will be $1/n^{th}$ that of the largest settlement. In other words the second largest settlement will have a population half that of the largest. When graphed, this simple rule produces an inverted J-shaped curve between rank and population.

Ranking of Towns: Many attempts have been made to devise a hierarchy of towns based on the belief that one could be found. One approach is to rate cities by the number of retail outlets, but this is difficult since 'shop' might be a large jewellers or corner store. One refinement of this is to give different weightings to different functions and then calculate a total for each settlement. Multivariate analysis has also been used, but the data are not always available and are difficult to collect. Other methods emphasize the interaction between a settlement and its field and rank settlement by the size of their spheres of influence; yet this method, too, has its difficulties. Graph theory has also been used to rank town.

Refugee: A refugee is defined by the UN convention relating to the status of Refugees 1951, and the UN protocol 1976, as a person who, owing to a well-founded fear of persecution for reasons of race, religion, nationality, membership of a particular social group or political opinion, is outside his or her country of nationality and who is unable or unwilling to return.

Region: An area that is differentiated from other areas according to specified criteria.

Regional Consciousness: A feeling held by the inhabitants of an area that 'their territory' is distinctive from other places.

Regional Geography: The study in geography of regions and of their distinctive qualities. A precondition of this study is the recognition of a region, and the delimitation of its boundaries. One approach has been to identify 'natural' regions while another was to establish economic regions based on agriculture and/or industry. Often there was an intimation of a link between the two types of region. Once the keystone of geography, the status of regional geography has been in decline since the 1950s, but, areal differentiation, which may be seen as a branch of regional geography, has grown in importance in recent years.

Regional Inequality: A disparity between the standards of living within a nation. It is difficult to quantify the prosperity or poverty of a region, but there are two basic indicators. The first is unemployment, the second indicator is per capita income. Other factors indicating disparity include the type of industry and its growth or decline, numbers of young people in further education, housing standards, health and quality of environment. Some would assert that economic development brings about regional inequality.

Relative Location: A position on the earth that is indicated in terms of other phenomena. For example, Delhi is located on the banks of Yamuna.

Religion: The recognition of God as object of worship, love and obedience.

Religion Geography of: The study of the spatial distribution of organized religions and their territorial development over time.

Replacement Level: The population growth rate at which the size of one generation is the same as that of the previous generation.

Residential Segregation: The tendency of an ethnic group to cluster in a particular residential areas.

Resource: Some component which fulfills people's need. Resources may be man-made—labour, skills, finance, capital, and technology—or natural ores, water, soil, natural vegetation, or even climate. The perception of a resource may vary through time; coal was of little significance to Neolithic man, while flint was of great importance. Such resources depend on relevant technology. Other resources like landscapes and ecosystems, may be permanently valued whatever the technology. Resource can be renewable (*flow resource*) or non-renewable (*stock resource*).

Ribbon Development: A built up area along a main road running outwards from the city centre.

Ring City: A city created along the outer side of a circular routeway while there is open space at the centre of the ring.

Room Density: Also known as the occupancy rate, this is the number of persons in a house per unit habitable room (kitchen and bathrooms are excluded). It is a widely used index since it is an easily calculated and sensitive indication of housing provision, where any density of over one person per room indicates overcrowding.

Rural Community: A group of people living in the same rural place who have common ties with each other and with their location. It is the smallest social group which caters for the daily social life of the inhabitants.

Rural Depopulation: The decrease in population of rural areas, whether by migration or falling birth rates as young people move away. It has been argued that the mechanization of agriculture leads to rural unemployment, and hence depopulation. Some writers argue that the move to the cities took place before mechanization and that machines were, therefore, needed to supplement the dwindling workforce.

Rural Planning: The management of rural areas with regard to some objective such as the maintenance and improvement of rural living standard, or easy access to jobs and social, economic and welfare services. The most usual aspects of rural planning are maintenance of rural landscape, developing the recreational use of countryside, and the planning of population, settlements, and amenities. Also studied are rural problems such as lack of access to amenities and services, rural unemployment, substandard housing, provisions of electricity, roads and drinking water.

Rural Urban Continuum: A typological approach to the differentiation of rural from urban communities. In its original form, the typology merely distinguished the extremes, but more recent interpretations have emphasized the gradual transformation which occurs from one pole to the other.

Rural Urban Fringe: The transition zone between the city and its suburbs, and the countryside. Certain type of land use are the characteristic of this zone: greenbelt, garden centres, country parks, riding-stables, golf-courses, sewage works and airports are common, and these are neither truly urban nor truly rural uses. They do, however, give an urban air to the countryside, an air which can be cited as an argument for further development—since the zone is not really 'countryside', it need to be preserved.

Satellite Town: A town designed to house the overspill population of a major city, but located well beyond the limits of that city, and operating as a discrete, self-contained entity. Most of the early new towns, like, Faridabad, Ballabgarh, Gurgaon, Ghaziabad, Noida, Dadri, Hapur, Moradnagar, Modinagar, Panipat, Narela, etc. are the satellite towns of Delhi.

Satisficer: A decision-maker whose aim is to make a choice which is acceptable rather than optimal, possibly because it is impossible to locate on the maximum-profit site (because of prior occupancy, planning restrictions). Logically impossible to pin point that site, or because the decision-maker lacks the necessary knowledge and/or ability.

It may be that maximization of profit is not the only goal. The decision-maker may be inclined to maximize psychic income, that is the satisfaction which may come to managers from causes other than financial ones, such as amenity. Other satisficers are loath to undertake the somewhat hazardous process of seeking the optimum because they are averse to uncertainty. Others may be guided by moral concerns.

Secondary Activity: Economic activities that produce goods by remaking materials such as through manufacturing, processing, industries and construction.

Secondary Industry (Secondary Sector): The creation of finished products from raw materials, that is, manufacturing.

Secondary Urbanization: Urbanization which results from forces which are external to a country, such as foreign, colonial settlement. For example, pre-colonial Australia had no urban settlement.

Sector Principle: The principle on which claims to territory in the Arctic and Antarctic are made. The territory is shared out in the form of arbitrary sectors, each one having an apex at the poles including outer area bounded by the coast. The principle has worked well in establishing control of the Arctic ice, but there are disputed claims over the land mass of the Antarctic which may possibly yield mineral resources.

Sector Theory: This theory was advanced by Homer Hoyt (1939) as an alternative to Burgess' concentric model, and was based on residential rent pattern in the USA. The view that housing areas in city develop in sectors along the lines of communication, from the CBD outwards. High quality areas are along the roads and also reflect the incidence of higher ground. Industrial sectors develop around railways and canals, away from high quality housing. Thus a high status residential area will spread out along the lines of the sector by the addition of new belts of housing beyond the outer arc of the city. Once contrasts in land use have developed in a sector near the city, the contrast will be perpetuated as the city grows.

Sedentary: Fixed, not moving, as a sedentary agriculture where the farmer and the fields are permanently settled. Compare with shifting cultivation.

Segregation: The separation of the subgroups of a large population, particularly into distinct residential areas. This segregation may be based on grounds of income, race, ethnicity, religion, or language.

Selective Migration: This may be spontaneous as when a particular age-sex group may constitute most of the migration. Most countries now practice selective migration by demanding qualifications, such as skills, youth and health for immigrants.

Sense of Place: Either the intrinsic character of a place, or the meaning people give to it, but, more often, a mixture of both.

Sequent Occupance: This concept, developed by D. Whittlesey (1929) owes much to human ecology. Each stage is seen as being established by its predecessor, although the sequence will almost certainly be interrupted by outside forces.

This concept is more complex, because it envisages interruptions and transformations, so there is no suggestion of an ideal sequence, or succession, as in ecology, instead, the idea is that each stage contains within it the seeds of the next.

Settlement: In human geography, any form of human habitation from a single house to largest city.

Settlement Hierarchy: A division of settlements into ranks, usually according to the size of population. In central place theory, large, high-order settlements provide high-order goods and services. In theory, the greater the rank of a settlement the more goods and services it provides.

Settlement Pattern: The nature of the distribution of settlements. Some settlement patterns may be seen as a reflection of cultural traditions. For example, the isolated farmstead is typical in the Greater Himalayas, and the Thar Desert, while the nucleated settlements are found in the Northern Plains of India.

Share Cropping: A type of farming whereby the tenant pays his rent to the landowner in produce rather than in cash. The landlord often provides seeds, stock, and in return for fixed proportion of the output.

Share-cropping usually shows low yields in comparison with owner-occupied farms or cash tenancies since the incentives are less.

Site: The internal characteristics of a place, such as topography, soil types, land use and buildings.

Situation: The position of a place relative to the location of external features; for example, a town may be situated in the centre of a plain and be located midway between two major cities.

Slum: An area of poor housing, often characterized by multi-occupance, inadequate sanitation. Slum population shows a high concentration of drug abusers, alcoholics, criminals and vandals and over-crowding.

Social Anthropology: The study of people in a social context with a strong historical bias. It tends to be concerned with the cultures and societies of the, as yet, non-industrial world.

Social Darwinism: The application of the concept of evolution to the development of human societies over time. It is an idea which emphasizes the struggle of existence of each society, and the survival of the fittest of them. Such ideas have been used to justify naked capitalism, and have been extended to defend power politics, imperialism and war.

Social Geography: Originally this was defined as the study of the spatial patterns of social, as distinct from political and economic factors. The subject now may be subdivided into three categories. The first lies in the spatial expression of capitalism; the city has a focal structure as an expression of class structures which are reflected in its morphology. Another aspect stresses the 'alternative' view of human geography which studies the response of economically disadvantaged rather than the successful. A third category emphasizes welfare geography.

Social Justice: The distribution of the benefits and the hardships in society, together with the way they are allocated. Geographers are particularly concerned with the spatial expression of social justice; where do the advantaged and disadvantaged groups live, why do they live there, and what is the connection between their place of residence and their future advantage or disadvantage. Such questions can be asked on local, regional, national and global scales. Also of interest to geographers is the way in which moral systems vary spatially and whether, with the rise of post-modernism, we can look for universal principles for social justice.

Social Mobility: The process by which individuals move from one position to another in society, with the positions generally considered to have relatively high or low status in the social system.

Social Network: The cluster of relatives, family and neighbours to which an individual or family is connected. Such groupings often share the same values and goals.

Social Physics: A view of human society which seeks analogies from the world of physics to aggregate human behaviour. Perhaps the best-known example of this is the gravity model which sees the attraction of town for the surrounding population being proportional to its population and inversely proportional to the distance away from the town. This is analogous to the gravitational attraction of a physical body in Newtonian physics.

Social Space: The combined use and perception of space by distinct social groups, as opposed to personal space. Social space provides an environmental framework for the behaviour of the group, such as a community, or natural area.

Social Well-being: A state of affairs where the basic needs of the populace are met. This is a society where income levels are high enough to cover basic wants, where there is no poverty, where unemployment is insignificant, where there is easy access to social, medical, and educational services, and where everyone is treated with dignity and consideration. Many attempts have been made to quantify social well-being.

Sociofacts: The institutions, such as family structure and educational system that bind a society together.

Space: The extent of an area, usually expressed in terms of the earth's surface. From this meaning, derives the term spatial; and spatial relationships are at the heart of geography. It is important to distinguish between *absolute space*, which refers to clearly distinct, real and objective space, and *relative space*, which is space as perceived by a person or society and concerns the relationship between events and between aspects of events.

Spatial Interaction: The exchange that occurs between places which are separated from each other.

Spatial Perspective: An emphasis on the locations, patterns or arrangements of phenomena.

Specific Evolution: The process of diversification and development of an individual culture as it adapts to its environment.

Sphere of Influence: Initially a region influence by a colonial power but distinctly colonized; this is an outmoded concept. The term is increasingly used as a synonym for the umland, catchment area, tributary area, or urban field of a city.

Sprawl: Development of new housing sites are relatively of low density and at locations that are not contiguous to the existing built-up area.

Spring-line Villages: A series of villages at the foot of a scarp through which water percolates until it emerges as a spring at the point where the rock type changes to an impermeable layer. Such springs attracted early settlements.

Squatter Settlements: An area of usually unauthorized, makeshift housing, generally at the edge of a Third World city. Local terms include *Barrio* (Spanish—Latin America), *Favela* (Portuguese—Brazil), *Jhuggi-Jhopri* (Delhi), *Jhopad-Patti* (Mumbai), *Bustee* (Kolkata), *Kampong* (South East Asia).

Squatter settlements grow because demand for cheap housing outstrips supply. Houses are made from available cheap materials such as packing cases, metal canes, plywood, and cardboard. Sanitation is grossly inadequate, electricity may not be available, and roads are not metalled. In addition, educational and medical facilities are severely limited.

Stable Population: A population where fertility and mortality are constant. This type of population will show an unvarying age distribution and will grow at a constant rate. Where fertility and mortality are equal, the stable population is stationary.

Stepwise Migration: A type of migration which occurs in a series of movements, for example, from a hamlet (Purva/Majra) to a village, from village to a town, and from a town to a city, i.e. up the hierarchy.

Stochastic Model (Stochastic Process): A stochastic process is one that develops in time according to probability theory. This means that the future behaviour of the process cannot be predicted with any certainty. In other words, it is a chance or random element which has a direct influence on the process or direction of change. Awareness of stochastic process is, therefore, crucial in many aspects of human geography, ranging from evolution of settlement patterns to the spatial diffusion of innovation, from industrial location to residential choice. The use of stochastic models to replicate the spatial organization of human activity has been a recent influential development in human geography.

Street Village: The German *Strabendorf*; a settlement of linear form strung out along a route-way. Most street villages grew up as Dark Age German colonists moved eastwards into the forests of Central Europe, cutting roads as they went.

Systematic Geography: The study of a particular element in geography, such as agriculture, population or settlement, seeking to understand the processes which influence it and the spatial patterns which it causes.

Taboo: A restriction imposed by social custom.

Telemobility: The transfer of information electronically, which reduces the need for movement of people and the transportation of goods.

Teleology: This is, in literal terms, the study of purposes, goals, or ends. In other words, the theory that acts, objects, states of affairs, and so on can only be justified in terms of the ends towards which they are directed, or by the functions they fulfill. 'Rightness' is not intrinsic in an action or process, but is dependent on consequences of the action or process. As a very simple example, the significance of a lioness stalking an antelope only becomes apparent when it is seen as directed to some other purpose, that is, of killing the antelope. In geography, Ritter was a teleologist and Rostows's theory is teleological.

Territorial Production Complex (TPC): A type of large scale industrial complex identified by Soviet planners. It is a part of the of the industrial hierarchy of a centrally planned economy, ranking in size and importance below the national economy and the major industrial unit but above specific industrial centres. Soviet planners used mathematical models to indicate the optimum location for territorial production complexes especially in the industrialization of Siberia. The term refers to planned, rather than spontaneously arising industrial complexes.

Territorial Seas, Territorial Water: The coastal waters together with the sea bed beneath the air space above them, over which a state claims sovereignty.

Threshold Population: The minimum population needed to justify the provision of a certain goods or service. The concept of threshold can be demonstrated in terms of the provision of medical services, which will have an increasing threshold population as the sophistication of the service provided increases; from a single GP, to a group of practice, to a hospital with consultants, and finally a specialist hospital such as the All India Institute of Medical Sciences, Escort Heart Institute, Apollo, Cancer Institute and Batra Hospitals in New Delhi.

Tertiary Activities: Economic activities that produce value through the performance of services.

Three-field System Agriculture: An agricultural practice in which arable land of a community is divided into three open fields with two fields being cultivated and one field, in rotation, left fallow.

Time Space Geography: An approach to geography developed at the University of Lund by Torsten Hagerstrand and his associates.

Time and space provide the room needed for sequences of events and Hagerstrand expressed this as a web model. This is based on four proportions: that space and time are scarce resources which individuals draw on to achieve their aims: that achieving an aim is subject to capability constraints, coupling constraints, and authority constraints; that these constraints interact to demarcate a series of probability boundaries; and that choices are made within these boundaries. A time space prism is a representation of the constraints limiting the time within which the individual can act.

Time geography provides a method of mapping spatial movements through time.

Toponym: The name of a place; a place name.

Topophilia: The feeling of affection which individuals have for particular places, a term introduced by Y.F. Tuan (19861). Places in this sense may vary in scale from a single room to a nation or continent. Topophilia is an important aspect of the symbolic meaning and significance of landscapes (see iconography).

Town: A relatively small urban place. No limiting figures of population or areal extent are agreed upon. Towns may be regarded as central places providing goods and services to their surroundings but without the degree of economic specialization to be found in a city.

Townscape: In urban geography, the objective, visible scene of the urban area or the subjective image of the city. The townscape has three separate, but closely related parts; the street plan or layout, the architectural style, and the land use. Townscape is the urban counterpart of landscape.

Traffic Density: A measure of the average number of vehicles per unit length of roadway over a given period of time. The critical density occurs when the volume of traffic is at the possible capacity on a given roadway.

Transhumance: A seasonal movement of men and animals between different seasonal grazing grounds. Shepherds leave their low altitude places and move to upland in the summer season and vice versa. A farmer practicing transhumance is not a nomad.

Transport Geography: A branch of human geography concentrating on the movement of people and goods, the patterns of such movements, the volume of people and goods carried, the cost of transport, and the role of transport in economic, political, and social development.

Umland: The area served by a city. The umland is known as a sphere of influence, catchment area, tributary area, or urban field.

Untouchable: The lowest social class in India, now dissolved; so named because to touch such person was considered to make a person of a higher caste impure.

Urban Containment: The policy of, limiting urban sprawl by restricting out-of-town development.

Urban Ecology: The application of the principles of ecology to a study of urban environment. (see human ecology).

Urban-field: The area surrounding the city which is influenced by it. The inhabitants of the urban field depend on city for services such as hospitals, higher education, employment, retailing, marketing, and finance and the city is served in its turn by labour, milk, vegetables, fruits and grains.

Urban Geography: The study of site, evolution, morphology, spatial pattern and classification of towns.

Urban Morphology: The form, function, and layout of city, and the study of these features, including their development over time.

Urban Planning: An attempt to manage the city, often in order to avoid, or alleviate, common urban problems such as inner city decay, overcrowding, traffic, pollution, and other forms of congestion.

Urban Renewal: The attempt to reinvigorate a run-down urban area, such as inner city. Slum clearance is redevelopment after wholesale clearance of the site, but this has effect of destroying communities and has become unacceptable.

Urban Sprawl: The extension of the city into the countryside, particularly associated with improvements in mass transport.

Urban System: Any network of town and cities, and their hinterlands, which can be seen as a system, since it depends on the movements of labour, goods and services, ideas and capital through the network. Crucial to the interactions within the system are efficient systems of transport and communication. With improved technology, it is possible to see urban systems which transcend national boundaries.

Urbanism: A way of life associated with urban dwelling. Urban dwellers follow a distinctly different way of life from rural dwellers. Physical and social stimuli are high and urban residents may react by becoming aloof and indifferent in their relationship with others. Stress is higher than in country, and this is said to account for higher levels of mental illness and crime in city

Urbanization: This is the migration of rural population into towns and cities; an increasing proportion of the world's population resides in towns. Urbanization indicates a change of employment structure from agriculture and cottage industries to mass production and service industries. This backs up the view that urbanization results from, rather than causes social change. This is most notable in the development of capitalism and its attendant industrialization. It is said that the development of the landless labourer and the concentration of wealth into a few hands encourages urbanization.

von Thunen Model: J.H. von Thunen had two basic models. Both were located in an isotropic plain where there was one market—the city—for surplus agricultural production. One form of transport was available and transport costs increased in direct proportion to distance. No external trade took place, and farmers acted as economic man. All farmers received the same price for a particular crop at any one time.

The first model postulates that the intensity of production of a crop declines with distance from the market since transport costs increase with distance from the market and the location rent is therefore lower. Intensive farming—which demands costly inputs—is only profitable where locational rent is higher to cover costs, so intensive farming takes place only near the city.

von Thunen's second model is concerned with land use patterns. Transport costs vary with the bulkiness and perishability of product. The highly perishable and bulky products are therefore produced near/around the city.

Weber's Theory of Industrial Location: A model of industrial location proposed by A. Weber (1909 and 1929), which assumes that industrialists choose a least cost location for the development of new industry. The theory is based on a number of assumptions, among them that markets are fixed at certain specific points, that transport costs are proportional to the weight of the goods and the distance covered by raw material or a finished product, that perfect competition exists, and decisions are made by economic man.

Weber postulated that raw materials and markets would exert a 'pull' on the location of an industry through transport costs. Industries with a high material index would be pulled towards the raw material. Industries with a low material index would be pulled towards the market.

Once a least cost location has been established, Weber goes on to consider the deflecting effect of labour costs. To determine whether the savings provided by moving to a location of cheaper or more efficient labour would more than offset the increase in transport costs, isodapanes are constructed around the point of production at the point of minimum transport costs. The extra price of the wage bill is calculated for the point of production. If the source of cheap labour lies within the isodapane which has the value of higher wages differential (the critical isodapane), it would be more profitable to choose the site with low labour costs rather than the least transport location.

Industrial location may be swayed by agglomeration economies. The saving which would be made if, three firms were to be located together, are calculated for each plant. The isodapane with that value is drawn around the three least-cost locations. If these isodapanes overlap, it would be profitable for all three to locate together in the area overlap

Welfare Geography: An approach in human geography which features inequalities in social well-being and social justice, looking at the real differentiation and spatial organization of human activity from the point of view of welfare of the people involved: who gets what, where, and why?

Welfare geography developed, in part, in response to what was seen as the rather 'bloodless' narrowly economic preoccupations of the quantitative revolution, and focuses on those factors which affect the quality of human life: crime/lack of crime, poverty/wealth, housing/homelessness, and the provision/lack of education, health, leisure, and social services.

There are three broad lines of enquiry. The first is to identify inequality in the distribution of any of the welfare indicators listed above. This provides a base from which to evaluate the impact of past or proposed changes. From this, the second aim is an attempt to explain these inequalities. Current interpretations include explanations which link inequality with the mode of production; it is argued that uneven development is the inevitable consequence of capitalism, while recognizing that inequalities of a less extreme spatial expression also arise under socialism. Other explanations show how the location of services benefits some and disadvantageous others. The third aim is to propose measures which will bring a fairer distribution of resources and opportunities.

Zero Population Growth: A condition in which the rate of annual increase of society is nil, attained when the crude birth rate and death rate are the same or a situation in which the number of births and deaths is equal over a sustained period of time. This would require an average number of 2.3 children per family.

Zone of Assimilation: A transition zone created by the advancing front of a moving Central Business District usually involving the invasion of residential areas by various types of central business (notably small offices and retailing firms).

Zonal Model: Another term for concentric zone model.

Zone of Discard: The area from which a Central Business District is migrating involving a shift away from slums, railways and rivers. The zone of discard is characterized by family clothing stores, second hand furniture stores, low grade restaurant, and bus stops. The process may be exacerbated where central business district retails sale are declining in the face of competition from outlying shopping centres.

Zone of Assimilation: The area which increasingly develops the functions of the CBD; the CBD of the future, characterized by whole scale redevelopment of shops, offices, and hotels.

Zone of Overlap: An area served by more than one urban centre, i.e. within two or more different urban fields.

References

Barthclomew, H., and Wood, J., 1955, *Land Use in American Cities*, Cambridge, Mass, Harvard University Press.

Forde, C.D. 1925, *Habitat, Economy and Society*, London, Methuen.

Gottmann, J., and Harper, R.A. (Eds.), 1972, *Metropolis on the Move*, London, Longman.

Husain, M., 2002, *Human Geography*, third edn, Jaipur, Rawat Publications.

Kirby, D.A., 1979, *Slum Housing and Residential Renewal, the Case in Urban Britain*, London, Longman.

Lebon, J.H. 1969, *An Introduction to Human Geography,* London, Hutchinson.

Mansion, A.M. 1991, *Global Environment Change*, New York, Longman.

Mayhew, S. 1997, *Oxford Dictionary of Geography*, Indian eds. Delhi, Oxford University Press.

Murphy, R.E., 1974, *The American City: An Urban Geography*, New York, McGraw Hill.

Robson, B.T., 1970, *Urban Social Areas,* London, Oxford University Press.

UNESCO, 1952, *What is Race*?

Yeates, M.H., and Garner, B., 1980, *The North American City* (3rd Edn) New York, Harper and Row.

Appendices

Lowest Life Expectancy in the World—2005-2006

Rank	*Country*	*Average Life Span in Years*
1.	Zambia	32.4
2.	Sierra Leone	34.2
3.	Lesotho	35.1
4.	Malawi	37.5
5.	Mozambique	38.1
6.	Rwanda	39.3
7.	Botswana	39.7
8.	Angola	40.1
9.	Burundi	40.9
10.	Cote de Ivoire	41.0
11.	Congo Dem. Republic	41.8
12.	Afghanistan	43.1
13.	Tanzania	43.3
14.	Namibia	44.3
15.	Kenya	44.6
16.	Central African Republic	44.7
17.	Ethiopia	45.5
18.	Djibouti	45.7
19.	Niger	46.2
20.	Cameroon	46.2
21.	South Africa	47.7
22.	Congo Republic	48.0
23.	Mali	48.6
24.	Guinea	49.1
25.	Equatorial Guinea	49.1

Source: The World Guide-A View from the South: 2005/2006, Oxford, New Internationalist™ Publications Ltd.

The Highest Life Expectancy in the World—2005-2006

Rank	*Country*	*Average Life Span in Years*
1.	Japan	81.6
2.	Sweden	80.1
3.	Iceland	78.8
4.	Canada	79.3
5.	Spain	79.3
6.	Australia	79.2
7.	New Zealand	78.9
8.	Norway	78.9
9.	Italy	78.7
10.	Austria	78.5
11.	U.K.	78.2
12.	Germany	78.0
13.	Hungary	71.9
14.	USA	77.1
15	Belarus	71.0
16.	China	71.0
17.	Russia	66.8

Source: The World Guide-A View from the South: 2005-2006, Oxford, New Internationalist TM Publications Ltd.

Per Capita Income
The Lowest Per Capita Income—2005-2006

Rank	*Country*	*Per Capita Income (GNI) in $ US*
1.	Congo Dem. Republic	99.0
2.	Burundi	100.0
3.	Ethiopia	100.0
4.	Sierra Leone	140.0
5.	Malawi	160.0
6.	Niger	170.0
7.	Mozambique	210.0
8.	Burkina Faso	220.0
9.	Chad	220.0
10.	Nepal	230.0
11.	Rwanda	230.0
12	Mali	230.0
13.	Madagascar	240.0
14.	Uganda	250.0
15.	Central African Republic	260.0
16.	Ghana	270.0
17.	Kyrgyzstan	290.0
18.	Nigeria	290.0
19.	Laos	310.0
20.	Zambia	330.0

Source: The World Guide-A View from the South: 2005-2006, Oxford, The Internationalist™ Publications Ltd.

Countries with the highest Per Capita Income (GNI): 2005-2006

Rank	*Country*	*Per capita income in $ US*
1.	Switzerland	37930.0
2.	Norway	37850.0
3.	USA	35060.0
4.	Japan	33560.0
5.	Denmark	30290.0
6.	Iceland	27970.0
7.	Sweden	24820.0
8.	Brunei	24000.0
9.	Netherlands	23960.0
10.	Finland	23510.0
11.	Austria	23390.0
12.	Belgium	23250.0
13.	Germany	22670.0
14.	Canada	22300.0
15.	France	22010.0
16.	Australia	19740.0
17.	Italy	18960.0
18.	New Zealand	13710.0
19.	Portugal	10840.0
20.	Russia	2140.0

Source: *The World Guide-A View from the South: 2005-2006,* Oxford, New Internationalist™ Publications Ltd.

The Largest Cities of the World-2000

Figures are for the urban agglomeration, defined as the population contained within the contours of a contiguous territory inhabited at urban levels without regard to administrative boundaries. They incorporate the population within the city plus the suburban fringe lying outside of built adjacent to the city boundaries.

Rank	*City*	*Population*
1.	Tokyo (Japan)	28,025,000
2.	Mexico (Mexico)	18,131,000
3.	Mumbai (India)	18,042,000
4.	Sao Paulo (Brazil)	17,711,000
5.	New York (USA)	16,626,000
6.	Shanghai (China)	14,173,000
7.	Lagos (Nigeria)	13,488,000
8.	Los Angles (USA)	13,129,000
9.	Kolkata (India)	12,900,000
10.	Buenos Aires (Argentina)	12,431,000
11.	Soul (South Korea)	12,215,000

Contd...

Rank	City	Population
12.	Beijing (China)	12,033,000
13.	Karachi (Pakistan)	11,774,000
14.	Delhi (India)	11,680,000
15.	Dhaka (Bangladesh)	10,980,000
16.	Manila (Philippines)	10,818,000
17.	Cairo (Egypt)	10,772,000
18.	Osaka (Japan)	10,609,000
19.	Rio de Janeiro (Brazil)	10,556,000
20.	Tianjin (China)	10,239,000
21.	Jakarta (Indonesia)	9,815,000
22.	Paris (France)	9,638,000
23.	Istanbul (Turkey)	9,413,000
24.	Moskava (Moscow) (Russia)	9,299,000
25.	London (U.K.)	7,640,000
26.	Lima (Peru)	7,443,000
27.	Tehran (Iran)	7,380,000
28.	Bangkok (Thailand)	7,221,000
29.	Chicago (USA)	6,945,000
30.	Bogota (Colombia)	6,834,000
31.	Hyderabad (India)	6,833,000
32.	Chennai (India)	6,639,000
33.	Essan (Germany)	6,550,000
34.	Hangzhou (China)	6,389,000
35.	Hong Kong (China)	6,097,000
36.	Lahore (Pakistan)	6,030,000
37.	Shenyang (China)	5,618,000
38.	Changchu (China)	5,556,000
39.	Bangalore (India)	5,544,000
40.	Harbin (China)	5,475,000
41.	Chendu (China)	5,930,000
42.	Santiago (Chile)	5,261,000
43.	Guangzhou (China)	5,162,000
44.	St. Petersburg (Russia)	5,068,000
45.	Kinshasha (Dem. Republic Congo)	5,068,000
46.	Baghdad (Iraq)	4,796,000
47.	Jinan (China)	4,789,000
48.	Wuhan (China)	4,750,000
49.	Toronto (Canada)	4,657,000
50.	Yangon (Myanmar)	4,458,000
51.	Algier (Algeria)	4,447,000
52.	Philadelphia (USA)	4,398,000
53.	Qinddao (China)	4,376,000
54.	Milano (Italy)	4,251,000
55.	Pusan (S. Korea)	4,239,000
56.	Belo Horizonte (Brazil)	4,160,000

Contd...

Rank	City	Population
57.	Ahmedabad (India)	4,154,000
58.	Madrid (Spain)	4,072,000
59.	San Francisco (USA)	4,051,000
60.	Alexandria (Egypt)	3,995,000
61.	Washington, D.C. (USA)	3,927,000
62.	Dallas (USA)	3,912,000
63.	Guadalajara (Mexico)	3,908,000
64.	Chongqing (China)	3,896,000
65.	Medellin (Colombia)	3,831,000
66.	Detroit (USA)	3,785,000
67.	Handan (China)	3,763,000
68.	Frankfurt (Germany)	3,700,000
69.	Porto Alegre (Brazil)	3,699,000
70.	Na Noi (Vietnam)	3,678,000
71.	Sydney (Australia)	3,665,000
72.	Santo Domingo (Dominican Republic)	3,601,000
73.	Singapore (Singapore)	3,587,000
74.	Casablanca (Morocco)	3,5,35,000
75.	Katowice (Poland)	3,488,000
76.	Pune (India)	3,485,000
77.	Bandung (Indonesia)	3,420,000
78.	Monterrey (Mexico)	3,416,000
79.	Montreal (Canada)	3,401,000
80.	Nagoya (Japan)	3,377,000
81.	Nanjing (China)	3,375,000
82.	Houston (USA)	3,365,000
83.	Abidjan (Cote dÍvore)	3,359,000
84.	Xian (China)	3,352,000
85.	Berlin (Germany)	3,337,000
86.	Riyadh (Saudi Arabia)	3,328,000
87.	Recife (Brazil)	3,307,000
88.	Dusseldorf (Germany)	3,251,000
89.	Ankara (Turkey)	3,190,000
90.	Melbourne (Australia)	3,198,000
91.	Salvador (Brazil)	3,180,000
92.	Dalian (China)	3,153,000
93.	Caracas (Venezuela)	3,153,000
94.	Adis Ababa (Ethiopia)	3,112,000
95.	Athens (Greece)	3,103,000
96.	Cape Town (S. Africa)	3,092,000
97.	Koln (Germany)	3,067,000
98.	Maputo (Mozambique)	3,017,000
99.	Napoli (Italy)	3,012,000
100.	Fortaleza (Brazil)	3,007,000

Source: The Times Concise Atlas of the World, 2001, Times Books, London.

World's Top Ten Energy Producers (Million tons of oil equivalent)

1.	USA	1835
2.	Russian Federation	1034
3.	China	835
4.	Saudi Arabia	529
5.	Canada	433
6.	United Kingdom	293
7.	India	251
8.	Iran	249
9.	Mexico	234
10.	Australia	209

World's Top Ten Energy Consumers

(in kw)

1.	USA	20389
2.	China	855
3.	Russian Federation	655
4.	Japan	536
5.	Germany	349
6.	India	315
7.	Canada	300
8.	France	252
9.	United Kingdom	246
10.	Brazil	204

Source: The Times Concise Atlas of the World, 2002, Times Books, London.

Continents—Mountains

Europe:

Peak	*Metre*	*Feet*	*Country*
Elbrus	5642	18510	Russian Federation
Dora-Dykh-Tau	5204	17073	Russian Federation
Shkhara	5201	17063	Georgia
Mount Blanc	4808	15764	France/Italy
Dufourspitze	4634	15203	Italy/Switzerland

Source: The Times Concise Atlas of the World, 2002, London.

Africa:

Peak	*Metre*	*Feet*	*Country*
Kilimanjaro	5892	19331	Tanzania
Kirinyaga (Mt. Kenya)	5199	17057	Kenya
Marghaerita	5110	16765	De.Rep.Congo/Uganda
Meru	4564	14977	Tanzania
Mt. Karismbi	4510	14796	Rwanda

Source: The Times Concise Atlas of the World, 2002, London.

Antarctica:

Peak/Mt.	*Metre*	*Feet*
Vinsopn Massif	4897	16 066
Mt. Tyree	4852	15 918
Mt.Kirkpatrick	4528	14 855
Mt. Markhan	4351	14 275
Mt. Jackson	4190	13 747

Source: The Times Concise Atlas of the World, 2002, London.

Asia:

Peak	*Metre*	*Feet*	*Country*
Mt. Everest	8848	29 028	China/Nepal
K2 (Oogr Feng)	8611	28 251	China/Jammu & Kashmir
Kanchenjunga	8586	28 169	India/Nepal
Makalu	8463	27 765	China/Nepal
Cho Oyu	8201	26 906	China/Nepal

Source: The Times Concise Atlas of the World, 2002, London.

Australia and New Zealand:

Peak	*Metre*	*Country*
Mt. Cook	3800	New Zealand
Mt. Tasman	3530	New Zealand
Mt. Kosisco	2228	Australia

North America:

Peak/Mt.	*Metre*	*Feet*	*Country*
Mt. McKinley	6194	20 321	USA
Mt. Logan	5959	19 550	Canada
Pico de Orizaba	5747	18 855	Mexico
Mt. St. Elias	5489	18 008	USA
Volcan Popcatepetl	5452	17 887	Mexico

Source: The Times Concise Atlas of the World, 2000, London.

South America:

Peak/Mt.	Metre	Feet	Country
Acancagua	6960	22 834	Argentina
Nevado Ojos del Salado	6908	22 664	Argentia/Chile
Cerro Bonete	6872	22 546	Argentina
Cerro Pissis	6858	22 500	Argentina
Cerro Tupungato	6800	22 309	Argentia/Chile

Source: The Times Concise Atlas of the World, 2000, London

1st Six Rivers in Length

Africa

Nile	6695 km	4160 miles
Congo	4667 km	2900 miles
Niger	4184 km	2599 miles
Zembezi	2736 km	1700 miles
Webi Shbbeelle	2490 km	1547 miles
Ubangi	2250 km	1398 miles

Asia

Chang Jiang (Yangtze)	6380 km	3964 miles
Ob-Irtysh	5568 km	3459 miles
Yenisey-Angara	5550 km	3448 miles
Huang He (Yellow River)	5464 km	3396 miles
Mekong	4425 km	2749 miles
Amur (Heilong Jiang,-Argun)	4416 km	2744 miles

Australia

Murray-Darling	3750 km	2330 miles
Darling	2739 km	1702 miles
Murray	2589 km	1608 miles
Murrumbidgee	1890 km	1050 miles
Lachtan	1480 km	919 miles
Macquarie	950 km	590 miles

Europe

Volga	3688 km	2291 miles
Danube	2850 km	1770 miles
Dnieper	2285 km	1419 miles
Kama	2028 km	1260 miles
Don	1931 km	1199 miles
Pecora	1802 km	1119 miles

North America

Mississippi-Missouri	5969 km	3709 miles
Mackenzie-Peace-Finlay	4241 km	2635 miles
Missouri	4086 km	2539 miles
Mississippi	3765 km	2339 miles
Yukon	3185 km	1979 miles
Rio-Grande	3057 km	1899 miles

South America

Amazon	6516 km	4049 miles
Rio de Laplata-Parana	4500 km	2796 miles
Purus	3218 km	1999 miles
Madeira	3200 km	1988 miles
Sao Francisco	2900 km	1802 miles
Tocantins	2750 km	1708 miles

Nations of the World: Area, Population and Capital

	Country	*Area (km²)*	*Population*	*Capital*
1.	Afghanistan	652,090	25,971,254	Kabul
2.	Albania	28,750	3,220,164	Tirana (Tirane)
3.	Algeria	2,381,740	32,877,042	Algiers (Alger)
4.	Andorra	453	75,379	Andodorra laVella
5.	Angola	1,246,700	14,532,929	Luanda
6.	Anguilla	90	12,246	The Valley
7.	Antigua and Barbuda	440	73,615	St. John's
8.	Argentina	2,780,400	39,310,826	Buenos Aires
9.	Falkland Islands	11,410	3,063	Port Stanley (Puerto Argentino)
10.	Armenia	29,800	3,042,663	Yerevan
11.	Aruba	190	103,205	Oranjestad
12.	Australia	7,741,220	20,092,317	Canberra
13.	Cocos Islands	14	630	West Island
14.	Christmas Island	135	1,853	Kingston
15.	Austria	83,860	8,120,243	Vienna (Wien)
16.	Azerbaijan	86,600	8,527,280	Baku
17.	Bahamas	13,880	320,650	Nassau
18.	Bahrain	710	754,331	Manama
19.	Bangladesh	144,000	152,592,662	Dhaka (Dacca)
20.	Barbados	430	272,214	Bridgetown
21.	Belarus	207,600	9,808,724	Minsk (Mensk)
22.	Belgium	33,100	10,359,127	Brussels (Bruxelles)
23.	Belize	22,960	266,260	Belmopan
24.	Benin	112,620	7,103,140	Porto Nova
25.	Bermuda	50	82,729	Hamilton
26.	Bhutan	47,000	2,391,782	Thimbu

Contd...

	Country	Area (km²)	Population	Capital
27.	Bolivia	1,098,580	9,138,490	La Paz
28.	Bosnia-Herzegovina	51,130	4,208,630	Sarajevo
29.	Botswana	581,730	1,800,577	Gaborone
30.	Brazil	8,547,400	182,797,708	Brasilia
31.	Brunei	5,770	374,242	Bandar Seri Begawan
32.	Bulgaria	110,910	7,763,459	Sofia (Sofija)
33.	Burkina Faso	274,000	13,797,527	Ouagadougou
34.	Burundi	27,830	7,381,712	Bujumbura
35.	Cambodia	181,040	14,825,135	Phnom Penh
36.	Cameroon	475,440	16,564,191	Yaounde
37.	Canada	9,970,610	31,972,474	Ottawa
38.	Cape Verde	4,030	481,873	Praia
39.	Cayman Islands	260	42,785	George Town
40.	Central African Rep.	622,980	3,961,946	Bengui
41.	Chad (Tashad)	1,284,000	9,117,376	N'djamena
42.	Chile	756,630	16,185,450	Santiago
43.	China	9,598,050	1,322,273,282	Beijing
44.	Taiwan	36,960	22,370,000	Taipei
45.	Colombia	1,138,910	45,600,244	Bogota
47.	Comoros	2,230	812,310	Moroni
48.	Congo Dem.Rep.	2,344,860	56,079,226	Kinshasa
49.	Congo	342,000	3,920,514	Brazzaville
50.	Costa Rica	51,100	4,327,072	San Jose
51.	Cote d'Ivoire	322,460	17,164,505	Abidjan (Yamoussoukro)
52.	Croatia	56,540	4.405,173	Zagreb
53.	Cuba	110,860	11,353,321	Havana (La Habana)
54.	Cyprus	9,250	813,331	Nicosia (Lefkosia)
55.	Czech Republic	78,870	10,215,797	Prague
56.	Denmark	43,090	5,385,540	Copenhagen
57.	Djibouti	23,200	720,992	Djibouti
58.	Dominica	750	78,966	Roseau
59.	Dominican Rep.	48,730	8,998,112	Santo Domingo
60.	Ecuador	283,560	13,378,641	Quito
61.	Egypt (Misr)	1,001,450	74,878,313	Cairo (Qahirah)
62.	El Salvador	21,040	6,709,339	San Salvador
63.	Equatorial Guinea	28,050	520,833	Malabo
64.	Eritrea	117,600	4,456,086	Asmara
65.	Estonia	45,100	1,294,143	Tallinn
66.	Ethiopia	1,104,300	74,188,932	Addis Ababa
67.	Fiji	18,270	854,471	Suva
68.	Finland	338,150	5,223,605	Helsinki
69.	France	551,500	60,711,094	Paris
70.	French Guiana	90,000	186,508	Cayenne

Contd...

	Country	Area (km²)	Population	Capital
71.	French Polynesia	4,000	651,647	Papeete
72.	Gabon	267,670	1,375,223	Libreville
73.	Gambia	11,300	1,499,176	Benjul
74.	Georgia	69,700	5,025,885	Tbilisi
75.	Germany	357,030	82,559,636	Berlin
76.	Ghana	238,540	21,832,963	Accra
77.	Greece	131,960	10,978,487	Athens
78.	Grenada	340	79,952	St. George's
79.	Guadeloupe	1,705	446,486	Basse-Terre
80.	Guam	550	167,834	Agana
81.	Guatemala	108,890	12,977,540	Quetzal
82.	Guinea	245,860	8,788,030	Conakry
83.	Guinea-Bissau	36,120	1,583,546	Bissau
84.	Guyana	214,970	768,003	George Town
85.	Haiti	27,750	8,549,254	Port-au-Prince
86.	Honduras	112,090	7,257,281	Tegucigalpa
87.	Hungary	93,030	9,784,352	Budapest
88.	Iceland	103,000	293,702	Reyjavik
89.	India	3,287,260	1,096,917,184	New Delhi
90.	Indonesia	1,904,570	225,313,459	Jakarta
91.	Iran	1,648,200	70,675,076	Tehran
92.	Iraq	438,320	26,555,227	Baghdad
93.	Ireland	70,720	4,040,185	Dublin
94.	Israel	21,060	6,684,817	Jerusalem/(Tel-Aviv)
95.	Italy	301,340	57,252,557	Rome
96.	Jamaica	10,990	2,700,868	Kingston
97.	Japan	377,800	127,914,376	Tokyo
98.	Jordan	89,210	5,750,139	Amman
99.	Kanaky-New Caledonia	18,580	236,979	Noumea
100.	Kazakhstan	2,724,900	15,364,445	Astana
101.	Kenya	580,370	32,849,169	Nairobi
102.	Kiribati	730	90,159	Bairiki
103.	Korea, North	120,540	22,876,149	Pyongyang
104.	Korea, South	99,260	48,182,450	Seoul
105.	Kuwait	17,820	2,671,421	Kuwait City
106.	Kyrgyzstan	199,900	5,277,887	Bishkek
107.	Laos (Lao)	236,800	5,918,323	Vientiane
108.	Latvia (Latvija)	64,600	2,264,965	Riga
109.	Lebanon	10,400	3,760,703	Beirut
110.	Lesotho	30,350	1,797,120	Maseru
111.	Liberia	111,370	3,602,847	Monrovia
112.	Libya	1,759,540	5,768,469	Tripoli
113.	Liechtenstein	160	34,077	Vaduz
114.	Lithuania	62,200	3,401,459	Vilnius

Contd...

	Country	Area (km²)	Population	Capital
115.	Luxembourg	2,600	464,904	Luxembourg
116.	Macedonia (Tfyr)	25,710	2,075,805	Skopje
117.	Madagascar	587,040	18,408,559	Antananarivo
118.	Malawi	118,480	12,572,372	Lilongwe
119.	Malaysia	329,750	25,324,815	Kuala Lumpur
120.	Maldives	300	337,722	Male
121.	Mali	1,240,190	13,829,090	Bamako
122.	Malta	320	397,334	Valletta
123.	Marshal Islands	200	54,200	Majuro
124.	Martinique	1,102	396,589	Fort-de-France
125.	Mauritania	1,025,520	3,068,742	Nouakchott
126.	Mauritius	2,040	1,243,871	Port Louis
127.	Mexico	1,958,200	106,384,786	Mexico City
128.	Micronesia	700	111,465	Palikir
129.	Moldova	4,258,650	33,850	Chisinau
130.	Monaco	2	35,017	Monaco
131.	Mongolia	1,566,500	2,666,950	Ulan Bator
132.	Montserrat	102	3,958	Playmouth
133.	Morocco (Al-Maghrib)	446,550	31,564,452	Rabat
134.	Mozambique	801,590	19,495,185	Moputo
135.	Myanmar (Burma)	676,580	50,695,720	Yangon (Rangoon)
136.	Namibia	824,290	2,032,315	Windhoek
137.	Naura	21	13,660	No official cap.
138.	Nepal	147,180	26,289,096	Kathmandu
139.	Netherlands	41,530	16,299,745	Amsterdam
140.	Netherlands Antilles	800	224,407	Willemstad
141.	New Zealand (Aotearoa)	270,530	3,931,823	Wellington
142.	Tokelau	12	1,521	Fakaofo
143.	Nicaragua	130,000	5,727,367	Managua
144.	Niger	1,267,000	12,872,815	Niamey
145.	Nigeria	923,770	130,235,642	Abuja
146.	Northern Marianas	500	85,693	Saipan
147.	Norway (Norge)	323,880	4,569,598	Oslo
149.	Oman	309,500	3,020,264	Muscat (Musqat)
150.	Pakistan	796,100	161,150,580	Islamabad
151.	Palau	460	21,354	Koror
152.	Palestine	6,220	3,815,250	Jerusalem (Al-Qus)
153.	Panama	75,520	3,234,654	Panama City
154.	Papua New Guinea	462,840	5,959,257	Port Moresby
155.	Paraguay	406, 750	6,159,680	Asuncion
156.	Peru	1,285,220	27,968,365	Lima
157.	Philippines	300,000	82,808,513	Manila
158.	Pitcairn	5	47	Adamstown
159.	Poland	323,250	38,515,955	Warsaw (Warzawa)

Contd...

	Country	Area (km²)	Population	Capital
160.	Portugal	91,980	10,080,174	Lisbon
161.	Puerto Rico	8,950	3,915,379	San Juan
162.	Qatar	11,000	627,567	Doha
163.	Reunion	2,510	776,888	Saint Denis
164.	Romania	238,390	22,227,815	Bucharest
165.	Russia	17,075,400	141,552,786	Moscow (Moskva)
166.	Rwanda	26,340	8,606,604	Kigali
167.	St. Helena	122	5,049	Jamestown
168.	St. Kitts-Nevis	380	41,578	Basseterre
169.	St. Lucia	620	151,509	Castries
170.	St. Vincent & Grenadines	390	121,215	Kingstown
171.	Samoa	2,840	181,618	Apia
172.	American samoa	200	64,820	Pago Pago
173.	San Marino	61	28,108	San Marino
174.	Sao Tome & Principe	960	168,732	Sao Tome
175.	Saudi Arabia	2,149,690	25,625,687	Riyadh
176.	Senegal	196,720	10,587,234	Dakar
177.	Serbia & Montenegro	120,000	10,513,058	Belgrade
178.	Seychelles	450	82,394	Victoria
179.	Sierra Leone	71,740	5,339,564	Freetown
180.	Singapore	620	4,371,515	Singapore
181.	Slovakia	49,010	5,411,393	Brastislava
182.	Slovenia	20,250	1,979,244	Ljubljana
183.	Soloman Islands	28,900	504,405	Honiara
184.	Somalia	637,660	10,741,665	Mogadishu
185.	South Africa	1,221,040	45,323,008	Tshwane (Pretoria)
186.	Spain (Espana)	505,990	41,184,085	Madrid
187.	Sri Lanka	65,610	19,366,334	Colombo
188.	Sudan	2,505,810	35,039,802	Khartoum
189.	Surinam	163,270	442,351	Paramaribo
190.	Swaziland	17,360	1,086,815	Mbabane
191.	Sweden	449,960	8,894,851	Stockholm
192.	Switzerland	41,290	7,156,665	Bern
193.	Syria (Suriyah)	185,180	18,650,334	Damascus
194.	Tajikistan	143,100	6,356,090	Dushanbe
195.	Tanzania	945,090	38,364,837	Dodoma
196.	Thailand	513,120	64,081,371	Bangkok
197.	Timor-Leste	14,870	857,405	Dili
198.	Togo	56,790	5,128,902	Lome
199.	Tonga	750	105,797	Nuku'alofa
200.	Trinidad & Tobago	5,130	1,311,259	Port of Spain
201.	Tunisia	163,610	10,041,690	Tunis
202.	Turkey (Turkiye)	774,820	73,301,553	Ankara
203.	Turkmenistan	488,100	5,014,531	Ashkabad

Contd...

	Country	Area (km²)	Population	Capital
205.	Turks & Caicos	430	22,028	Cockburn Town
206.	Tuvalu	26	10,851	Fongalfale
207.	Uganda	241,040	27,623,190	Kampala
208.	Ukraine	603,700	47,782,268	Kiev (Kijef)
209.	United Arab Emirates	83,600	3,106,458	Abu-Dhabi
210.	United Kingdom	242,910	59,598,039	London
211.	United States	9,629,090	300,037,902	Washington D.C.
212.	Uruguay	176,220	3,463,202	Montevideo
213.	Uzbekistan	447,400	26,868,064	Tashkent
214.	Vanuatu	12,190	222,237	Vila
215.	Vatican City	1	440	Vatican City
216.	Venezuela	912,050	26,639,527	Caracas
217.	Vietnam	331,690	83,584,512	Hanoi
218.	Virgin Islands	340	113,466	Charlotte Amalie
219.	Virgin Islands	153	21,697	Road Town
220.	Western Sahara	266,000	323,961	L'ayoun
221.	Western Papua (Papua)	418,000	1,800,000	Jayapura
222.	Yemen	527,970	21,480,302	Sana'a
223.	Zambia	752,610	11,043,312	Lusaka
224.	Zimbabwe	390,760	12,963,496	Harare

Selected Countries: Export of Goods and Services (Current US million $)

Country	Export
Indonesia	61,210
Brazil	67,141
Denmark	73,654
Saudi Arabia	78,214
Australia	88,780
Malaysia	108,261
Switzerland	112,355
Mexico	170,588
Spain	174,076
Mexico	185,154
South Korea	190,741
Belgium	193,831
Netherlands	227,048
Canada	304,276
Italy	307,805
China	332,367
France	365,625
United Kingdom	386,216
Japan	432,547
Germany	645,542
USA	1,103,100

South Asian Countries: Export of Goods

Bhutan	229
Maldives	545
Nepal	1,610
Sri Lanka	5,825
Bangladesh	6,850
Pakistan	10,235
India	78,156

Selected Countries: Import of goods and Services (Current Million $)

Iran	19,066
Saudi Arabia	45,589
Brazil	72,339
Thailand	72,709
Australia	88,371
Malaysia	91,696
Switzerland	101,620
Spain	182,681
South Korea	183,885
Mexico	185,154
Belgium	193,831
Netherlands	227,048
Canada	268,355
Italy	290,358
China	313,802
France	345,072
Japan	406,428
United Kingdom	416,944
Germany	610,572
U.S.A.	1,466,900

South Asian Countries: Import of Goods

Bhutan	122
Nepal	1,610
Bangladesh	9,061
Sri Lank	6,964
Pakistan	10,900
India	82,900

Did you know

if the world were a village of 100 people ...

14 would not have enough to eat.
31 would not have any electricity.
17 would not have clean water to drink.
29 would be aged under fifteen.

Did you know:

If world were a village of 100 persons:

The richest person in the village would have more wealth than the poorest 57 combined.

If the world were a village of 100 people:

22 would be Chinese.
17 would be Indian.
5 would be American
About 3.75 would have been Indonesians.
3 would have been Brazilians
2.7 would have been Pakistani.
2.5 would have been Bangladeshi
Total = 52.95 per cent

Did you know:

If the world were a village of 100 people...

At least 18 would have mobile phone.

Coffee Production in the World (2004)

Brazil	32.0%
Colombia	11.0%
Vietnam	10.0%
Philippines	6.0%
India	3.5%
Mexico	3.4%

Top 10 Coffee Drinking Countries, 2004

	Country	*kg/person*		*Country*	*kg/person*
1.	Finland	10.00	6.	Netherlands	6.74
2.	Norway	8.85	7.	Germany	6.73
3.	Denmark	8.58	8.	Austria	5.46
4.	Sweden	8.0	9.	France	5.44
5.	Switzerland	6.95	10.	Italy	

Index